Pharmaceutical Formulations with Antimicrobial Properties

Pharmaceutical Formulations with Antimicrobial Properties

Guest Editors

Marcel Popa
Anca Niculina Cadinoiu

MDPI

Basel • Beijing • Wuhan • Barcelona • Belgrade • Novi Sad • Cluj • Manchester

Guest Editors
Marcel Popa
"Cristofor Simionescu"
Faculty of Chemical
Engineering and
Environment Protection
"Gheorghe Asachi" Technical
University
Iasi
Romania

Anca Niculina Cadinoiu
Department of Biomaterials
"Apollonia" University of Iasi
Iasi
Romania

Editorial Office
MDPI AG
Grosspeteranlage 5
4052 Basel, Switzerland

This is a reprint of the Special Issue, published open access by the journal *Pharmaceutics* (ISSN 1999-4923), freely accessible at: https://www.mdpi.com/journal/pharmaceutics/special_issues/antimicrobial_properties.

For citation purposes, cite each article independently as indicated on the article page online and as indicated below:

Lastname, A.A.; Lastname, B.B. Article Title. *Journal Name* **Year**, *Volume Number*, Page Range.

ISBN 978-3-7258-7931-1 (Hbk)
ISBN 978-3-7258-7932-8 (PDF)
https://doi.org/10.3390/books978-3-7258-7932-8

Contents

About the Editors

Marcel Popa

Marcel Popa is a Professor at the "Gheorghe Asachi" Technical University of Iași, Romania, "Cristofor Simionescu" Faculty of Chemical Engineering and Environmental Protection. He also collaborates with the "Apollonia" University of Iași, Romania. His research interests include polysaccharide modification, bioactive polymers, biomaterials, hydrogels, interpenetrated networks, micro- and nanoparticles (spheres and capsules), polymer–drug conjugates, with a focus on drug delivery systems and antimicrobial formulations. He has authored and co-authored numerous publications in the field and has participated in national and international research projects.

Anca Niculina Cadinoiu

Anca Niculina Cadinoiu is an Associate Professor at "Apollonia" University in Iasi, Romania. His research interests include drug delivery, biomaterials, polymer nanoparticles, liposomes, and hydrogels. Her work focuses on the development and characterization of micro/nanoscale carriers as well as the incorporation of bioactive compounds into advanced delivery systems for biomedical applications. She is the author and co-author of several scientific publications and has been involved in national and international research projects.

Preface

The articles collected in this Reprint are from the Special Issue "Pharmaceutical Formulations with Antimicrobial Properties", published in *Pharmaceutics*, and represent an ever-evolving and increasingly important field. Researchers are challenged to find new methods and therapeutic agents to help combat antimicrobial resistance.

What stands out in this collection is a common effort to collaborate and find solutions. Instead of focusing only on antimicrobial agents, many works explore how formulation strategies such as hydrogels, nano/microcarriers, or lipid-based systems can significantly influence therapeutic outcomes. These approaches aim not only to improve efficacy but also to address practical challenges, such as targeted administration, reduced toxicity, and increased patient compliance.

We hope that this Reprint will serve not only as a useful reference but also as a source of inspiration for further research and collaboration. Addressing antimicrobial resistance is a collective effort, and each contribution is a step forward in this ongoing effort.

Marcel Popa and Anca Niculina Cadinoiu
Guest Editors

pharmaceutics

MDPI

Editorial

Editorial on Special Issue "Pharmaceutical Formulations with Antimicrobial Properties"

Anca N. Cadinoiu [1,*] and Marcel Popa [1,2]

1 Faculty of Medical Dentistry, Apollonia University of Iasi, 700511 Iasi, Romania
2 Academy of Romanian Scientists, Ilfov Street, 050044 Bucharest, Romania
* Correspondence: jancaniculina@yahoo.com

Even though numerous studies on the systemic administration of antimicrobial drugs can be found in the literature, they still have many shortcomings related to the site-specific drug delivery, unwanted side effects and even potential toxicity. To overcome the above limitations, it is necessary to explore new approaches for the administration of active principles with antimicrobial properties.

In this Special Issue, original papers have been published that bring new perspectives in the development of novel pharmaceutical formulations for antimicrobial applications. Active principles with antimicrobial properties, such as antibiotics [1–5], antifungals [6,7], nutraceuticals [8], silver or ZnO nanoparticles [9,10], plant materials [11,12] and biocides [13], have been used to obtain mainly controlled drug delivery systems.

A first study, in which the Chapman University School of Pharmacy, Chapman University Irvine, USA and the Department of Ophthalmology School of Medicine, University of Pittsburgh School of Medicine, USA collaborated, was designed to evaluate the stability of vancomycin eye drops in normal saline solution, phosphate-buffered saline (PBS) and balanced salt solution (BSS), while being stored for 7, 14 and 28 days at room temperature or refrigerated [1]. The study detected no difference in the stability of vancomycin over 28 days of storage at either room temperature or refrigeration, but a difference in pH and turbidity was observed.

Two other research teams from the Interdisciplinary Excellence Centre, Faculty of Pharmacy, Institute of Pharmaceutical Technology and Regulatory Affairs, University of Szeged, Hungary and the School of Engineering, Institute for Materials and Processes, University of Edinburgh, UK studied the preformulation of poly(vinylpyrrolidone)-based nanofiber mats loaded with ciprofloxacin (CIP) for oral administration [2]. The aim of this study was to increase the water solubility and diffusion of the active pharmaceutical ingredient and to study the in vitro drug release and its kinetics. CIP-loaded nanofibers with a small, uniform fiber diameter and a smooth surface morphology were obtained by electrospinning.

Another study involving researchers from Turkey, UK and Romania presents the fabrication of bacterial cellulose (BC)/polycaprolactone (PCL) patches loaded with different antibiotics (amoxicillin (AMX), ampicillin (AMP) and kanamycin (KAN)) for transdermal delivery [3]. The obtained antibiotic-loaded patches displayed micron-scale fibers and an interfibrous pore size. In addition, the microfibers presented a porous structure, which imparted improved biocompatibility and increased drug loading capacity. Moreover, from the BC/PCL patches produced, different types of antibiotics could be successfully loaded and released, regardless of their type. The obtained antibiotic patches can meet the characteristic requirements needed in local transdermal applications with accelerated release characteristics, adjustability and versatility for dressing trials, and lower the cost of pharmaceutical industries.

A multidisciplinary research team from Hungary and Croatia managed to prepare and test several new formulations based on elastic liposomes loaded with azithromycin

(AZT), which differ in surface charge (cationic, neutral and anionic) [4]. Their antibacterial potential was evaluated against the *C. trachomatis* serovar D laboratory strain and the clinical isolate *C. trachomatis* serovar F. All tested liposomes efficiently delivered AZT to HeLa 229 cells infected with the laboratory *Chlamydia* strain, showing minimum inhibitory concentrations (MICs) and minimum bactericidal concentrations (MBCs) of AZT even 4–8 times lower than those obtained with free AZT. Also, AZT-loaded liposomes were effective against the clinical *Chlamydia* strain by decreasing the MIC values by two-fold compared to free AZT.

Three Research Institutes from Măgurele, Romania and three other Departments from the Faculty of Biology, University of Bucharest, Romania developed composite materials based on bioglass/ciprofloxacin/poly(methyl methacrylate) as multifunctional thin films for hard tissue implants [5]. Briefly, a thin layer of PMMA material was deposited by matrix-assisted pulsed laser evaporation (MAPLE) onto Ti substrates. A second layer consisting of bioglass + ciprofloxacin was applied by MAPLE to the initial thin film. The results of this study demonstrated that the laser-deposited coatings are biocompatible and resistant to microbial colonization and biofilm formation.

A stable and highly skin-permeable topical delivery system for itraconazole (ITZ) was obtained and tested by a multidisciplinary research team in Korea [6]. This topical delivery system aimed to improve the solubility and skin penetration of ITZ, as well as increase the storage duration at which it remains stable. The in vivo antifungal efficacy of this new system was compared with a commercially available oral formulation. The permeabilities of the obtained ITZ formulations through the artificial membrane, human skin and human nail were also evaluated. After analyzing the permeability results, the optimal formulation was tested in vivo to demonstrate its antifungal efficacy at a dose identical to that of commercially available oral ITZ, using a BALB/c mouse model of cutaneous *Candida albicans* infection.

Another study that sought to obtain pharmaceutical formulations based on antifungal drugs was published following collaboration between researchers from Saudi Arabia and Ethiopia. Butenafine hydrochloride nano lipid carrier (NLC) formulations were obtained and optimized [7]. The optimized BF-NLCopt exhibited a nanometer size with high drug entrapment efficiency. Further, the optimized BF-NLCopt was immobilized in a topical gel system for topical application for the treatment of fungal infections. The drug release pattern of BF-NLCopt was improved compared to BF-NLCopt-gel. The results obtained in the permeability tests demonstrated a good permeability of BF-NLCopt and BF-NLCopt-gel through the tested membrane. The conclusions of this study highlighted that the butenafine nano lipid carrier gel system acts as a potential delivery system in the treatment of antifungal diseases.

As is known, modulation of the immune system refers to a certain alteration in the immune response in the form of stimulation, amplification, expression or inactivation of some stage of the immune response. A material which exhibits a modifying immune system response to a threat is an immunomodulatory and this concept currently underlies the development of functional foods. The study of a group of researchers from Romania [8] aimed to evaluate the immunomodulatory properties of two novel nutraceutical formulations, one based on curcuminoids from different natural sources and another based on probiotics (*Lactobacillus acidophilus, Bifidobacterium animalis*) combined with actives from *Helianthus tuberosus* and *Plantago ovate* as herbal prebiotics. The results obtained are encouraging and attest to the great potential of the obtained nutraceutical formulations for their designated purposes, recommending them for further in vivo studies.

Another group of Romanian researchers proposes a new type of full-IPN hydrogel, based on chitosan and poly(vinyl alcohol), obtained by combining two cross-linking methods: covalent and physical. In fact, the authors obtained a biocomposite consisting of silver nanoparticles dispersed in the matrix of IPN, designed to exert a synergistic antimicrobial activity through the action of both chitosan and silver [9]. The biocomposite obtained is of hydrogel type, with pores whose sizes vary between 40 and 60 μm, with swelling capacity in water that depends on different factors, among which the chemical nature of the epoxy crosslinker is important. The tests showed that the obtained materials present a high inhibitory activity against *S. aureus* and *K. pneumonia*, but low activity for *P. aeruginosa* and *E. coli*, and may be considered non-cytotoxic. The swelling behavior, the mechanical properties, and the antimicrobial activity recommend the use of this biocomposite in biomedical applications as wound or oral dressings.

The same type of matrix (with hydrogel character), but covalently cross-linked with glutaric aldehyde, was used to disperse ZnO particles in order to obtain a biocomposite proposed for the treatment of bacterial infections [10]. The increased concentration of covalent cross-linker led to increased mechanical properties, but when ZnO nanoparticles were incorporated into the hydrogels, the tensile strength and elongation at break decreased compared with the sample without metal oxide NPs. The degree of covalent crosslinking and the amount of ZnO nanoparticles incorporated influence the morphology and swelling degree of hydrogels. The biocomposite is practically non-toxic and shows significant antimicrobial activity against *S. aureus*, *E. coli*, and *K. pneumonia*.

A review discusses the possibility of using plant materials in the management of many inflammatory diseases, including periodontitis—one of the most serious oral diseases which affects tissues, being caused by pathogenic bacteria and environmental factors [11]. It is well known that a series of products extracted from plants exhibit antibacterial, anti-inflammatory and antioxidant activities, and affect the periodontium structure. The review is based on the consultation of 180 bibliographic references, most of them recent, and analyzes the use and applicability of plant materials in periodontitis. Their benefits concern the biocompatibility, the lower cost and the better safety profile.

In another paper of this issue, harmala alkaloid-rich fraction (HARF) (which is hydrophobic) was included in 2-hydroxy propyl-β-cyclodextrin to obtain a host–guest complex; by doing so, the water solubility and bioavailability of HARF was improved [12]. The complex was co-encapsulated with ascorbic acid into poly(lactide-glicolide)-based nanoparticles coated with poly(ethylene glycol), the entrapment efficiency being very high (more than 80%). The obtained polymer–drug system exhibited the highest antibacterial activity against *Staphylococcus aureus* and *Escherichia coli*, and high selective antiviral activity against the H1N1 influenza virus, with minimum host toxic effects.

The last paper of the issue concerns the preparation of ligand brush nanocapsules, based on poly(vinyl pyrrolidone), coated with SiO_2 and grafted with poly(methalic acid sodium salt) [13]. The ligand-grafted nanocapsule diameter was 888 nm, and they manifested a self-controlled antibacterial response in the pH and humidity conditions needed for medical applications. Ligand-brush NCs containing an anionic antimicrobial drug had a rapid release effect because of the repellent electrostatic force and swelling properties of the ligand brushes. The experiments proved that *Escherichia coli* was eradicated at pH 6–7 and humidity 45–100%, conditions in which most bacterial and fungal strains thrive.

The editors of this Special Issue of *Pharmaceutics*, titled *Pharmaceutical Formulations with Antimicrobial Properties*, are researchers with decades of experience in the field of controlled drug release, and they consider these articles reporting recent results concerning the obtaining of polymer–drug systems, their properties and their biomedical applications justified and useful.

The editors express their gratitude for the kindness and cooperation of the contributors, who are reputable researchers in the field and have responded to our call to contribute to this Special Issue of *Pharmaceutics* with the results of their recent research.

Author Contributions: Conceptualization, A.N.C. and M.P.; writing—original draft preparation, A.N.C. and M.P.; writing—review and editing, A.N.C. and M.P.; visualization, A.N.C. and M.P. All authors have read and agreed to the published version of the manuscript.

Funding: A.N.C. and M.P. were supported by the NO Grants 2014-2021, under Project contract number 15/2020, and a grant of the Romanian Ministry of Education and Research, CNCS-UEFISCDI, project number PN-III-P4-ID-PCE-2020-2009, within PNCDI III.

Conflicts of Interest: The authors declare no conflict of interest.

References

1. Ross, C.; Syed, B.; Pak, J.; Jhanji, V.; Yamaki, J.; Sharma, A. Stability Evaluation of Extemporaneously Compounded Vancomycin Ophthalmic Drops: Effect of Solvents and Storage Conditions. *Pharmaceutics* **2021**, *13*, 289. [CrossRef] [PubMed]
2. Uhljar, L.É.; Kan, S.Y.; Radacsi, N.; Koutsos, V.; Szabó-Révész, P.; Ambrus, R. In Vitro Drug Release, Permeability, and Structural Test of Ciprofloxacin-Loaded Nanofibers. *Pharmaceutics* **2021**, *13*, 556. [CrossRef] [PubMed]
3. Altun, E.; Yuca, E.; Ekren, N.; Kalaskar, D.M.; Ficai, D.; Dolete, G.; Ficai, A.; Gunduz, O. Kinetic Release Studies of Antibiotic Patches for Local Transdermal Delivery. *Pharmaceutics* **2021**, *13*, 613. [CrossRef] [PubMed]
4. Bogdanov, A.; Janovák, L.; Vraneš, J.; Meštrović, T.; Ljubin-Sternak, S.; Cseh, Z.; Endrész, V.; Burián, K.; Vanić, Ž.; Virok, D.P. Liposomal Encapsulation Increases the Efficacy of Azithromycin against *Chlamydia trachomatis*. *Pharmaceutics* **2022**, *14*, 36. [CrossRef] [PubMed]
5. Negut, I.; Ristoscu, C.; Tozar, T.; Dinu, M.; Parau, A.C.; Grumezescu, V.; Hapenciuc, C.; Popa, M.; Stan, M.S.; Marutescu, L.; et al. Implant Surfaces Containing Bioglasses and Ciprofloxacin as Platforms for Bone Repair and Improved Resistance to Microbial Colonization. *Pharmaceutics* **2022**, *14*, 1175. [CrossRef] [PubMed]
6. Subedi, L.; Song, S.-Y.; Jha, S.K.; Lee, S.-H.; Pangeni, R.; Koo, K.-T.; Kim, B.J.; Cho, S.-S.; Park, J.W. Preparation of Topical Itraconazole with Enhanced Skin/Nail Permeability and In Vivo Antifungal Efficacy against Superficial Mycosis. *Pharmaceutics* **2021**, *13*, 622. [CrossRef] [PubMed]
7. Mahdi, W.A.; Bukhari, S.I.; Imam, S.S.; Alshehri, S.; Zafar, A.; Yasir, M. Formulation and Optimization of Butenafine-Loaded Topical Nano Lipid Carrier-Based Gel: Characterization, Irritation Study, and Anti-Fungal Activity. *Pharmaceutics* **2021**, *13*, 1087. [CrossRef]
8. Raducanu, A.E.; Tihauan, B.-M.; Marinas, I.C.; Ciuperca, O.T.; Tebrencu, C.E.; Ionescu, E.; Onisei, T. The Biological Effects of Novel Nutraceuticals with Curcuminoids and Other Plant-Derived Immunomodulators and Pre-Probiotics. *Pharmaceutics* **2021**, *13*, 666. [CrossRef]
9. Suflet, D.M.; Popescu, I.; Pelin, I.M.; Ichim, D.L.; Daraba, O.M.; Constantin, M.; Fundueanu, G. Dual Cross-Linked Chitosan/PVA Hydrogels Containing Silver Nanoparticles with Antimicrobial Properties. *Pharmaceutics* **2021**, *13*, 1461. [CrossRef] [PubMed]
10. Rata, D.M.; Cadinoiu, A.N.; Popa, M.; Atanase, L.I.; Daraba, O.M.; Popescu, I.; Romila, L.E.; Ichim, D.L. Biocomposite Hydrogels for the Treatment of Bacterial Infections: Physicochemical Characterization and In Vitro Assessment. *Pharmaceutics* **2021**, *13*, 2079. [CrossRef] [PubMed]
11. Gośsciniak, A.; Paczkowska-Walendowska, M.; Skotnicka, A.; Ruchała, M.A.; Cielecka-Piontek, J. Can Plant Materials Be Valuable in the Treatment of Periodontal Diseases? Practical Review. *Pharmaceutics* **2021**, *13*, 2185. [CrossRef] [PubMed]
12. Fahmy, S.A.; Mahdy, N.K.; Al Mulla, H.; ElMeshad, A.N.; Issa, M.Y.; Azzazy, H.M.E.-S. PLGA/PEG Nanoparticles Loaded with Cyclodextrin-Peganum harmala Alkaloid Complex and Ascorbic Acid with Promising Antimicrobial Activities. *Pharmaceutics* **2022**, *14*, 142. [CrossRef] [PubMed]
13. Koh, E.; Lee, Y.T. Preparation of Ligand Brush Nanocapsules for Robust Self-Controlled Antimicrobial Activity with Low Cytotoxicity at Target pH and Humidity. *Pharmaceutics* **2022**, *14*, 280. [CrossRef] [PubMed]

pharmaceutics MDPI

Article

In Vitro Drug Release, Permeability, and Structural Test of Ciprofloxacin-Loaded Nanofibers

Luca Éva Uhljar [1], Sheng Yuan Kan [2], Norbert Radacsi [2], Vasileios Koutsos [2], Piroska Szabó-Révész [1] and Rita Ambrus [1,*]

[1] Interdisciplinary Excellence Centre, Faculty of Pharmacy, Institute of Pharmaceutical Technology and Regulatory Affairs, University of Szeged, Eötvös Street 6, H-6720 Szeged, Hungary; uhljar.luca.eva@szte.hu (L.É.U.); ReveszPiroska@szte.hu (P.S.-R.)

[2] School of Engineering, Institute for Materials and Processes, The University of Edinburgh, King's Buildings, Edinburgh EH9 3FB, UK; skyshengyuan@hotmail.com (S.Y.K.); n.radacsi@ed.ac.uk (N.R.); vasileios.koutsos@ed.ac.uk (V.K.)

* Correspondence: ambrus.rita@szte.hu

Abstract: Nanofibers of the poorly water-soluble antibiotic ciprofloxacin (CIP) were fabricated in the form of an amorphous solid dispersion by using poly(vinyl pyrrolidone) as a polymer matrix, by the low-cost electrospinning method. The solubility of the nanofibers as well as their in vitro diffusion were remarkably higher than those of the CIP powder or the physical mixture of the two components. The fiber size and morphology were optimized, and it was found that the addition of the CIP to the electrospinning solution decreased the nanofiber diameter, leading to an increased specific surface area. Structural characterization confirmed the interactions between the drug and the polymer and the amorphous state of CIP inside the nanofibers. Since the solubility of CIP is pH-dependent, the in vitro solubility and dissolution studies were executed at different pH levels. The nanofiber sample with the finest morphology demonstrated a significant increase in solubility both in water and pH 7.4 buffer. Single medium and two-stage biorelevant dissolution studies were performed, and the release mechanism was described by mathematical models. Besides, in vitro diffusion from pH 6.8 to pH 7.4 notably increased when compared with the pure drug and physical mixture. Ciprofloxacin-loaded poly(vinyl pyrrolidone) (PVP) nanofibers can be considered as fast-dissolving formulations with improved physicochemical properties.

Keywords: amorphous solid dispersion; ciprofloxacin; electrospinning; nanofibers; poorly water-soluble drug; povidone

1. Introduction

Nowadays, one of the main challenges in pharmaceutical technology is dealing with poorly water-soluble drugs. The number of Biopharmaceutics Classification System (BCS) Class II and Class IV drugs is estimated to make up ~40% of currently marketed drugs and ~90% of compounds currently under development [1–3]. One of the BCS Class II/IV drugs is ciprofloxacin (CIP), which is a worldwide-used, broad-spectrum, second-generation fluoroquinolone antibiotic. Fluoroquinolones inhibit DNA replication in bacteria because their mechanism of action is the inhibition of bacterial DNA gyrase and topoisomerase IV. The lack of cross-resistance between fluoroquinolones and other classes of antibiotics makes this antibiotic family important. CIP and its derivatives can effectively treat antibiotic-resistant infections (e.g., nosocomial pneumonia), so they are usually reserved as "drugs of last resort" [4]. Since CIP is active against many Gram-negative and Gram-positive bacteria, it is directed to treat several bacterial infections. CIP is indicated for a broad spectrum of diseases, from bacterial conjunctivitis through bone and joint infections to complicated urinary tract infections. Accordingly, the available dosage forms on the market are ophthalmic and otic solutions, film-coated tablets, oral suspensions, and solutions for infusions. Thus, various routes for drug administration, both local and systemic, are applied. Despite common use, CIP has only 56–77% bioavailability [4]. The solubility of the drug is pH-dependent, as CIP is highly soluble below pH 5 and above pH 10 but almost insoluble around neutral pH level [5]. According to the BCS, CIP belongs to Class IV, with low water solubility (0.067 mg/mL at 25 °C, pH 7.5; [6]) and poor permeability [4].

In most cases, nanocarrier-based drug delivery systems can offer a solution to the low solubility problem of a drug. Nanostructures are innovative formulations with nanometer-scale in at least one dimension and a large surface-area-to-volume ratio. The latter ensures that a large amount of the drug can come into contact with the surrounding medium. In this way, the dissolution rate and even solubility can increase [7–9]. Furthermore, numerous drug delivery systems contain the active pharmaceutical ingredient (API) in an amorphous state and therefore cause increased solubility of the drug [10,11]. With the increase of solubility and dissolution rate, the bioavailability of the API increases [12–14]. Currently, intensive research is underway to formulate CIP into different nanocarriers to treat several diseases. Some examples are collected in Table 1.

Table 1. Ciprofloxacin-loaded nanocarriers under investigation with the aimed indication.

Type of Nanocarrier	Excipients Applied in Nanocarrier	Aimed Indications	Targeted Organs	References
Polymeric nanoparticles (NP) in nanofibers (NF)	PLGA and PCL—NP PEOT/PBT—NF	tissue engineering	middle ear	[15]
Nanoparticles and coated nanoparticles	PLGA and chitosan (coat)	root canal infection	tooth	[16]
Composite nanoparticles	synthetic nano-HA and sodium alginate	tissue engineering	bone	[17]
Microspheres	PLGA	osteomyelitis, orthopedic infections	bone	[18]
Microparticles	calcium carbonate, sodium hyaluronate	lung infections	lungs	[19]
Nanocrystals inside liposomes	HSPC, cholesterol	lung infections	lungs	[20]
Amorphous nanoparticle complex	dextran sulfate	non-cystic fibrosis bronchiectasis	lungs	[21]
Lipid-core nanocapsules	PCL, sorbitan monostearate, oleic acid, polysorbate 80	cystic fibrosis	lungs	[22]
Nanofibers	PVP	wound infections	skin	[23]
Nanofibers	PVA, chitosan, graphene oxide	wound infections	skin	[24]

Abbreviations: HA—hydroxyapatite; HSPC—hydrogenated soy phosphatidylcholine; PCL—poly(ε-caprolactone); PEOT/PBT—Poly(ethylene oxide terephthalate)/poly(butylene terephthalate) copolymer; PLGA—poly(D,L-lactide-*co*-glycolide); PVA—poly(vinyl alcohol); PVP—poly(vinyl pyrrolidone).

Polymeric nanofibers are considered solid dispersions with a large specific surface area that can stabilize drugs in their amorphous state. Thus, polymer-based nanofibers provide an attractive approach for the development of dosage forms due to the enhanced solubility and dissolution rate [25,26]. Nanofibers have a variety of uses currently under investigation in different fields of science. In pharmaceutical and medical fields, at least four applications have to be mentioned: wound dressings, filtration, drug delivery systems, and tissue engineering scaffolds [27–31]. Electrospinning is an effective and inexpensive method for the fabrication of polymeric nanofibers; hence, it is the most widely used technique for nanofiber production in the industry. During the electrospinning procedure, which is generated by a high voltage power supply, the ejection and travel of the polymer-drug fluid jet are induced by the large potential difference between a needle and a collector [32]. As the jet travels towards the collector, the solvent evaporates, and the jet solidifies into nano-sized fibers. It is possible to control the mean fiber diameter and morphology by changing the polymer-drug solution, the applied voltage, the needle–collector distance, the flow rate, the collector speed, or the environmental temperature and humidity [33,34].

Poly(vinyl pyrrolidone) (PVP) is a water-soluble and biocompatible polymer, which is commonly used as a pharmaceutical excipient and food additive [35]. PVP is also widely used as a carrier polymer that enables electrospinning [23,36–45].

Ciprofloxacin-loaded PVP nanofibers as potential wound dressings have been recently investigated [23,41,44]. However, to the best of our knowledge, the development of a potential per os formulation has not been published yet.

In this article, the preformulation studies of PVP-based nanofiber mats loaded with CIP for per os administration are investigated. The aim of the present study was to increase the water solubility and diffusion of the API and study the in vitro drug release and its kinetics. Additionally, the study aimed to investigate the effect of the polymer–drug solution and the flow rate on the fiber diameter and morphology. Special attention was paid to structural characterization, drug entrapment efficiency, solubility, in vitro dissolution, and in vitro diffusion of the CIP-loaded nanofibrous samples when producing immediate-release nanofibrous formulation, and therefore achieving effective antibiotic therapy.

2. Materials and Methods

2.1. Materials

Ciprofloxacin base (CIP; Mw = 331.35; purity >98%) was gifted by Teva Pharmaceutical Works Ltd. (Debrecen, Hungary). Polyvinylpyrrolidone (PVP) (Mw = 1,300,000) was obtained from Alfa Aesar (Lancashire, United Kingdom). Ethanol (99.99% purity) and chloroform (99.8% purity) were purchased from Fisher Scientific (Loughborough, United Kingdom).

Phosphate buffer solutions (PBS, pH 7.4 and 6.8) were prepared in-house as follows. To prepare 1 L of the pH 7.4 PBS, 1.44 g disodium phosphate dihydrate ($Na_2HPO_4 \times 2$ H_2O), 0.12 g potassium dihydrogen phosphate (KH_2PO_4), 8.00 g sodium chloride (NaCl), and 0.20 g potassium chloride (KCl) were dissolved in approximately 0.9 L of distilled water. Then the pH was adjusted to 7.4 using aqueous solutions of NaOH, bringing the buffer up to a volume of 1 L. For the preparation of 1 L of the pH 6.8 PBS, 6.8 g KH_2PO_4 was dissolved in 900 mL of distilled water before mixing 77.00 mL of aqueous solutions of NaOH (0.2 M). The pH was adjusted to 6.8 using the NaOH solution. All used chemicals were purchased from Sigma-Aldrich (Budapest, Hungary).

All other chemicals were analytical grade, and distilled water was used.

2.2. Methods

2.2.1. Preparation of the Solutions for Electrospinning

As the first step of the nanofiber preparation, the electrospinning solutions were prepared. PVP powder was dissolved in ethanol, whereas ciprofloxacin was dissolved in chloroform in a separate vessel. Constant stirring was applied to both at room temperature for 24 h to generate homogenous solutions using a magnetic stirrer at 700 rpm. The nominal concentration of the CIP solutions was 1 mg/mL, while the polymer concentrations varied between 5–20% w/v. The PVP and the drug solutions were mixed in 1:1, 1:2, 1:3 volume ratio, respectively, to yield the requested concentrations. The full details of the samples prepared are listed in Table 2.

Table 2. The composition of each sample and their preparation procedures.

Sample	PVP Solution (*w/v* %)	PVP:CIP (*v/v*)	Flow Rate (mL/h)
NF1	5	1:0	2
NF2	20	1:3	2
NF3	15	1:2	2
NF4	10	1:1	0.5
NF5	10	1:1	1
NF6	10	1:1	2
NF7	10	1:1	3
NF8	10	1:1	4

2.2.2. Electrospinning Procedure

The prepared solutions were filled in 2 mL syringes fitted with stainless steel 20 G needles. The electrospinning process was carried out using a commercially available electrospinning device (IME Medical Electrospinning, Waalre, The Netherlands; Figure 1). The applied potential difference was 24 kV, and the needle-collector distance was maintained at 10 cm. Different flow rates (0.5, 1, 2, 3, and 4 mL/h) were provided by a syringe pump (Table 2). All the experiments were performed at ambient conditions (23 °C temperature and 36–42% relative humidity). Additionally, a control sample containing no CIP was prepared using the same conditions as were used for the other samples (a constant 2 mL/h pump flow rate was used for the pure PVP sample).

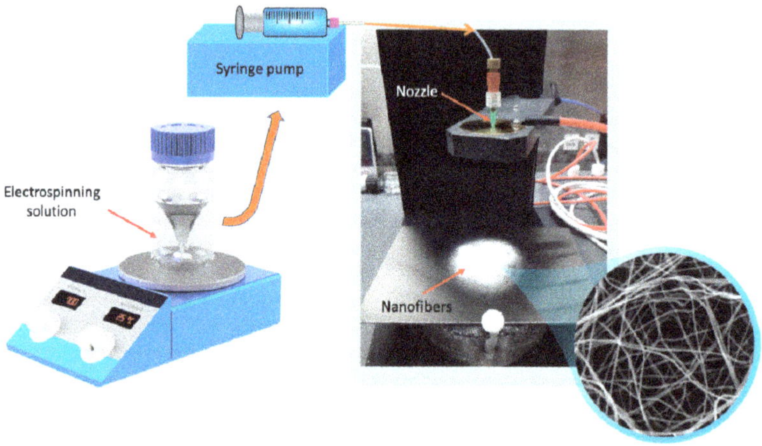

Figure 1. Preparation of ciprofloxacin-loaded nanofibers.

2.2.3. Preparation of the Physical Mixture

For the structural characterization and in vitro studies, physical mixtures were used as reference samples, prepared by a shaker mixer (Turbula System Schatz; Willy A. Bachofen AG Maschinenfabrik, Basel, Switzerland). PVP and CIP were homogenized under controlled conditions (50 rpm, 10 min) in the same weight ratio as most nanofibrous samples.

2.2.4. Size and Morphological Measurements

The morphology of the electrospun nanofibers was observed by scanning electron microscopy (SEM; Hitachi S4700, Hitachi Scientific Ltd., Tokyo, Japan) at 10 kV. Specimens of the fiber mats (0.4 × 0.3 cm) were cut and coated with approximately 10 nm thin film of gold–palladium using a sputter coater (Bio-Rad SC 502, VG Microtech, Uckfield, UK) before the SEM imaging was performed. One hundred fibers from each formulation were randomly selected to carry out the fiber diameter measurement (ImageJ 1.44p software; Bethesda, MD, USA).

2.2.5. Structural Characterization

CIP, physical mixture, and CIP-loaded nanofibers were characterized by X-ray powder diffraction (XRPD; D8 Advance, Bruker AXS GmbH, Karlsruhe, Germany). The samples were measured with Cu K λ_I radiation (40kV/40mA, λ = 1.5406 Å) in between diffraction angle 3–40° for their structural properties.

Differential scanning calorimeter (DSC; Mettler Toledo DSC821e; Mettler Inc., Schwerzenbach, Switzerland) was applied to evaluate the thermal behavior of the samples using 25–300 °C temperature range at a heating rate of 5 °C min^{-1}.

Fourier transform infrared spectroscopy (FTIR; Thermo Nicolet AVATAR 330, Madison, WI, USA) was performed after discs of a KBr and nanofibers had been made by a compression molding technique. The discs were scanned 128 times over the range 4000–400 cm^{-1} and with a resolution of 4 cm^{-1}.

2.2.6. Drug Loading and Drug Entrapment Efficiency

The drug loading and the amount of CIP entrapped in the optimized nanofibrous mat was quantified by UV spectrophotometry (ABL&E-Jasco UV/VIS Spectrophotometer V-730, Budapest, Hungary). A known mass of NF6 sample (about 160 mg) was dissolved in 50 mL of 0.1 M hydrochloric acid solution. The amount of CIP in the solution was calculated by UV spectroscopy analysis at a λ_{max} of 277 nm.

The amount of the loaded CIP of the NF6 formulation was calculated by comparing the released mass of CIP (w_{UV}) with the mass of the dissolved nanofiber mat (w_{UV}). The drug loading (DL%) was calculated by the following Equation (1).

$$DL\ (\%) = (w_{UV}/w_{MAT}) \times 100. \quad (1)$$

The drug entrapment efficiency of the nanofiber (EE) was calculated by the following Equation (2).

$$EE\ (\%) = (w_{UV}/w_{ES}) \times 100, \quad (2)$$

where w_{UV} is the calculated mass of CIP released from the nanofibers, and w_{ES} is the mass of CIP dissolved before the electrospinning procedure.

Each experiment was performed in four parallel measurements, and the average values and standard deviations are reported.

2.2.7. Solubility Tests

Solubility tests were carried out both in distilled water and PBS (pH 7.4) as CIP is a poorly water-soluble drug and has a U-shaped pH-dependent solubility with a minimum around pH 7.4. The first step was to place the pure drug, the physical mixture, and the drug-loaded nanofibrous samples in 3 mL of distilled water (pH 6.3) or PBS (pH 7.4), which were then stirred for 24 h at room temperature with a magnetic bar. The samples were then filtered (0.22 mm, FilterBio PES Syringe Filter; Labex Ltd., Budapest, Hungary), suitably diluted, and measured by UV spectrophotometry (ABL&E-Jasco UV/VIS Spectrophotometer V-730, Budapest, Hungary) at 275 nm and 271 nm in the case of water and PBS, respectively. The solubility tests of each sample were performed in triplicates, and the average values with the standard deviations are reported.

Additionally, the solubility of the CIP was measured with the same methodology in Fasted State Simulating Gastric Fluid (FaSSGF; Biorelevant.com accessed on 6 April 2021, London, UK) and Fasted State Simulating Intestinal Fluid (FaSSIF; Biorelevant.com accessed on 6 April 2021, London, UK) to get data for the in vitro two-stage drug release study. For the UV spectroscopy, 277 nm and 272 nm in the case of the FaSSGF and FaSSIF media were set as the wavelength of light, respectively. The results of the measurements are reported in Section 3.5.

2.2.8. In Vitro Drug Release Studies

Two types of in vitro drug release studies were executed to get a complete picture of the CIP release. Firstly, single medium dissolution (pH 7.4 PBS) with a modified paddle method (Hanson Research SR8-Plus Dissolution Test Station, Chatsworth, CA, USA) was used to measure the drug release from NF6 nanofibers (containing 25 mg CIP) compared with its corresponding physical mixture, and also with 25 mg raw CIP powder. The release studies were carried out in 100 mL of the PBS medium at 37 °C. The paddle was rotated at 100 rpm. Then, 5 mL samples were taken manually from the buffer solution after 5, 10, 15, 30, 60, and 90 min. After sampling, the volume was replaced with fresh PBS. The amount of drug present in the aliquots was determined with UV-Vis spectrophotometry (ABL&E-Jasco UV/VIS Spectrophotometer V-730, Budapest, Hungary) at a λ_{max} of 275 nm. The cumulative CIP release (~25 mg = 100%) was calculated using the calibration curve of CIP in pH 7.4 PBS. Each experiment was performed in triplicate, and the average values and standard deviations are reported.

Besides, to mimic the in vivo conditions better, a two-stage release test as a biorelevant gastrointestinal transfer protocol was also studied by the Hanson dissolution tester mentioned above. A portion of 25 mg CIP powder was compared with a 100-fold larger mass of NF6 electrospun sample as the nanofibers were containing 1 w/w % CIP. The samples were first added to 25 mL of FaSSGF, and 2 mL aliquots were taken at the time points: 1, 3, 5, 10, 15, and 30 min. Right after the last sampling, 25 mL of FaSSIF Concentrate, as a "bolus", was quickly added to the gastric medium. The FaSSIF Concentrate's surfactant concentration and buffer strength are doubled compared with those of regular FaSSIF. In this way, after the addition of 25 mL of FaSSIF Concentrate to 25 mL of FaSSGF, the final dissolution medium will be the regular FaSSIF. The methodology and the preparation of the FaSSIF Concentrate are reported by J. Mann et al. [46]. Next, 2 mL aliquots were taken 1, 3, 5, 10, 15, 30, and 60 min after the medium change. Each taken volume was replaced with fresh FaSSIF media. The dissolution vessels were thermostated at 37 °C, and the paddle speed was set to 100 rpm during the experiment. Similar to the first release study, the cumulative drug release (~25 mg = 100%) was calculated from the absorbance values measured by UV-Vis spectrophotometry at a λ_{max} of 277 nm and 272 nm in the case of the FaSSGF and FaSSIF media, respectively. Each experiment was performed in triplicate, and the average values and standard deviations are reported.

2.2.9. Study of Drug-Release Kinetics and Mechanism

The release kinetics of CIP from the electrospun nanofiber and physical mixture was compared with the dissolution kinetics of the CIP powder. Five different mathematical models (zero order, first order, Hixson–Crowell, Higuchi, and Korsmeyer–Peppas model) were fitted with the obtained cumulative drug release vs. time curves to describe the kinetics. To evaluate which model was followed, the value of the regression coefficient (R^2) was determined and compared.

2.2.10. In Vitro Diffusion Study

For the estimation of the passive diffusion of CIP through biological membranes, the Side-Bi-Side™ (Crown Glass, Somerville, NJ, USA) diffusion test was carried out. The cellulose ester membrane (pore diameter = 0.45 μm) was dipped in isopropyl myristate before use. The donor phase was pH 6.8 PBS, and the acceptor phase was pH 7.4 PBS; both were kept at 37 °C temperature. The diffused drug content was measured in real-time at 272 nm by AvaLight DH-S-BAL spectrophotometer (AVANTES, Apeldoorn, The Netherlands) connected to an AvaSpec-2048L transmission immersion probe (AVANTES, Apeldoorn, The Netherlands). The optical path length was 1 cm. The diffusion tests of each sample were performed in triplicates, and the average values with the standard deviations are reported.

The flux, J, was calculated from the linear part of the graph using the following Equation (3):

$$J = \partial m/(A\ \partial t), \tag{3}$$

where m is the cumulative amount of API transported in t time, and A is the surface area of the membrane (0.875 cm^2). The permeability coefficient (K_p) was determined by normalizing the flux to the donor concentration (C_d), according to the Equation (4):

$$K_p = J/C_d, \tag{4}$$

2.2.11. Statistical Analysis

The significance levels of the differences between the measured fiber diameters of the samples produced under different parameters were examined by one-way analysis of variance (ANOVA) with post hoc Tukey HSD test. All the experimental results of solubility and single medium dissolution tests are expressed as mean ± standard deviation and statistically compared by a two-sample t-test. The experimental results with p values <0.05, <0.01, and <0.001 were assumed to be statistically significant.

3. Results and Discussion

3.1. Optimization of the Electrospinning Parameters

Reviewing the works focusing on the electrospinning of drug-loaded PVP nanofibers, the solution flow rate was found to range between 0.2 and 2 mL/h [23,36–45]. The average fiber diameter of the different formulations was largely varying. One of the aims of this research was to investigate the effect of flow rate on the diameter of PVP nanofibers, keeping the other preparation parameters constant. Additionally, nanofibers with different PVP:CIP volume ratios were prepared and studied. SEM was used to visualize the nanofibers, and then their morphology was observed, and the average fiber diameter was measured. The narrowest nanofiber formulation with the finest morphology was targeted during the optimization.

The morphology of drug-free and various drug-loaded nanofibers are represented in Table 3 and Figure 2. Continuous, smooth-surfaced nanofibers were successfully prepared from all the solutions studied except for sample NF3. PVP:CIP 1:2 volume ratio eventuated discontinuous, worm-like nanofibers with largely varying diameters (889 ± 265 nm). The average diameter of the pure PVP fibers was 815 ± 216 nm. Considering the continuous fibers, the addition of the CIP decreased the fiber diameter. The decrease was significant except for sample NF2 (Table 4). This is expected to be related to the increase in the conductivity of the electrospinning solutions by adding the CIP [47]. Higher solution conductivity could facilitate the elongation of the jet and generate thinner fibers [48]. Additionally, the PVP nanofibers had some narrowing that slightly resembled the formation of beads, as shown in Figure 2. This was probably due to the low PVP concentration [49]. As the flow rate was increased, the average fiber diameter increased, which was expected from the literature [49]. However, the difference was not significant in every case (Table 4). Besides, the 3 mL/h flow rate (NF7) caused incomplete solvent evaporation and merged fibers, while the sample produced by 4 mL/h flow rate (NF8) showed some bead-like structures on top of the merged fibers (Figure 2). According to the results of this study, two formulations—namely, NF5 and NF6—had the most optimal morphology and diameter distribution. Thus, the 1:1 ratio between the PVP and CIP was found optimal, similar to our previous study [50]. The higher flow rate is desirable for the preparation of samples because of the larger number of nanofibers produced during the same period of time. Since the applied flow rate of NF6 was 2 times higher than the flow rate of NF5, finally, NF6 was selected for further studies.

Table 3. Fiber morphology and average diameter for different formulations.

Sample	Fiber Morphology		Average Diameter (nm)
NF1	continuous	smooth	815 ± 216
NF2	continuous	smooth	787 ± 140
NF3	discontinuous	worm-like	889 ± 265
NF4	continuous	smooth	542 ± 103
NF5	continuous	smooth	601 ± 87
NF6	continuous	smooth	645 ± 82
NF7	continuous	smooth	663 ± 79
NF8	continuous	rough	718 ± 105

Table 4. Q statistic values of the different nanofibrous formulations as pairs of treatments in one-way ANOVA with post hoc Tukey HSD test. The significantly different (** $p < 0.01$; * $p < 0.05$) pairs of treatments are marked.

Sample	NF1	NF2	NF3	NF4	NF5	NF6	NF7	NF8
NF1	-	1.77	4.54 *	15.47 **	12.54 **	9.91 **	7.88 **	4.31 *
NF2	1.77	-	6.45 **	14.24 **	11.21 **	8.51 **	6.58 **	3.12
NF3	4.54 *	6.45 **	-	19.67 **	16.88 **	14.21 **	11.69 **	7.58 **
NF4	15.47 **	14.24 **	19.67 **	-	3.19	5.53 **	5.85 **	7.46 **
NF5	12.54 **	11.21 **	16.88 **	3.19	-	2.43	3.07	5.05 **
NF6	9.91 **	8.51 **	14.21 **	5.53 **	2.43	-	0.88	3.15
NF7	7.88 **	6.58 **	11.69 **	5.85 **	3.07	0.88	-	2.23
NF8	4.31 *	3.12	7.58 **	7.46 **	5.05 **	3.15	2.23	-

3.2. Structural Characterization

The change in the crystallinity of CIP was studied by XRPD and DSC (Figure 3A,B). The XRPD of the CIP powder showed high crystallinity, represented by several smaller and three longer sharp peaks at around 2-Theta = 14.5, 20.9, and 25.4°. The diffractogram of the physical mixture also had the same peaks along with a broad peak between 2-Theta = 7–15°. The broad peak that appeared in the diffractogram of PVP, physical mixture, and the half of nanofibrous samples prove the polymer's amorphous nature. The three characteristic peaks of the CIP were missing from the spectra of the nanofibrous samples, which indicates physicochemical interactions between the drug and the polymer matrix and the amorphous form of CIP created by the fast evaporation of the solvent during electrospinning.

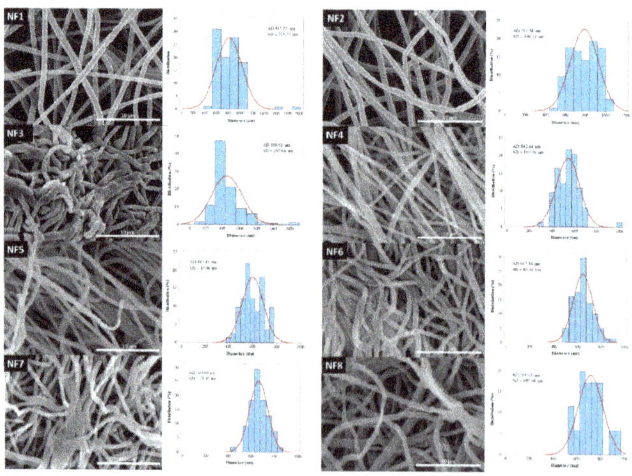

Figure 2. SEM images and diameter distributions of nanofibrous samples (5000× magnification).

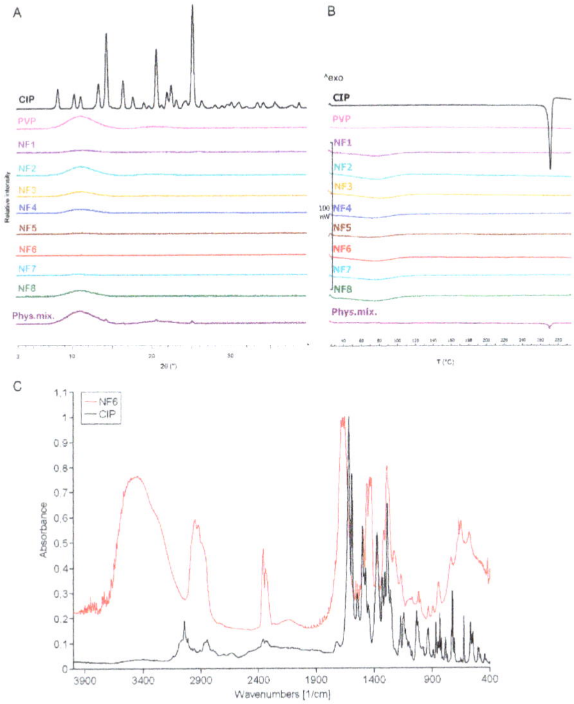

Figure 3. XRPD diffractogram (**A**) and DSC thermograms (**B**) of ciprofloxacin (CIP), PVP, physical mixture, and NF1–NF8 nanofibrous samples. All the electrospun samples are amorphous solid dispersions. FTIR spectra (**C**) of CIP and NF6 nanofibrous sample. The observed shifts and widenings confirm the successful incorporation of the CIP into the polymeric fibers.

Figure 3B presents the DSC thermograms of CIP, PVP, physical mixture, and all the samples listed in Table 2. The melting point of the pure drug is well-defined by a large endothermic peak at 275 °C, also appearing on the thermogram of the physical mixture. However, the flat PVP and NF1-NF8 thermograms suggest amorphous components. It can be concluded from both results that the drug was incorporated in amorphous form into the nanofibers.

Main drug–polymer interactions were observed by FTIR, and the spectra are given in Figure 3C. One prominent characteristic CIP peak at 1732 cm^{-1} (νC=O) shifted toward the higher wavenumbers and stretched due to the interaction with the hydroxyl groups of the PVP [51]. The wide band between 3050–3750 cm^{-1} (νC-OH) was enlarged by the hydrogen bonding between the CIP and PVP in the nanofibers. Moreover, the bands between 3050–2750 cm^{-1} (alkene and aromatic νC-H), between 2380–2275 cm^{-1} (νC-N), at 1500 cm^{-1} (quinolone), and at 1290 cm^{-1} (νC-O) appeared wider in case of the nanofiber [52]. These shifts and widenings confirm the successful incorporation of the CIP into the polymeric fibers.

3.3. *Drug Loading and Drug Entrapment Efficiency*

Nanofibers, in general, exhibit very high entrapment efficiency because the electrospinning procedure can be considered as an in situ solidification of a polymer solution. Since both CIP and PVP are non-volatile in nature, high entrapment efficiency was expected under the consideration of complete miscibility.

Entrapment efficiency, as well as drug loading, was calculated by UV spectrophotometry. Solvent with low pH (0.1 M HCl) was used to earn complete dissolution. The NF6 nanofibrous sample met our expectations with 92 ± 8% entrapment efficiency. Furthermore, the theoretical drug loading of the NF6 formulation was 0.99%, while the calculated drug loading was 0.92 ± 0.08%. These data are consistent with the high entrapment efficiency.

3.4. Solubility Test

The optimization study revealed that the sample NF6 had the most promising size and morphology for the solubility and dissolution studies. Solubility tests of the raw CIP powder, the physical mixture, and the drug-loaded nanofibrous samples were carried out in distilled water and PBS, as shown in Table 5. CIP has a pH-dependent, U-shaped solubility profile showing high solubility at pH < 5 and pH > 10 and poor solubility around neutral pH level as its isoelectric point is 7.42 [5]. Thus, distilled water and pH 7.4 PBS were chosen as solvents to observe the effect of the polymer nanocarrier on the solubility. The pH was measured at the beginning and the end of the tests. The pH values of the solvents before adding the CIP were 6.3 and 7.4 in the case of the distilled water and PBS, respectively. During the 24 h solubility study, the pH of the CIP solution in water increased to 7.1 because CIP as a weak base slightly alkalized the solution. With the increase of the pH level, the solubility of the drug decreases, which reduces the dissolution of the remaining solid CIP in the system. The buffer capacity of the PBS was sufficient to prevent a notable pH shift.

Table 5. Solubility data of the samples in distilled water and phosphate buffer. The dissolution of the ciprofloxacin (CIP) shifted the pH from 6.3 to 7.1 in distilled water. The nanofiber NF6 has significantly higher (** $p < 0.01$; * $p < 0.05$) solubility than the physical mixture and the raw CIP powder. Statistical analysis: Two-sample *t*-test.

Sample	Solubility [mg/mL] in Water (pH 6.3 → pH 7.1)		Solubility [mg/mL] in PBS (pH 7.4)	
CIP	0.071 ± 0.001	** (vs. NF6)	0.099 ± 0.001	* (vs. NF6)
Physical mixture	0.182 ± 0.035	** (vs. NF6)	0.123 ± 0.001	* (vs. NF6)
Nanofiber (NF6)	0.862 ± 0.074		0.629 ± 0.186	

Studies with both solvents suggested that the incorporation of CIP into nanofibers caused a significant increase ($p < 0.01$ in the case of water and $p < 0.05$ in the case of PBS) in the drug solubility. In distilled water, the final CIP concentration showed a 12-fold increase in the case of nanofibers compared with the raw CIP, while in the PBS, the nanofibers were approx. 6.4 times more soluble. Furthermore, the solubility of the drug in the physical mixture was not significantly higher in either solvent. Thus, it can be seen that only the presence of PVP could enhance solubility by increasing wettability. However, this enhancement was much lower than in the case of nanofibers, where solid molecular dispersion of CIP formed during the electrospinning procedure. Solid molecular dispersions can guarantee increased solubility by decreasing the particle size and improving wettability. This finding may be correlated with the characterization results because inside the nanofiber the CIP is in its amorphous form.

3.5. In Vitro Drug Release

The in vitro release profiles of electrospun nanofibers and raw CIP powder were investigated both in single medium dissolution and two-stage biorelevant release tests. On the one hand, pH 7.4 PBS was chosen as a medium, while in this pH level, CIP has a minimum solubility value, and the pH of the terminal ileum [53], as well as the blood, is around 7.4. The pH-dependent solubility of a drug can cause incomplete dissolution or precipitation, which leads to suboptimal bioavailability. This might generate a problem with CIP because it occurs mostly at high administered doses [4]. On the other hand, any pH shift that occurs during the transfer from the stomach to the small intestine may affect the solubility and the bioavailability of CIP [54]. To investigate this effect, conditions similar to in vivo (biosimilar media, change in pH after 30 min, body temperature) were used by a two-stage release study [46].

Figure 4 presents the cumulative drug release vs. time curves of CIP powder, physical mixture, and electrospun sample (NF6) measured in PBS (pH 7.4). In the case of the CIP powder, the dissolution was not complete within 90 min, most probably due to the poor solubility of the drug at this pH. Until 90 min, only 41 ± 3% of the drug was liberated. Similarly, the drug release from the physical mixture was 67 ± 12% at the end of the measurement (90 min). However, NF6 showed a significantly higher dissolution rate than CIP powder ($p < 0.001$) at every measured point. Additionally, the nanofibrous sample demonstrated significantly higher drug release than the physical mixture ($p < 0.05$) in time points at 5 and 10 min. Besides, while all the samples showed fast dissolution behavior, the release was the fastest from the nanofibers. The improved dissolution rate resulted in 94 ± 6% dissolved CIP within only 5 min. The distinct difference between the raw CIP and the nanofibers could be caused by the high surface-to-volume ratio, the high wettability, and the amorphous drug inside the nanofibers.

In addition, a two-stage biorelevant release study was executed to mimic the in vivo conditions since the solubility of CIP is sensitive to changes in pH of the gastrointestinal environment. The raw CIP powder and the NF6 nanofibrous sample were placed into 25 mL of FaSSGF medium, and the drug release was studied for 30 min. From the raw CIP, the release was fast and complete as the whole amount of the added powder was dissolved within 10 min (Figure 5). This was expected since Hansmann et al. published a similar result with Ciprobay® 500 mg IR tablet [54], and the 24 h solubility test in FaSSGF (described in Section 2.2.7) showed 7.794 ± 0.675 mg/mL average CIP concentration. However, in the case of the electrospun sample, the drug release was slower. It barely reached 90% until the last sampling of the FaSSGF medium. The difference could have been caused by the strong gelation of the high molecular weight PVP (Mw = 1,300,000) in low pH levels forming a viscous matrix around the CIP, which could remarkably slow down its diffusion to the dissolution medium.

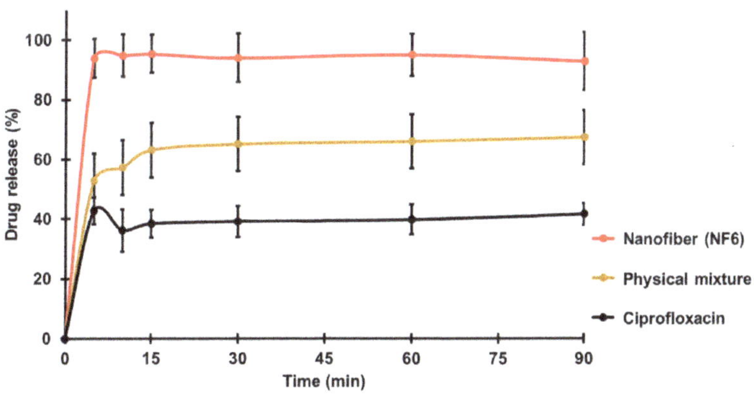

Figure 4. In vitro dissolution of ciprofloxacin (CIP) from NF6 nanofiber, physical mixture, and CIP powder in pH 7.4 phosphate buffer solution. All measured NF6 values were significantly higher than the raw CIP values ($p < 0.001$), and the first two were significantly higher than the physical mixture values ($p < 0.05$). Statistical analysis: Two-sample t-test.

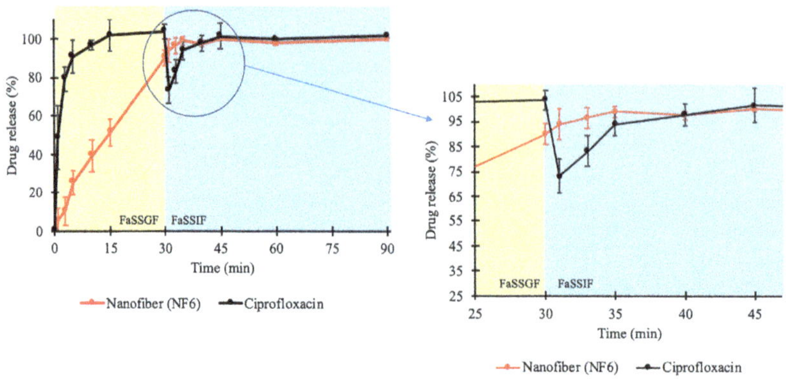

Figure 5. In vitro dissolution of ciprofloxacin (CIP) from NF6 nanofiber and CIP powder during the two-stage biorelevant release study. The dissolution medium (FaSSGF) simulated gastric condition until 30 min when the medium was turned into FaSSIF biorelevant fluid. The time range around the change of the medium (25–45 min) is magnified for better visualization.

As the second phase of the two-stage study, at 30 min, FaSSIF Concentrate was added to the FaSSGF dissolution medium to create a 1:1 volume ratio mixture (FaSSIF medium). With this step, the emptying of the stomach into the small intestine is modeled easily [54,55]. Due to the change of pH, the solubility of CIP decreased. As a result of the 24 h solubility test, 0.205 ± 0.002 mg/mL drug concentration was detected in FaSSIF (described in Section 2.2.7). In the case of the raw CIP samples, this 38-fold decrease caused the precipitation of the pre-dissolved CIP, which was visible in the vessels. Additionally, a slope can be observed at the 30 min time point in Figure 5. Interestingly, the curve of the nanofibrous sample has no slope, while the dissolution of the drug continued even after the change of the medium. Controlled release without precipitation was observed in the case of the electrospun sample. From the time point at 40 min, the two curves run together at around 100%. The time range between 25 and 45 min is magnified in Figure 5.

3.6. Drug-Release Kinetics and Mechanism

To describe the release kinetics from nanofibers, five different mathematical models are usually used—namely, zero order, first order, Hixson–Crowell model, Higuchi model, and Korsmeyer–Peppas model. The regression coefficient (R^2) values of the different drug release models are listed in Table 6.

Table 6. Regression coefficient values of the different drug release models.

Release Study	Single Medium (pH 7.4 PBS)			Two-Stage Biorelevant (FaSSGF)		Two-Stage Biorelevant (FaSSIF)	
Sample	CIP	Physical Mixture	Nanofiber (NF6)	CIP	Nanofiber (NF6)	CIP	Nanofiber (NF6)
Zero order	0.3270	0.2955	0.6122	0.5820	0.9810	0.2687	0.2344
Fist order	0.8463	0.6565	0.9807	0.9085	0.9602	0.9276	0.9268
Hixson–Crowell	0.3449	0.3540	0.6431	0.9204	0.9873	0.6452	0.5282
Higuchi	0.8908	0.8632	0.9577	0.9502	0.9616	0.8273	0.8304
Korsmeyer–Peppas	0.8967	0.8684	0.9993	0.8895	0.9794	0.8717	0.9229

In the case of the single medium (pH 7.4 PBS) dissolution study, the Korsmeyer–Peppas model showed superiority over the other models studied for describing the release kinetic of the raw CIP, the physical mixture, and the NF6 electrospun sample. However, the Korsmeyer–Peppas model could not be fitted very well to the curves of CIP powder

(R^2 = 0.8967) and physical mixture (R^2 = 0.8684), while the drug release of nanofibers was almost perfectly described by the model (R^2 = 0.9993). This is reasonable since the Korsmeyer–Peppas model describes drug release from a polymeric system. It takes into account several mechanisms simultaneously, such as the diffusion of water into the polymer matrix, the swelling, and the dissolution of the polymer [52,56].

Furthermore, the Korsmeyer–Peppas model showed a high R^2 value both in FaSSGF (R^2 = 0.9794) and FaSSIF (R^2 = 0.9229), considering the release kinetics of the two-stage biorelevant dissolution study. However, in the case of the latter, the first order kinetics could be fitted to the curve even more precisely (R^2 = 0.9268). First order was found to be the release kinetics of the raw CIP in FaSSIF medium as well, which means that the dissolution rate of CIP from the PVP matrix was dependent on the drug concentration. This can be explained by the change of pH caused by the change of the medium at 30 min. The solution was supersaturated, and the higher pH caused the decrease of the solubility of the drug. Near to its solubility limit, the concentration of the CIP could affect the release kinetics.

In terms of the other medium, the release of the CIP powder followed the Higuchi model (R^2 = 0.9502) while the highest R^2 values of the nanofibrous sample were related to zero order (R^2 = 0.9810) and the Hixson–Crowell model (R^2 = 0.9873). The difference in the course of the curves can also be seen in Figure 5. The Higuchi model describes drug release from different matrix systems that contain water-soluble drugs [52]. The FaSSGF is a good solvent of the CIP and without any polymer in the system, it could dissolve freely. On the other hand, the PVP formed a viscous hemisphere gel at the bottom of the vessel, causing the Hixson–Crowell model. This release model considers the dissolution of a tablet or polymer matrix but with the maintenance of their geometrical characteristics.

3.7. In Vitro Diffusion Study

An in vitro diffusion study was executed to compare the capacity of CIP from different samples (NF6 nanofiber, physical mixture, CIP powder) for crossing biological barriers, e.g., the small intestine cells (Figure 6). In the literature, it is suggested that the CIP is absorbed from the duodenum and the proximal jejunum [57]. Thus, during the in vitro diffusion study, pH 6.8 PBS was used for the donor phase. The acceptor phase (pH 7.4 PBS) modeled the intracellular pH ("set-point" pHi 7.35) of intestinal absorptive cells [58]. Since the absorption of ciprofloxacin seems to be mainly mediated not by active but by passive diffusion [54], a synthetic membrane was used to separate the two phases.

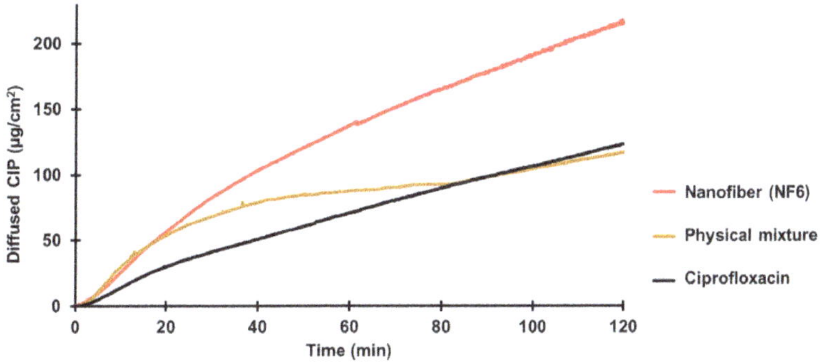

Figure 6. In vitro diffusion of ciprofloxacin (CIP) from the NF6 nanofibrous sample, physical mixture, and CIP powder (SD ± 2%).

According to the results, the diffusion of the CIP was remarkably higher from the nanofibrous sample (215 µg/cm^2) than the raw CIP (122 µg/cm^2) or the physical mixture (118 µg/cm^2). Figure 6 shows the diffusion profiles of CIP from the samples. The diffused CIP over time curve of the physical mixture interestingly runs with the curve of the nanofibrous sample for the first 20 min, but after that it becomes flatter as the CIP diffusion slows down. Finally, at the time range between 80 and 120 min, it overlaps with the curve of the pure CIP powder. This fascinating behavior could be caused by the presence of PVP. As written previously, PVP is a wetting agent that could increase the solubility of a drug. More solute CIP means a higher concentration gradient, which results in faster diffusion. However, the effect of this increase has a limit, and beyond that, the solution and diffusion of CIP slow down.

The calculated flux (J) and permeability coefficient (K_p) values are shown in Table 7. The diffused CIP amount was 1.7 times higher from the NF6 sample than from the nonfibrous samples. The presence of the PVP did not affect the diffusion since the results of the physical mixture were similar to the CIP powders. Moreover, the permeability coefficient was increased 1.9 times by the nanofiber formulation.

Table 7. Calculated flux (J) and permeability coefficient (K_p) values.

Sample	J (μg/(cm^2 h))	K_p (cm/h)
CIP	60.94	0.112
Physical mixture	58.80	0.105
Nanofiber (NF6)	107.71	0.217

4. Conclusions

Since CIP is a BSC Class IV drug, various pharmaceutical technological approaches are desirable for improving its bioavailability through the improvement in solubility, dissolution, and permeability. One such approach is the formulation of nanofibers via electrospinning, which is a simple and cost-effective production technique. Electrospun nanofibers, besides solid nanoparticles, are considered as amorphous solid dispersions and promising nanocarriers. Nanofibers are favorable drug carriers because of their high specific surface area, the wide variety of polymers and APIs that are spinnable, the ease of material combination, and the capability for mass production.

In the present study, CIP-loaded nanofibers were successfully fabricated by electrospinning. With the optimized process parameters, the nanofibers had a small, uniform fiber diameter with smooth surface morphology. The flow rates used in PVP-based nanofiber fabrication were compared. It was found that a higher flow rate produced thicker fibers, but in the case of too high a flow rate, the fibers were merged because the solvent evaporation was incomplete. According to the results of the XRPD and the DSC measurements, the CIP lost its crystallinity during the electrospinning procedure, and an amorphous form was produced. This form of the drug, along with the increased surface area, is the reason for the significantly higher solubility and in vitro dissolution rate in pH 7.4 PBS achieved with the nanofibrous samples. In the single medium release study (pH 7.4 PBS), the nanofibrous formulation demonstrated fast dissolution, and the release kinetics followed the Korsmeyer–Peppas model. In contrast, the raw CIP showed incomplete dissolution due to its poor solubility at this pH level. To mimic more precisely the in vivo conditions, a two-stage biorelevant release study was executed. Since the CIP is more soluble at low pH levels, a supersaturated solution was formed with FaSSGF medium. Then, with the change of the medium to FaSSIF, the solubility of the drug changed, and precipitation occurred. The precipitation could be prevented by nanofibers, since the PVP formed a viscous matrix around the CIP and released it with the dissolution kinetics described the best by Hixson–Crowell model and zero order kinetics. This study predicts a controlled release in the stomach for other API-loaded PVP-based nanofibers that might be useful in the development of local drug delivery systems. Moreover, incorporation of CIP into nanofibers could provide noticeably higher in vitro diffusion through the membrane. Therefore, our results show that CIP-loaded PVP-nanofibers could be considered as fast-dissolving formulations with improved physicochemical properties that may be suitable for further studies to keep in mind when developing an oral dosage form.

Author Contributions: Conceptualization, L.É.U., P.S.-R., and R.A.; methodology, L.É.U., S.Y.K., N.R., V.K., P.S.-R., and R.A.; validation, V.K. and P.S.-R.; formal analysis and investigation, L.É.U. and R.A.; writing—original draft preparation, L.É.U.; writing—review and editing, S.Y.K., N.R., V.K., P.S.-R., and R.A.; visualization, L.É.U. and N.R.; supervision, N.R., V.K., P.S.-R., and R.A.; funding acquisition, R.A. All authors have read and agreed to the published version of the manuscript.

Funding: The publication was funded by The University of Szeged Open Access Fund (FundRef, Grant No. 5249).

Institutional Review Board Statement: Not applicable.

Informed Consent Statement: Not applicable.

Data Availability Statement: The datasets used and/or analyzed are available from the corresponding author on reasonable request.

Acknowledgments: This project was supported by the Gedeon Richter's Talentum Foundation, Gedeon Richter Plc. In addition, the Ministry of Human Capacities, Hungary grant TKP 2020, and EFOP 3.6.3-VEKOP-16-2017-00009 are acknowledged. The authors would like to thank Areen Alshweiat of University of Szeged for her help in the sample preparation and Gábor Katona of University of Szeged and Gregory Myles of The University of Edinburgh for reviewing the paper.

Conflicts of Interest: The authors declare no conflict of interest.

References

1. Di, L.; Kerns, E.; Carter, G. Drug-Like Property Concepts in Pharmaceutical Design. *Curr. Pharm. Des.* **2009**, *15*, 2184–2194. [CrossRef]
2. Williams, H.D.; Trevaskis, N.L.; Charman, S.A.; Shanker, R.M.; Charman, W.N.; Pouton, C.W.; Porter, C.J.H. Strategies to Address Low Drug Solubility in Discovery and Development. *Pharmacol. Rev.* **2013**, *65*, 315–499. [CrossRef] [PubMed]
3. Loftsson, T.; Brewster, M.E. Pharmaceutical Applications of Cyclodextrins: Basic Science and Product Development: Pharmaceutical Applications of Cyclodextrins. *J. Pharm. Pharmacol.* **2010**, *62*, 1607–1621. [CrossRef] [PubMed]
4. Olivera, M.E.; Manzo, R.H.; Junginger, H.E.; Midha, K.K.; Shah, V.P.; Stavchansky, S.; Dressman, J.B.; Barends, D.M. Biowaiver Monographs for Immediate Release Solid Oral Dosage Forms: Ciprofloxacin Hydrochloride. *J. Pharm. Sci.* **2011**, *100*, 22–33. [CrossRef]
5. Ross, D.; Riley, C. Aqueous Solubilities of Some Variously Substituted Quinolone Antimicrobials. *Int. J. Pharm.* **1990**, *63*, 237–250. [CrossRef]
6. Caço, A.I.; Varanda, F.; Pratas de Melo, M.J.; Dias, A.M.A.; Dohrn, R.; Marrucho, I.M. Solubility of Antibiotics in Different Solvents. Part II. Non-Hydrochloride Forms of Tetracycline and Ciprofloxacin. *Ind. Eng. Chem. Res.* **2008**, *47*, 8083–8089. [CrossRef]
7. Radacsi, N.; Ambrus, R.; Szabó-Révész, P.; van der Heijden, A.; ter Horst, J.H. Atmospheric Pressure Cold Plasma Synthesis of Submicrometer-Sized Pharmaceuticals with Improved Physicochemical Properties. *Cryst. Growth Des.* **2012**, *12*, 5090–5095. [CrossRef]
8. Kariminia, S.; Shamsipur, A.; Shamsipur, M. Analytical Characteristics and Application of Novel Chitosan Coated Magnetic Nanoparticles as an Efficient Drug Delivery System for Ciprofloxacin. Enhanced Drug Release Kinetics by Low-Frequency Ultrasounds. *J. Pharm. Biomed. Anal.* **2016**, *129*, 450–457. [CrossRef]
9. Zattoni, A. Flow Field-Flow Fractionation for the Analysis of Nanoparticles Used in Drug Delivery. *J. Pharm. Biomed. Anal.* **2014**, *87*, 53–61. [CrossRef] [PubMed]
10. Jermain, S.V.; Brough, C.; Williams, R.O. Amorphous Solid Dispersions and Nanocrystal Technologies for Poorly Water-Soluble Drug Delivery—An Update. *Int. J. Pharm.* **2018**, *535*, 379–392. [CrossRef]
11. Ogawa, N.; Hiramatsu, T.; Suzuki, R.; Okamoto, R.; Shibagaki, K.; Fujita, K.; Takahashi, C.; Kawashima, Y.; Yamamoto, H. Improvement in the Water Solubility of Drugs with a Solid Dispersion System by Spray Drying and Hot-Melt Extrusion with Using the Amphiphilic Polyvinyl Caprolactam-Polyvinyl Acetate-Polyethylene Glycol Graft Copolymer and d-Mannitol. *Eur. J. Pharm. Sci.* **2018**, *111*, 205–214. [CrossRef] [PubMed]
12. Kumar, R. Nanotechnology Based Approaches to Enhance Aqueous Solubility and Bioavailability of Griseofulvin: A Literature Survey. *J. Drug Deliv. Sci. Technol.* **2019**, *53*, 101221. [CrossRef]
13. Murdande, S.B.; Pikal, M.J.; Shanker, R.M.; Bogner, R.H. Solubility Advantage of Amorphous Pharmaceuticals: II. Application of Quantitative Thermodynamic Relationships for Prediction of Solubility Enhancement in Structurally Diverse Insoluble Pharmaceuticals. *Pharm. Res.* **2010**, *27*, 2704–2714. [CrossRef] [PubMed]
14. Horter, D.; Dressman, J.B. Influence of Physicochemical Properties on Dissolution of Drugs in the Gastrointestinal Tractq. *Adv. Drug Deliv. Rev.* **2001**, *46*, 75–87. [PubMed]

15. Günday, C.; Anand, S.; Gencer, H.B.; Munafò, S.; Moroni, L.; Fusco, A.; Donnarumma, G.; Ricci, C.; Hatir, P.C.; Türeli, N.G.; et al. Ciprofloxacin-Loaded Polymeric Nanoparticles Incorporated Electrospun Fibers for Drug Delivery in Tissue Engineering Applications. *Drug Deliv. Transl. Res.* **2020**, *10*, 706–720. [CrossRef] [PubMed]
16. Arafa, M.G.; Mousa, H.A.; Afifi, N.N. Preparation of PLGA-Chitosan Based Nanocarriers for Enhancing Antibacterial Effect of Ciprofloxacin in Root Canal Infection. *Drug Deliv.* **2020**, *27*, 26–39. [CrossRef] [PubMed]
17. Benedini, L.; Laiuppa, J.; Santillán, G.; Baldini, M.; Messina, P. Antibacterial Alginate/Nano-Hydroxyapatite Composites for Bone Tissue Engineering: Assessment of Their Bioactivity, Biocompatibility, and Antibacterial Activity. *Mater. Sci. Eng. C* **2020**, *115*, 111101. [CrossRef]
18. Mocanu, A.-G.; Belu, I. Formulation and Characterization of Ciprofloxacin Loaded PLGA Microspheres for Applications in Orthopedic Infections. *Curr. Health Sci. J.* **2017**, *43*, 306–310. [CrossRef]
19. Lamy, B.; Tewes, F.; Serrano, D.R.; Lamarche, I.; Gobin, P.; Couet, W.; Healy, A.M.; Marchand, S. New Aerosol Formulation to Control Ciprofloxacin Pulmonary Concentration. *J. Control. Release* **2018**, *271*, 118–126. [CrossRef]
20. Khatib, I.; Tang, P.; Ruan, J.; Cipolla, D.; Dayton, F.; Blanchard, J.D.; Chan, H.-K. Formation of Ciprofloxacin Nanocrystals within Liposomes by Spray Drying for Controlled Release via Inhalation. *Int. J. Pharm.* **2020**, *578*, 119045. [CrossRef]
21. Torge, A. Ciprofloxacin-Loaded Lipid-Core Nanocapsules as Mucus Penetrating Drug Delivery System Intended for the Treatment of Bacterial Infections in Cystic Fibrosis. *Int. J. Pharm.* **2017**, *527*, 92–102. [CrossRef]
22. Tran, T.-T.; Vidaillac, C.; Yu, H.; Yong, V.F.L.; Roizman, D.; Chandrasekaran, R.; Lim, A.Y.H.; Low, T.B.; Tan, G.L.; Abisheganaden, J.A.; et al. A New Therapeutic Avenue for Bronchiectasis: Dry Powder Inhaler of Ciprofloxacin Nanoplex Exhibits Superior Ex Vivo Mucus Permeability and Antibacterial Efficacy to Its Native Ciprofloxacin Counterpart. *Int. J. Pharm.* **2018**, *547*, 368–376. [CrossRef]
23. Rancan, F.; Contardi, M.; Jurisch, J.; Blume-Peytavi, U.; Vogt, A.; Bayer, I.S.; Schaudinn, C. Schaudinn Evaluation of Drug Delivery and Efficacy of Ciprofloxacin-Loaded Povidone Foils and Nanofiber Mats in a Wound-Infection Model Based on Ex Vivo Human Skin. *Pharmaceutics* **2019**, *11*, 527. [CrossRef]
24. Yang, S.; Zhang, X.; Zhang, D. Electrospun Chitosan/Poly (Vinyl Alcohol)/Graphene Oxide Nanofibrous Membrane with Ciprofloxacin Antibiotic Drug for Potential Wound Dressing Application. *Int. J. Mol. Sci.* **2019**, *20*, 4395. [CrossRef] [PubMed]
25. Paaver, U.; Heinämäki, J.; Laidmäe, I.; Lust, A.; Kozlova, J.; Sillaste, E.; Kirsimäe, K.; Veski, P.; Kogermann, K. Electrospun Nanofibers as a Potential Controlled-Release Solid Dispersion System for Poorly Water-Soluble Drugs. *Int. J. Pharm.* **2015**, *479*, 252–260. [CrossRef] [PubMed]
26. Sipos, E.; Kósa, N.; Kazsoki, A.; Szabó, Z.-I.; Zelkó, R. Formulation and Characterization of Aceclofenac-Loaded Nanofiber Based Orally Dissolving Webs. *Pharmaceutics* **2019**, *11*, 417. [CrossRef] [PubMed]
27. Villarreal-Gómez, L.J.; Cornejo-Bravo, J.M.; Vera-Graziano, R.; Grande, D. Electrospinning as a Powerful Technique for Biomedical Applications: A Critically Selected Survey. *J. Biomater. Sci. Polym. Ed.* **2016**, *27*, 157–176. [CrossRef] [PubMed]
28. Ambekar, R.S.; Kandasubramanian, B. Advancements in Nanofibers for Wound Dressing: A Review. *Eur. Polym. J.* **2019**, *117*, 304–336. [CrossRef]
29. Barhoum, A.; Pal, K.; Rahier, H.; Uludag, H.; Kim, I.S.; Bechelany, M. Nanofibers as New-Generation Materials: From Spinning and Nano-Spinning Fabrication Techniques to Emerging Applications. *Appl. Mater. Today* **2019**, *17*, 1–35. [CrossRef]
30. Khodadadi, M.; Alijani, S.; Montazeri, M.; Esmaeilizadeh, N.; Sadeghi-Soureh, S.; Pilehvar-Soltanahmadi, Y. Recent Advances in Electrospun Nanofiber -mediated Drug Delivery Strategies for Localized Cancer Chemotherapy. *J. Biomed. Mater. Res. A* **2020**, *108*, 1444–1458. [CrossRef]
31. Udomluck, N.; Koh, W.-G.; Lim, D.-J.; Park, H. Recent Developments in Nanofiber Fabrication and Modification for Bone Tissue Engineering. *Int. J. Mol. Sci.* **2019**, *21*, 99. [CrossRef]
32. Park, S.; Park, K.; Yoon, H.; Son, J.; Min, T.; Kim, G. Apparatus for Preparing Electrospun Nanofibers: Designing an Electrospinning Process for Nanofiber Fabrication. *Polym. Int.* **2007**, *56*, 1361–1366. [CrossRef]
33. Motamedi, A.S.; Mirzadeh, H.; Hajiesmaeilbaigi, F.; Bagheri-Khoulenjani, S.; Shokrgozar, M. Effect of Electrospinning Parameters on Morphological Properties of PVDF Nanofibrous Scaffolds. *Prog. Biomater.* **2017**, *6*, 113–123. [CrossRef]
34. Pelipenko, J.; Kristl, J.; Janković, B.; Baumgartner, S.; Kocbek, P. The Impact of Relative Humidity during Electrospinning on the Morphology and Mechanical Properties of Nanofibers. *Int. J. Pharm.* **2013**, *456*, 125–134. [CrossRef] [PubMed]
35. Sun, Y.; Zhu, L.; Wu, T.; Cai, T.; Gunn, E.M.; Yu, L. Stability of Amorphous Pharmaceutical Solids: Crystal Growth Mechanisms and Effect of Polymer Additives. *AAPS J.* **2012**, *14*, 380–388. [CrossRef]
36. Yu, D.-G.; Zhang, X.-F.; Shen, X.-X.; Brandford-White, C.; Zhu, L.-M. Ultrafine Ibuprofen-Loaded Polyvinylpyrrolidone Fiber Mats Using Electrospinning: Ultrafine Ibuprofen-Loaded PVP Fibers. *Polym. Int.* **2009**, *58*, 1010–1013. [CrossRef]
37. Suganya, S.; Senthil Ram, T.; Lakshmi, B.S.; Giridev, V.R. Herbal Drug Incorporated Antibacterial Nanofibrous Mat Fabricated by Electrospinning: An Excellent Matrix for Wound Dressings. *J. Appl. Polym. Sci.* **2011**, *121*, 2893–2899. [CrossRef]
38. Quan, J.; Yu, Y.; Branford-White, C.; Williams, G.R.; Yu, D.-G.; Nie, W.; Zhu, L.-M. Preparation of Ultrafine Fast-Dissolving Feruloyl-Oleyl-Glycerol-Loaded Polyvinylpyrrolidone Fiber Mats via Electrospinning. *Colloids Surf. B Biointerfaces* **2011**, *88*, 304–309. [CrossRef] [PubMed]
39. Dai, X.-Y.; Nie, W.; Wang, Y.-C.; Shen, Y.; Li, Y.; Gan, S.-J. Electrospun Emodin Polyvinylpyrrolidone Blended Nanofibrous Membrane: A Novel Medicated Biomaterial for Drug Delivery and Accelerated Wound Healing. *J. Mater. Sci. Mater. Med.* **2012**, *23*, 2709–2716. [CrossRef]

40. Jiang, Y.-N.; Mo, H.-Y.; Yu, D.-G. Electrospun Drug-Loaded Core–Sheath PVP/Zein Nanofibers for Biphasic Drug Release. *Int. J. Pharm.* **2012**, *438*, 232–239. [CrossRef]
41. Contardi, M.; Heredia-Guerrero, J.A.; Perotto, G.; Valentini, P.; Pompa, P.P.; Spanò, R.; Goldoni, L.; Bertorelli, R.; Athanassiou, A.; Bayer, I.S. Transparent Ciprofloxacin-Povidone Antibiotic Films and Nanofiber Mats as Potential Skin and Wound Care Dressings. *Eur. J. Pharm. Sci.* **2017**, *104*, 133–144. [CrossRef]
42. Radacsi, N.; Campos, F.D.; Chisholm, C.R.I.; Giapis, K.P. Spontaneous Formation of Nanoparticles on Electrospun Nanofibres. *Nat. Commun.* **2018**, *9*, 4740. [CrossRef] [PubMed]
43. He, P.; Zhong, Q.; Ge, Y.; Guo, Z.; Tian, J.; Zhou, Y.; Ding, S.; Li, H.; Zhou, C. Dual Drug Loaded Coaxial Electrospun PLGA/PVP Fiber for Guided Tissue Regeneration under Control of Infection. *Mater. Sci. Eng. C* **2018**, *90*, 549–556. [CrossRef] [PubMed]
44. Li, H.; Zhang, Z.; Godakanda, V.U.; Chiu, Y.-J.; Angkawinitwong, U.; Patel, K.; Stapleton, P.G.; de Silva, R.M.; de Silva, K.M.N.; Zhu, L.-M.; et al. The Effect of Collection Substrate on Electrospun Ciprofloxacin-Loaded Poly(Vinylpyrrolidone) and Ethyl Cellulose Nanofibers as Potential Wound Dressing Materials. *Mater. Sci. Eng. C* **2019**, *104*, 109917. [CrossRef]
45. Franco, P.; De Marco, I. The Use of Poly(N-Vinyl Pyrrolidone) in the Delivery of Drugs: A Review. *Polymers* **2020**, *12*, 1114. [CrossRef] [PubMed]
46. Mann, J.; Dressman, J.; Rosenblatt, K.; Ashworth, L.; Muenster, U.; Frank, K.; Hutchins, P.; Williams, J.; Klumpp, L.; Wielockx, K.; et al. Validation of Dissolution Testing with Biorelevant Media: An OrBiTo Study. *Mol. Pharm.* **2017**, *14*, 4192–4201. [CrossRef]
47. Sang, Q.; Williams, G.R.; Wu, H.; Liu, K.; Li, H.; Zhu, L.-M. Electrospun Gelatin/Sodium Bicarbonate and Poly(Lactide-Co-ε-Caprolactone)/Sodium Bicarbonate Nanofibers as Drug Delivery Systems. *Mater. Sci. Eng. C* **2017**, *81*, 359–365. [CrossRef]
48. Yan, S.; Xiaoqiang, L.; Shuiping, L.; Xiumei, M.; Ramakrishna, S. Controlled Release of Dual Drugs from Emulsion Electrospun Nanofibrous Mats. *Colloids Surf. B Biointerfaces* **2009**, *73*, 376–381. [CrossRef]
49. Cleeton, C.; Keirouz, A.; Chen, X.; Radacsi, N. Electrospun Nanofibers for Drug Delivery and Biosensing. *ACS Biomater. Sci. Eng.* **2019**, *5*, 4183–4205. [CrossRef]
50. Radacsi, N.; Giapis, K.P.; Ovari, G.; Szabó-Révész, P.; Ambrus, R. Electrospun Nanofiber-Based Niflumic Acid Capsules with Superior Physicochemical Properties. *J. Pharm. Biomed. Anal.* **2019**, *166*, 371–378. [CrossRef]
51. Karimi, K.; Pallagi, E.; Szabó-Révész, P.; Csóka, I.; Ambrus, R. Development of a Microparticle-Based Dry Powder Inhalation Formulation of Ciprofloxacin Hydrochloride Applying the Quality by Design Approach. *Drug Des. Devel. Ther.* **2016**, *10*, 3331–3343. [CrossRef] [PubMed]
52. Kataria, K.; Gupta, A.; Rath, G.; Mathur, R.B.; Dhakate, S.R. In Vivo Wound Healing Performance of Drug Loaded Electrospun Composite Nanofibers Transdermal Patch. *Int. J. Pharm.* **2014**, *469*, 102–110. [CrossRef]
53. Fallingborg, J. Intraluminal PH of the Human Gastrointestinal Tract. *Dan. Med. Bull.* **1999**, *46*, 183–196. [PubMed]
54. Hansmann, S.; Miyaji, Y.; Dressman, J. An in Silico Approach to Determine Challenges in the Bioavailability of Ciprofloxacin, a Poorly Soluble Weak Base with Borderline Solubility and Permeability Characteristics. *Eur. J. Pharm. Biopharm.* **2018**, *122*, 186–196. [CrossRef] [PubMed]
55. Permanadewi, I.; Kumoro, A.C.; Wardhani, D.H.; Aryanti, N. Modelling of Controlled Drug Release in Gastrointestinal Tract Simulation. *J. Phys. Conf. Ser.* **2019**, *1295*, 012063. [CrossRef]
56. Supramaniam, J.; Adnan, R.; Mohd Kaus, N.H.; Bushra, R. Magnetic Nanocellulose Alginate Hydrogel Beads as Potential Drug Delivery System. *Int. J. Biol. Macromol.* **2018**, *118*, 640–648. [CrossRef] [PubMed]
57. Harder, S.; Fuhr, U.; Beermann, D.; Staib, A. Ciprofloxacin Absorption in Different Regions of the Human Gastrointestinal Tract. Investigations with the Hf-Capsule. *Br. J. Clin. Pharmacol.* **1990**, *30*, 35–39. [CrossRef] [PubMed]
58. Calonge, M.L.; Ilundáin, A. Na+-H+ Exchange Activity and Cellular PH Regulation in Enterocytes Isolated from Chick Small Intestine. *Biochim. Biophys. Acta BBA Biomembr.* **1990**, *1029*, 201–210. [CrossRef]

pharmaceutics

MDPI

Article

Dual Cross-Linked Chitosan/PVA Hydrogels Containing Silver Nanoparticles with Antimicrobial Properties

Dana M. Suflet [1,*], Irina Popescu [1], Irina M. Pelin [1], Daniela L. Ichim [2], Oana M. Daraba [2], Marieta Constantin [1,*] and Gheorghe Fundueanu [1]

[1] Petru Poni Institute of Macromolecular Chemistry, Aleea Grigore Ghica Voda 41A, 700487 Iasi, Romania; ipopescu@icmpp.ro (I.P.); impelin@icmpp.ro (I.M.P.); ghefun@icmpp.ro (G.F.)

[2] Faculty of Medical Dentistry, "Apollonia" University of Iasi, 2 Muzicii Str., 700511 Iasi, Romania; danielaluminitaichim@yahoo.com (D.L.I.); oana_daraba@yahoo.com (O.M.D.)

* Correspondence: dsuflet@icmpp.ro (D.M.S.); marieta@icmpp.ro (M.C.)

Abstract: Stable chitosan/PVA-based hydrogels were obtained by combining covalent and physical cross-linking methods. As covalent cross-linkers, epoxy agents with different chain lengths were used, while freeze–thaw cycles were applied for additional physical cross-linking. The chemical structure of the hydrogel was examined by FTIR spectroscopy whereas the morphology was analyzed by SEM, showing well-defined pores with dimensions of around 50 μm in diameter. It was proved that gel fraction and the network morphology were deeply influenced by the synthesis conditions. Chitosan/PVA hydrogel showed a relative high swelling rate, reaching equilibrium in the first hour. The values obtained for the elastic modulus were relatively low (3–30 kPa); as a result, these hydrogels are soft and very flexible, and are ideal candidates for medical applications as wound or oral dressings. In addition, the natural antimicrobial activity of chitosan was enhanced by in situ generation of silver nanoparticles (AgNPs) under UV irradiation. The total amount of Ag from hydrogel was determined by elemental analyses and its crystalline state was confirmed by XRD. The CS/PVA hydrogels entrapped with AgNPs exhibited high inhibitory activity against *S. aureus* and *K. pneumonia*. The vitality tests confirmed the lack of cytotoxicity of CS/PVA hydrogels without and with AgNPs.

Keywords: chitosan; poly(vinyl alcohol); hydrogels; silver nanoparticles

1. Introduction

Bacterial infections caused by various pathogens, for example *Staphylococcus aureus* (*S. aureus*), *Escherichia coli* (*E. coli*), *Klebsiella pneumonia* (*K. pneumonia*) or *Pseudomonas aeruginosa* (*P. aeruginosa*), still represent a major health problem. The healing time can be influenced by different factors such as genetic vulnerability, nutritional status, age, coexisting illnesses, the time of organism exposure to bacterial agents, excessive drugs intake, and air pollution [1]. Effective treatment consists of antibiotics administration, but an increasing body resistance or unwanted secondary effects have been observed, and for this reason, new therapeutic devices or drug delivery systems are being developed [2]. For treating infected skin wounds, oral or intestinal mucosal diseases, polysaccharide-based hydrogels with prolonged antimicrobial activity are frequently used [3]. Hydrogels can be impregnated with antibiotics, antimicrobial agents, and silver or gold nanoparticles, acting through non-stereospecific mechanisms that involve bacterial membrane disruption or biofilm generation [3,4].

Hydrogels are usually polymeric networks, extensively developed as intelligent systems for biomedical and pharmaceutical applications, including contact lenses [5], vascular prostheses and coatings for stents and catheters [6], tissue engineering [5], diagnostics [7], and drug delivery systems [8]. Hydrogels are hydrophilic networks capable of retaining large amounts of water yet remaining insoluble and maintaining their three-dimensional structure [9]. The polymers used to prepare hydrogels need to be physically and/or chemically cross-linked to prevent their dissolution. By chemical cross-linking, the polymer chains are strongly bonded via a covalent linkage, while the hydrogels which are physically cross-linked possess weak linkages (physical domain junctions, hydrogen bondings, hydrophobic interactions, ionic interactions, van der Waals interactions, or combinations of these [10–12]).

Natural and synthetic polymers offer many possibilities for the design and development of hydrogels that can be used in biomedical applications. However, their combination allows us to obtain ideal materials since they often cumulate synergistically in the properties of the two components. Among natural polymers, polysaccharides are frequently used in the manufacture of hydrogels, due to their non-toxic, biocompatible, and biodegradable properties. Many studies regarding the use of polysaccharide-based hydrogels as drug delivery systems [13], scaffolds [14], cartilage substitutes [15], and sorbent materials [16] have been reported in the literature. The most widely used polysaccharides for obtaining hydrogels with applications in medicine and pharmacy are cellulose, alginate, chitosan, starch, dextran, pullulan, and their derivatives [17,18].

Among them, chitosan (CS) is a cationic polysaccharide widely used in pharmaceutical formulations or for other biomedical applications [14–16,19–23]. Many researchers have highlighted its attractive proprieties, such as biodegradability, biocompatibility, cellular binding capability, antimicrobial [24], antifungal [25], antioxidant [26], antitumor activity [27], and wound healing properties [28]. These and other positive features, such as hydrophilic and cationic charges, make chitosan a suitable polymer for the delivery of active ingredients such as drugs, growth factors, stem cells, and peptides.

Regarding synthetic polymers, poly(vinyl alcohol) (PVA) is the most used of them for the synthesis the hydrogels with biomedical and pharmaceutical applications due to its high biocompatibility, non-toxicity, and non-carcinogenicity [29,30]. Studies on using PVA-based hydrogels or mixtures with natural/synthetic polymers are frequently reported in the literature with applications such as wound dressings [29], artificial articular cartilage [31], and drug delivery systems [32]. PVA-based hydrogels can be prepared by chemical or physical cross-linking. Physically cross-linked hydrogels (obtained by the freeze–thawing (F–T) technique) have drawn greater research attention [33,34]. Single or more cycles of the F–T for a concentrated PVA solution leads to the formation of a network without cross-linking agents. However, PVA-based hydrogels obtained by the F–T method have a low resistance when they are stored for a long time at room temperature [29]. In addition, PVA-based hydrogels possess insufficient elasticity and limited hydrophilicity which restrict their use for biomedical applications. Therefore, combining PVA with natural polymers is a solution to induce specific mechanical properties, to increase the biocompatibility and bioactivity of the materials [29].

Several recent investigations have focused on the combination of CS with PVA in order to obtain soft, flexible, and swellable hydrogels by different methods for medical applications [35,36]. As has been stated, chitosan is known for its weak activity against a wide range of microorganisms, in which the most acceptable antimicrobial mechanism is found to include the presence of charged groups in the polymer backbone and their ionic interactions with bacteria wall constituents. Therefore, the addition of a supplementary antimicrobial agent in the cross-linked network of the hydrogel could substantially improve its antibacterial activity. However, the choice of the antibacterial agent is a very important factor. Various antibiotics are often used as antibacterial agents, but a common problem is the possibility to develop allergies or resistance, especially in long-term use. As a viable alternative, silver nanoparticles (AgNPs) exhibit strong inhibitory effects against a wide variety of microorganisms, such as Gram-positive and Gram-negative bacteria [37]. The advantages regarding the use of AgNPs compared with other formulations include the possibility of using lower concentrations of AgNPs to achieve similar antibacterial effects as other antimicrobial agents [38]. Reduction of silver ions in solution is one of the general methods for AgNP preparation. However, the presence of a residual reducing reagent used for nanoparticle syntheses such as citric acid, sodium borohydride, or other organic compounds may exhibit biological hazards. Therefore, the use of UV irradiation to reduce Ag^+ ions embedded in a polymer matrix seems to be a more appropriate approach for the preparation of nanosystems for bioapplications [39].

Most studies reporting the synthesis of CS/PVA-based hydrogels, by various methods, use concentrated solutions of PVA (10% *w/v* or more) [40,41] and few reports present the production of hydrogels with semi-diluted PVA solutions (6–10%) [42,43]. Furthermore, from our knowledge, data concerning the synthesis of CS/PVA hydrogel combining two cross-linking procedures (physically by freeze–thawing method and covalently using an epoxy cross-linking agent) have not been reported in the literature.

In this context, this paper presents original results regarding a novel method for the synthesis of stable CS/PVA-based hydrogels by a double cross-linking procedure: physically by a freeze–thawing method and covalently using an epoxy cross-linking agent. The morphology of new CS/PVA hydrogels was analyzed by scanning electron microscopy (SEM) and their chemical structure was confirmed by FTIR spectroscopy. The main characteristics of these hydrogels were studied, such as the gel fraction, the degree of swelling in pure water or phosphate buffers, and the mechanical behavior. Then, the native antimicrobial activity of chitosan was enhanced by the in situ generation of AgNPs under UV irradiation. The total amount of Ag from CS/PVA-based hydrogels was determined by elemental analyses and the crystalline form of Ag was confirmed by X-ray diffraction (XRD). The cytotoxicity and in vitro antibacterial activity of the hydrogels with or without AgNPs were also investigated.

2. Materials and Methods

2.1. Materials

Chitosan (CS) and poly(vinyl alcohol) (PVA, Mowiol® 20–98 with M_w = 125.000 g/mol and 98% hydrolysis degree) were purchased from Sigma Aldrich Chemie GmbH (Steinheim, Germany). The deacetylation degree of CS was 82.79% determined by NMR spectroscopy, and the molecular weight, M_v = 233.6 kDa, was calculated from viscometric measurements according to the method described in the literature [44]. Ethylene glycol diglycidyl ether (EGDGE), 1,4-butanediol diglycidyl ether (BDDGE), hydrochloric acid (HCl), and sodium hydroxide (NaOH) were purchased from Merck (Darmstadt, Germany), and silver nitrate ($AgNO_3$) from Honeywell Fluka (Seelze, Germany). All reagents were used without further purification and all experiments were performed using distilled water.

2.2. Methods

2.2.1. CS/PVA Hydrogel Preparation

First, the CS solutions (1 or 2%, *w/v*) were prepared by dissolving a proper amount of CS in 0.1 M HCl solution at room temperature (RT) while stirring overnight. Then, the pH of the CS solutions was adjusted to 5.0 by gradually adding 0.1–10 M of NaOH solution. PVA solutions (1 or 2%, *w/v*) were prepared by dissolving the polymer in water under continuous stirring at 80 °C for 4 h. Hydrogels (CS/PVA) were prepared by mixing CS and PVA solutions in different ratios so as to obtain 0.5:1, 1:1, and 1:0.5 weight ratios between partners in hydrogels. Then, a volume of cross-linking agent (EGDGE or BDDGE) was slowly added, to make a 1:1 or 1:2 stoichiometric ratio between the amine groups of CS and the epoxy groups of the cross-linking agent ([NH_2]:[epoxy]). The mixtures were poured into Petri dishes (d = 5 cm) and stored for 24 or 48 h at RT, until the chemical cross-linking reaction of the polymer chains occurred. Then, the samples were subjected to a freezing process for 24 h at −20 °C until the physical cross-linking of the PVA chains took place. The time of the thawing process (t_T) was maintained the same as the initial storing time: 24 h or 48 h at RT. The F–T process was repeated for 3 and 6 cycles in order to establish the optimum conditions for the synthesis of the highest performing hydrogel.

The overall encoding of the hydrogels is $CS_x/PVA_yR_z(t_T)$ wherein CS denotes chitosan, PVA means poly(vinyl alcohol), R represents cross-linking agent where E is EGDGE and B is BDDGE, x is the concentration of the CS in hydrogel, y shows the PVA concentration, z is the $[NH_2]$:[epoxy] ratio, and t_T is the time at RT, before each freeze step. For example, $CS_{0.5}/PVA_1E_2(24)$ means the hydrogel was obtained starting with 0.5% CS (w/v), 1% PVA (w/v), EGDGE as cross-linker with $[NH_2]$:[epoxy] = 1:2 and 24 h holding time (equal with the thawing time).

2.2.2. In Situ Generation of AgNPs within Hydrogels

Briefly, AgNPs were generated in situ by immersing circular discs of hydrogel overnight (diameter = 10 mm, thickness = 1.5 mm) in the silver nitrate solutions with various concentrations (1 mM, 5 mM, and 10 mM). The silver ion-loaded discs were rinsed with distilled water (2 × 20 mL) in order to remove the excess of silver ions, then were exposed for 15 min to ultraviolet light using a UV lamp (365 nm; 9 W) for reducing the Ag^+ to Ag^0. Finally, the discs were freeze-dried for further characterization or kept in hydrated form for antimicrobial assays.

2.2.3. Physico-Chemical Characterization of Hydrogels

Fourier-Transform Infrared Spectroscopy (FTIR)

FTIR analyses were used to ascertain the occurrence of covalent cross-links during the fabrication of the hydrogels. The FTIR spectra were recorded on KBr pellet using a Vertex 70 Bruker spectrometer (Bruker Optik GmbH, Ettlingen, Germany).

Scanning Electron Microscopy (SEM)

The surface and internal morphology of CS/PVA hydrogels were investigated using a Quanta 200 Scanning Electron Microscope (FEI Company, Bruno, Czech Republic) and a Verios G4 UC Scanning Electron Microscope (Thermo Scientific, Bruno, Czech Republic) equipped with an energy dispersive X-ray spectrometer (EDS, EDAX Octane Elect Super SDD detector, Ametek, Mahwah, NJ, USA) with a better magnification that allowed the visualization of AgNPs. The samples were coated with 10 nm platinum using a Leica EM ACE200 Sputter coater to provide electrical conductivity and to prevent charge buildup during exposure to the electron beam.

Gel Fraction

The gel fraction was used to evaluate the percentage of polymers (CS and PVA) involved in the covalent and physical cross-linking reaction. Briefly, after the last freezing cycle, the CS/PVA hydrogels were freeze-dried and weighed (W_i, g). Then, to remove the non-cross-linked polymer chains and unreacted cross-linking agent, the hydrogels were washed with distilled water until the conductivity of the distilled water was 5 μS/cm^2. Finally, the washed hydrogels were lyophilized again and weighed (W_p, g). The gel fraction (GF, %) that can be assimilated with a degree of cross-linking, was calculated according to Equation (1) [45]:

$$GF(\%) = \frac{W_p}{W_i} \times 100 \quad (1)$$

where W_p is the weight of purified sample in the dried state; W_i is the weight of dried sample before purification.

Swelling Behavior

The swelling kinetic in aqueous medium was studied by the gravimetric method. Briefly, freeze-dried and weighed hydrogels (W_d, g) were immersed in distilled water, in aqueous solutions with various pH (adjusted with 0.1 N NaOH or 0.1 N HCl) or in phosphate buffer solution (PBS) with pH = 6.8, simulating the oral environment. At selected time intervals, the hydrated hydrogels were carefully withdrawn, the excess of superficial water was removed with a wet filter paper, then the samples were weighed (W_t, g). This procedure was repeated until no further weight increment occurred and equilibrium was attained (W_e, g). The swelling ratio (SR) and equilibrium swelling ratio (SR_{eq}) were calculated according to Equations (2) and (3), respectively:

$$SR(\%) = \frac{(W_t - W_d)}{W_d} \times 100 \quad (2)$$

$$SR_{eq}(\%) = \frac{(W_e - W_d)}{W_d} \times 100 \quad (3)$$

Mechanical Properties

Uniaxial compression tests were performed using a Texture Analyser (Brookfield Texture PRO CT3®, Brookfield Engineering Laboratories Inc., Middleboro, MA, USA) at RT. The hydrogels in the hydrated state with cylindrical shape (10–15 mm diameter and 6–3 mm height) were compressed with 0.067 N trigger load between two parallel plates with a compression rate of 0.1 mm/s up to 70% strain. The elastic modulus (E) was calculated according to Equation (4) [46]:

$$E = \frac{\sigma}{\varepsilon} = \frac{\frac{F}{A}}{\frac{\Delta l}{l_0}} \quad (4)$$

where σ is the compressive stress (N/m^2), ε is the strain, F is the force (N), A is the cross-sectional area of the hydrogel (m^2), Δl is the change in length, and l_0 is the original length. The elastic modulus E was calculated from the slope of stress−strain curves between 5 and 15% compressions.

X-ray Diffraction (XRD)

The X-ray diffraction analysis of CS/PVA hydrogels with AgNPs was performed with a Rigaku Miniflex 600 diffractometer (Rigaku, Tokyo, Japan) using CuKα- emission (λ = 1.5406 Å) in the angular range (2θ) of 10° to 90°, at a scanning step of 0.01° and a recording rate of 5°/min. Background subtraction, smoothing and WPPF (Whole Powder Pattern Fitting) refinement of XRD data were made using the SmartLab II v.4 software package for powder X-ray diffraction analysis, while the diffraction peaks were identified using the Crystallography Open Database (COD).

Elemental Analyses

The amount of silver in hydrogels was determined using an atomic absorption spectrometer (ContrAA 800 D, Analytik Jena AG, Jena, Germany). The 5–10 mg of CS/PVA hydrogel loaded with AgNPs was digested in 1 mL of 65% nitric acid (HNO_3) at ambient temperature for 24 h. Later, 1 mL of 65% HNO_3 was added for a total digestion.

Antimicrobial Activity Assessments

The antimicrobial activities of the CS/PVA hydrogels without and with AgNPs were determined by the disk diffusion technique. The tests were conducted against typical reference strains of Gram-positive bacteria such as *S. aureus* ATCC 25923, and Gram-negative bacteria such as *E. coli* ATCC 11775, *K. pneumoniae* ATCC BAA–1705, and *P. aeruginosa* ATCC 10145. Disk diffusion test was performed on commercially available antimicrobial disks with MacConkey agar for *E. coli*, *P. aeruginosa*, and *K. pneumoniae*, and mannitol salt agar for *S. aureus*. A suspension of microorganisms of 0.5 McFarland density was inoculated on a plate with the substrate. The CS/PVA hydrogel discs with 12 mm diameter (d_s, mm) were firstly hydrated in sterile deionized water and then placed on the agar plate and incubated for 24 h at 37 °C. The antibacterial activities were assessed by measurement of the diameters of the inhibition (d_{in}, mm) zones according to the Kirby–Bauer method [47]. The experiments were repeated three times.

Cytotoxicity Assay

The cytotoxicity analysis of the hydrogels was performed using Human Dermal Fibroblasts adult (HDFa). The cells were grown in DMEM (Dulbecco's Modified Eagle Medium, Life Technologies Limited., Paisley, UK) supplemented with 10% FBS (Fetal Bovine Serum) and an antibiotic cocktail consisting of 1% (*v*/*v*) penicillin–streptomycin and 1% (*v*/*v*) non-essential amino acids. The medium was changed every day. The cell culture flasks (NuncTM EasYFlask TM, ThermoFisher Scientific, ThermoFisher Scientific, Roskilde, Denmark) were incubated at 37 °C in a 95% humidified atmosphere and 5% CO_2 (MCO-5AC CO_2 Incubator, Panasonic Healthcare Co., Ltd., Sakata Oizumi-Machi Ora-Gun Gunma, Japan). Cells were allowed to grow in four culture flasks to reach 80% confluence and then were trypsinized with 0.25% trypsin solution at 37 °C for 3 min, followed by the addition of fresh medium to neutralize trypsin. Concentration of cells was 1×10^4 cells/cm^2. After centrifugation (Rotofix-32A, Hettich, Andreas GmbH Hettich & Co.KG, Tuttlingen, Germany,) and re-suspension in fresh medium, the viable cells were plated in flasks and allowed to settle and attach for 24 h prior to treatment. The sterilized hydrogels were incubated for 24 h in direct contact with the subculture cells. After the incubation time, the cell viability was determined using 3-(4,5-Dimethyl-2-thia zolyl)-2,5-diphenyl-2H-tetrazolium bromide (Merck Millipore, Darmstadt, Germany) following the MTT assay. Each type of hydrogel was tested in triplicate.

Statistical Analysis

Data were presented as mean ±SD. The ANOVA single factor was performed for statistical analysis and $p < 0.05$ was considered to be statistically significant.

3. Results and Discussions

3.1. *Synthesis of CS/PVA Hydrogel*

The new CS/PVA-based hydrogels were successfully obtained, from diluted solutions of PVA, by combining two cross-linking techniques: (a) in the first stage, the covalent cross-linking of the CS chains was performed at RT with EGDGE or BDDGE, known to have low toxicity [48,49]; (b) in the second stage, the physical cross-linking of the PVA chains was achieved by the freeze–thaw cycles [33]. With the aim to obtain a hydrogel with optimal properties, the influence of synthesis parameters on the hydrogel characteristics was studied (Table 1). Thereby, the weight ratio between CS and PVA, the length of the cross-linking agent, the [NH_2]:[epoxy] ratio, the storage time at RT, and the number of freeze–thaw cycles were varied. The efficiency of the cross-linking process is well expressed by the gel fraction presented in Table 1.

Table 1. The influence of synthesis parameters on the main characteristics of CS/PVA hydrogels.

Sample Code	Time (h)	GF, %		SR_{eq} *, g/g		Elastic Modulus **	
		3c	6c	Water	PBS 6.8	(kPa)	R^2
$CS_{0.5}/PVA_1E_1$	(24)	52.8 (±1.3)	61.6 (±1.4)	72 (±3)	67 (±4)	2.9 (±0.4)	0.9895
$CS_{0.5}/PVA_1E_2$		58 (±2)	66.8 (±1.2)	64 (±2)	62 (±2)	6.1 (±0.3)	0.9906
$CS_{0.5}/PVA_1B_1$		38 (±3)	45.8 (±1.3)	82 (±3)	75 (±2)	2.6 (±0.2)	0.9528
$CS_{0.5}/PVA_1B_2$		43.3 (±1.2)	52.2 (±1.1)	79 (±3)	73 (±2)	9.8 (±0.4)	0.9874
CS_1/PVA_1E_1		54.1 (±0.9)	68.5 (±1.2)	61 (±3)	48.1 (±1.1)	20.7 (±1.5)	0.9919
CS_1/PVA_1E_2		58.2 (±1.2)	69 (±3)	48 (±3)	45.3 (±0.5)	27.4 (±0.5)	0.9942
CS_1/PVA_1B_1		41.2 (±0.8)	57 (±2)	75 (±2)	65 (±2)	10.8 (±0.1)	0.9383
CS_1/PVA_1B_2		56 (±2)	67 (±3)	69 (±2)	49 (±2)	15.1 (±0.6)	0.9908
$CS_1/PVA_{0.5}E_1$		64 (±3)	78 (±5)	49 (±3)	44.9 (±1.4)	17.1 (±1.3)	0.9907
$CS_1/PVA_{0.5}E_2$		69 (±4)	81 (±5)	46 (±3)	42.3 (±1.1)	18.7 (±1.5)	0.9954
$CS_1/PVA_{0.5}B_1$		62 (±3)	74 (±5)	91 (±5)	74.0 (±1.1)	14.7 (±1.2)	0.9885
$CS_1/PVA_{0.5}B_2$		62 (±3)	76 (±5)	69 (±3)	65.9 (±1.1)	14.6 (±1.4)	0.9278

Table 1. *Cont.*

Sample Code	Time (h)	GF, %		SR_{eq} *, g/g		Elastic Modulus **	
		3c	6c	Water	PBS 6.8	(kPa)	R^2
$CS_{0.5}/PVA_1E_1$	(48)	55 (±2)	70 (±4)	68 (±3)	65 (±3)	3.2 (±1.2)	0.9295
$CS_{0.5}/PVA_1E_2$		60 (±2)	73 (±4)	61 (±3)	49 (±2)	4.8 (±1.2)	0.9774
$CS_{0.5}/PVA_1B_1$		39.5 (±0.4)	71 (±4)	78 (±4)	83 (±3)	-	-
$CS_{0.5}/PVA_1B_2$		45.9 (±1.4)	74 (±2)	75 (±4)	82 (±3)	2.9 (±0.4)	0.9292
CS_1/PVA_1E_1		56 (±2)	74 (±2)	52 (±3)	48.1 (±1.1)	28.6 (±1.6)	0.9922
CS_1/PVA_1E_2		62 (±2)	75 (±2)	44 (±2)	43 (±2)	31.7 (±1.1)	0.9934
CS_1/PVA_1B_1		53.5 (±1.5)	63 (±2)	65 (±3)	61.7 (±1.2)	4.0 (±1.4)	0.9736
CS_1/PVA_1B_2		60.4 (±1.1)	70 (±2)	61 (±3)	59.3 (±1.1)	7.2 (±0.9)	0.9821
$CS_1/PVA_{0.5}E_1$		70 (±3)	77 (±3)	47 (±2)	42.5 (±1.4)	16.3 (±1.1)	0.9883
$CS_1/PVA_{0.5}E_2$		71 (±3)	84 (±4)	44 (±2)	41 (±2)	16.6 (±1.4)	0.9856
$CS_1/PVA_{0.5}B_1$		63 (±2)	76 (±4)	84 (±5)	69 (±3)	-	-
$CS_1/PVA_{0.5}B_2$		69 (±3)	76 (±4)	63 (±4)	58 (±3)	6.7 (±1.6)	0.9904

*, **-SR_{eq} and Elastic modulus were determined for the hydrogels obtained after six freeze–thaw cycles.

GF values between 38 and 84% were obtained (Table 1), values that were deeply influenced by the synthesis conditions. Therefore, as a first general rule, the *GF* values increase with increasing the number of F–T cycles; more microcrystalline domains between polymer chains are formed, resulting in the increase of cross-linking density [33,50,51]. It must be noticed that a number of fewer than three F–T cycles led to poorly stable hydrogels that were difficult to handle due to a low degree of cross-linking (a low number of chemical and physical bonds). As a second general rule, the *GF* values increase with increasing the holding time (t_T) at RT. In fact, maintaining the hydrogel for a longer period of time at room temperature inevitably leads to an increase in the degree of cross-linking by covalent bonds. Regarding the composition of the hydrogel, it can be observed that the maximum *GF* value of 84% was obtained for the hydrogels with the highest CS content and using EGDGE as a cross-linking agent. Covalent linkages between amino groups of CS and epoxy groups of cross-linker secure the chemical stability of the hydrogel. Moreover, EGDGE is more reactive than BDDGE in reaction with CS.

Figure 1 shows optical photographs of the CS/PVA hydrogels in hydrated state after six F–T cycles and a holding time of 24 h. It is known that during the F–T process in the structure of the PVA-based hydrogel, two regions are created. These regions include areas with high density of PVA chains and areas with low density. Repeating the F–T process leads to the formation of crystallites in the PVA-rich area, which induces a randomly distributed heterogeneity in the structure, creating cloudy/opaque PVA hydrogels [52]. In our case, the transparent CS/PVA hydrogels were obtained by the covalent cross-linking with BDDGE, while the EGDGE led to opaque hydrogels. The transparency is probably caused by the longer length and higher hydrophobicity of BDDGE chains, compared with EGDGE, which prevents the formation of PVA densely crystalline areas.

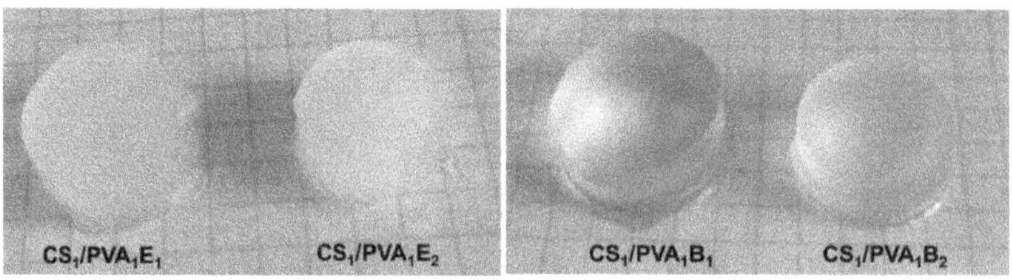

Figure 1. Optical photomicrographs of CS/PVA hydrogels after the last six freeze–thaw cycles and a holding time of 24 h.

3.2. Fourier-Transform Infrared Spectroscopy

The chemical structure of the CS/PVA hydrogels was assessed by FTIR spectrometry (Figure 2). FTIR spectra of hydrogels show both characteristic bands for parent polymers (CS and PVA) and the increase of intensity of some bands due to the presence of the cross-linking agent. Thus, the large band at 3437–3454 cm^{-1} is attributed to O–H and N–H stretching, as well as the intra- and inter-molecular hydrogen bonds, the bands at 2923–2873 cm^{-1} to C–H asymmetric and symmetric stretching from CH_2 and CH groups, the band at 1637–1644 cm^{-1} is assigned to the associated water, C–OH from the glycosidic units of polysaccharide chains and also the presence of residual N-acetyl groups (C=O stretching of amide I), and N-H in plane deformation coupled with C–N stretching of amide II (secondary amide) from CS. The bands from 1425–1444 cm^{-1} correspond to CH_2 deformation groups, at 1384 cm^{-1} to $-CH_3$ symmetric deformation, and at 1334–1338 cm^{-1} to C-N bond stretching of amide III. The band at 1154 cm^{-1} was attributed to the glycosidic linkage (asymmetric bridge stretch) and the absorption bands in the range 1083–883 cm^{-1} belong to the C–O–C glycosidic ring (skeletal vibration involving the C–O stretch).

The increase in the intensity of the bands from 2923 cm^{-1} (CH_2 groups), 1450–1437 cm^{-1} (CH_2 bonds) and at 854–853 cm^{-1} (NH groups) in the spectrum of the hydrogel confirms the covalent cross-linking. In the case of hydrogel with AgNPs, where neutral Ag atoms (Ag^0) form coordination bonds with NH_2 and OH electron-rich groups, the FTIR spectrum presents a decrease in intensity of the band from 3430 cm^{-1}, indicating that the N-H vibrations were affected by the attachment of silver to the nitrogen from CS. The shift of the bands from 1643-1004 cm^{-1} to higher or lower wavelengths or sometimes splitting (as the band from 1643 cm^{-1} was split in 1623 and 1621 cm^{-1}) also reflects the interactions between Ag, O, and N atoms [53,54].

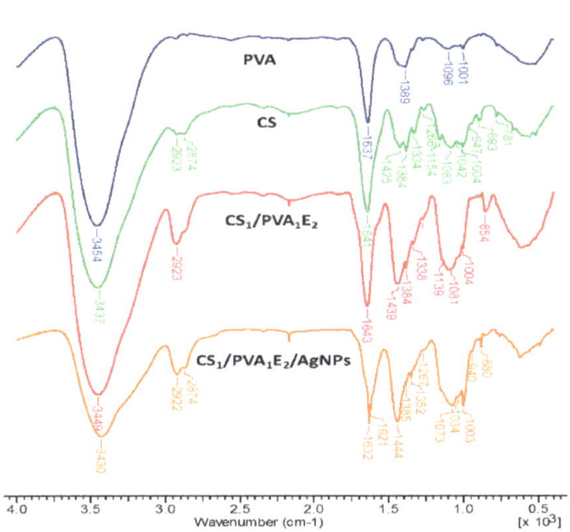

Figure 2. FT–IR spectra of chitosan (CS), poly(vinyl alcohol) (PVA), cross-linked hydrogels (CS_1/PVA_1E_2), and hydrogel loaded with silver nanoparticles (CS_1/PVA_1E_2/AgNPs).

3.3. SEM Analyses

Scanning electron micrographs of the hydrogels in cross section show a three-dimensional network with well-defined pores with a diameter ranging between 40 and 60 μm. As expected, the increase of [NH_2]:[epoxy] molar ratio led to a decrease of both size and size distribution, due to the increasing number of cross-links between polymeric chains (Figure 3a,b). Furthermore, the increase of the holding time at RT from 24 h (Figure 3b) to 48 h (Figure 3c) led to a structure with smaller pores and thicker walls, when EGDGE was used as cross-linker. If BDDGE was used as cross-linker, the morphology of the hydrogel was not significantly influenced by the holding time (Figure 3e,f).

An influence of the length of cross-linking agent on the pore size and pore size distribution was observed. Thus, EGDGE, which possesses a short chain between the epoxy groups (two C atoms) led to a structure with smaller pores (Figure 3a–c), while BDDGE, with four C atoms between the epoxy groups, led to a structure with larger but irregular pores (Figure 3d–f). It must also be mentioned that an increase of the CS:PVA weight ratio from 0.5:1 (Figure 3g) to 1:1 (Figure 3b), and also the CS content in hydrogel, 1:0.5 ratio (Figure 3h), determined the generation of larger pores that follow an orderly arrangement.

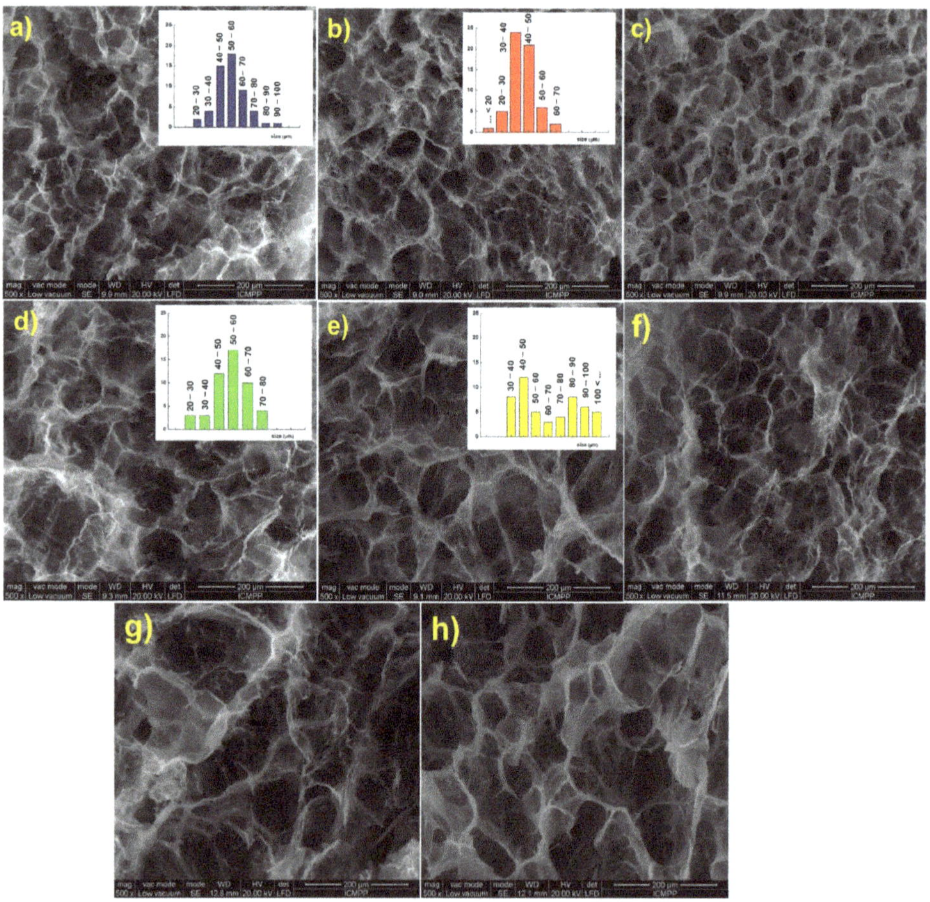

Figure 3. Scanning electron micrographs of the CS/PVA hydrogels in cross section: (**a**) $CS_1/PVA_1E_1(24)$; (**b**) $CS_1/PVA_1E_2(24)$; (**c**) $CS_1/PVA_1E_2(48)$; (**d**) $CS_1/PVA_1B_1(24)$; (**e**) $CS_1/PVA_1B_2(24)$; (**f**) $CS_1/PVA_1B_2(48)$; (**g**) $CS_{0.5}/PVA_1E_2(24)$; (**h**) $CS_1/PVA_{0.5}E_2(24)$.

3.4. Swelling Behaviour

If the hydrogel is synthesized as a biomaterial with potential applications in skin (or oral) dressing, it is absolutely necessary to determine the retention capacity and retention kinetics in water and simulated physiological conditions.

The retention capacity of water and PBS (pH = 6.8) at equilibrium is presented in Table 1. As is clearly evident, the hydrogels retain a large amount of water or PBS due to the hydrophilic nature of the two polymers. For example, the sample $CS_1/PVA_{0.5}B_1$ obtained at a holding time of 24 h retains about 92-times more water and 74-times more PBS than its own weight in the dried state. The lower retention capacity of PBS than water, and this is noted for all samples, is due to the salting-out effect of the ions present in buffer that hinder the hydration of polymeric chains. The water retention capacity also depends on the holding time; the hydrogels obtained after 48 h retained less water or PBS than those obtained after 24 h as a result of the increasing number of hydrogen and covalent bonds between polymer chains that makes the network less expandable. The length of the cross-linking agent as well as the $[NH_2]$:[epoxy] molar ratio deeply influenced the retention capacity of the CS/PVA hydrogels. Hydrogels cross-linked with BDDGE (4 C atoms between epoxy groups) have higher values of solvent retention than those obtained with EGDGE (2 C atoms between epoxy groups). A longer cross-linker leads to a hydrogel with larger pores and therefore with a higher water (PBS) retention capacity.

Concerning the $[NH_2]$:[epoxy] molar ratio, it is obvious that a higher ratio of cross-linker compared to the amino groups leads to a more efficient cross-linking reaction and therefore to a less expandable structure that retains less water (PBS). Because CS/PVA hydrogel contain ionizable amino groups with a $pK_a = 6.5$, the influence of pH on the swelling capacity was studied [55]. As shown in Figure 4a, the highest values of swelling ratios are obtained at the lowest values of pH due to the full protonation of the amine groups which induces electrostatic repulsions and structure expansion [56], and then the swelling ratio decreases. However, because a significant part of the primary amine groups is transformed into secondary amines during the covalent cross-linking, the influence of pH is not significant.

Besides the retention capacity, an important characteristic of a hydrogel used in biomedical applications is the swelling rate until it reaches equilibrium. Tanaka and co-workers proved that the promptness of the swelling response is inversely related to the size of the hydrogel [57]. Furthermore, a porous structure facilitates the migration of water through the pores ("convective" transport) of the hydrogels, increasing the swelling rate [58].

The hydrogels synthesized in this study have relatively small sizes and porous structures. As follows, they display relatively high swelling rates; in particular, after 1 h, equilibrium is reached (Figure 4b).

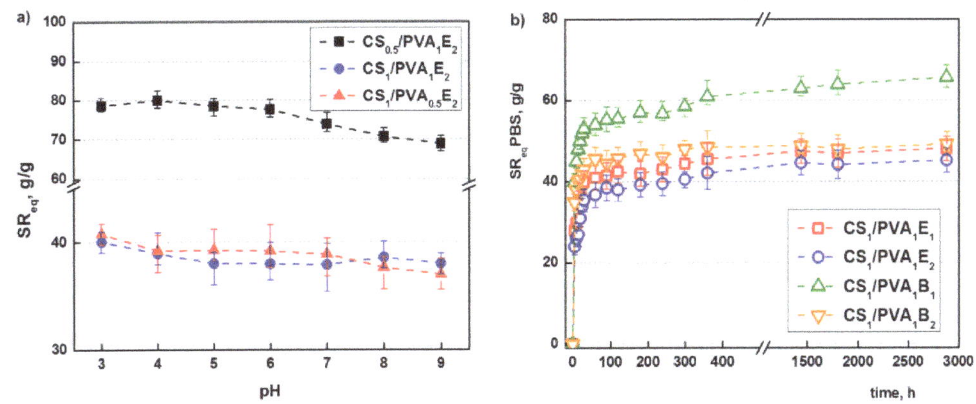

Figure 4. Influence of the pH of the swelling medium on the SR_{eq} (t_T = 24 h) (**a**). Swelling kinetics in simulated physiological conditions, PBS at pH = 6.8 (t_T = 24 h) (**b**).

3.5. Mechanical Properties

The mechanical properties of hydrogels are decisive factors for their future biomedical applications. Compression tests were performed to get information about the elastic behavior of the hydrogels (Figure 5). The values of elastic modulus were calculated between 5 and 15% compression, where the stress–strain curves are linear (Figure 5, inset), and the obtained values are presented in Table 1. Values of the elastic modulus cover a wide range from 2.58 kPa for sample $CS_{0.5}/PVA_1B_1$ to 31.70 kPa for sample CS_1/PVA_1E_2. As a general rule, the Young modulus increases with the increasing amount of cross-linker, due to the large number of the cross-links between polymeric chains and therefore a reduced mobility of the macromolecules. The increase of the number of carbon atoms of the cross-linker from 6 to 8 determined a decrease of the elastic modulus. Firstly, the longer-length cross-linker (BDDGE) has a lower reactivity than the shorter-length cross-linker (EGDGE) and gives rise to less cross-linked and therefore soft and more elastic three-dimensional structures. Secondly, a longer length of the cross-linker leads to an increased chain mobility and therefore elasticity. The increase of CS/PVA weight ratio from 0.5:1 to 1:1 ($CS_{0.5}/PVA_1$ to CS_1/PVA_1) determined a significant increase of the compression modulus due to the increased number of covalent cross-links, resulting a stiffening of the hydrogels. In contrast, the decrease of the PVA amount in hydrogels from CS_1/PVA_1 to $CS_1/PVA_{0.5}$ led to softer materials. These results are confirmed by the literature data showing that PVA can be used as a polymer for reinforcing hydrogels [29].

However, the values of the modulus of elasticity of hydrogels are relatively low (3–30 kPa) compared to other polymeric materials, which makes them very elastic. As a result, these hydrogels are ideal candidates for medical applications, as wound or oral dressings that must be strong, soft, and flexible materials [59,60]. Nevertheless, hydrogels with very low values of Young modulus are less resistant, and, therefore, for subsequent experiments samples with average values were chosen ($CS_1/PVA_{0.5}E_2$ and $CS_1/PVA_{0.5}B_2$ (t_T = 24 h)).

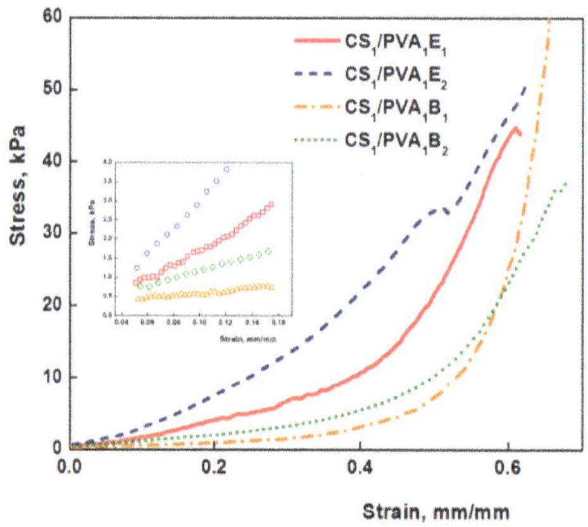

Figure 5. Stress–strain curve for CS/PVA hydrogel under compressive loading (t_T = 24 h). In detail the linear range for determining the elastic modulus is represented.

3.6. In Situ Synthesis and Immobilization of Silver Nanoparticles

Chitosan is known to exhibit antibacterial activity, but this activity may be diminished in mixtures with other compounds [61]. Thus, to improve the antimicrobial effect, CS/PVA-based hydrogels were immersed in $AgNO_3$ solutions of different concentrations. The reduction of Ag^+ ions to Ag^0 was performed in the presence of UV radiation as an ecological method that avoids the use of potentially toxic reducing agents. In addition, this method can allow the sterilization of hydrogels at the same time. In this context, the effect of $AgNO_3$ solution concentration on silver particle size and biological properties was studied.

The SEM images showed that the size of the Ag particles increases from 88 nm to 400 nm (average diameter, d_{av}) ($p < 0.05$) with the increase of $AgNO_3$ concentration from 1 mM to 10 mM (Figure 6a,b). This aspect was supported by the elemental analysis of Ag from hydrogel. Thus, we determined a content of 0.11, 1.1, and 2.3 wt.% Ag in hydrogel after immersion in 1 mM, 5 mM, and 10 mM $AgNO_3$, respectively.

The crystalline structure of the AgNPs was analyzed by XRD. The XRD spectrum of CS/PVA hydrogel (Figure 6c) exhibited a major peak at 20.2° which corresponds to the polymorph of the pattern [62] and certifies that a strong interaction occurred between CS and PVA molecules in the hydrogel [63]. The spectrum of CS/PVA hydrogel with AgNPs shows the diffraction peaks at around 38.1°, 44.3°, 64.4°, 77.3°, and 81.5° which were associated to the (111), (200), (220), (311), and (222) crystalline planes of face-centered cubic crystals of the metallic Ag structure in hydrogel, which confirms the successful synthesis of crystalline AgNPs (space group 225:Fm-3m, Card No. 9008459) [64]. In addition, the spectrum showed the presence, together with Ag^0, of AgCl nanoparticles resulting from the reaction of Ag^+ with the counterion Cl^- from the ionized form of amino groups of chitosan, due to solubilization of CS in HCl. Diffraction peaks observed at 28.2°, 32.6°, 46.9°, 55.63°, 58.3°, 68.5°, 75.6°, and 77.9° correspond to the face-centered cubic structures (111), (200), (220), (311), (222), (400), (331), and (400) planes of AgCl Nps, respectively [65]. These results were confirmed by energy dispersive spectroscopy (EDX); accordingly (Figure 6d), the elements C, O, N, Ag, and Cl were assigned to the sample and the Pt coating layer, respectively.

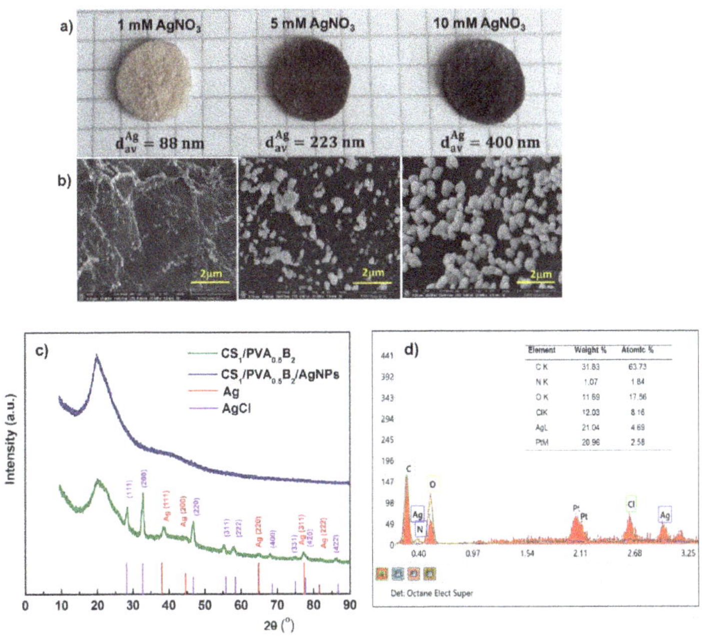

Figure 6. Optical microphotographs of CS/PVA hydrogel (sample $CS_1/PVA_{0.5}B_2$ (t_T = 24 h)), in situ loaded with silver nanoparticles by using different concentrations of $AgNO_3$ (**a**). Scanning electron micrographs (cross-sections) of the corresponding hydrogels loaded with Ag nanoparticles (**b**). XRD patterns of hydrogel without and with AgNPs after immersion in 10 mM $AgNO_3$ (**c**). Energy-dispersed X-ray spectrometry (EDS) patterns of the hydrogel with AgNPs after immersion in 10 mM $AgNO_3$ (**d**).

3.7. Antimicrobial Activity Assessments

The antibacterial activity of CS/PVA hydrogels with and without AgNPs was investigated using Gram-negative (*E. coli*, *P. aeruginosa*, and *K. pneumoniae*) and Gram-positive (*S. aureus*) bacteria by the disk diffusion method. Regardless of the Ag percentage or the size of AgNPs, all hydrogels with silver nanoparticles showed intense inhibitory activity against *S. aureus* (Table 2, Figure 7a,e). In this case, the inhibitory zone diameter (d_{iz}) was between 42 mm and 45 mm. We also observed an intense inhibitory activity against *K. pneumonia*. In this case, both the amount of Ag in the hydrogel and the AgNPs size had a slight influence on the inhibition of this bacterium, observing an increase in d_{iz} from 20 mm to 24 mm with increasing of Ag percentage and size (Figure 7b,f). The hydrogels loaded with AgNPs showed an insufficient activity against *P. aeruginosa* (Table 2, Figure 7c,g) and *E. coli* (Table 2, Figure 7d,h) when the d_{iz} was less than 15 mm. The hydrogels without AgNPs did not show inhibitory activity, probably due to the fact that the majority of amine groups in CS are involved in the cross-linking reactions. The results are not totally consistent with those obtained by Li et al. (2019) [66], since our AgNP-loaded hydrogels showed a strong antibacterial effect against *S. aureus* and *K. pneumoniae* and have almost no activity against *E. coli*. In fact, the antimicrobial activity was dependent on the content in AgNPs, increasing with the increase of Ag percentage in hydrogel, with this aspect largely demonstrated in the literature [24,29,38,54].

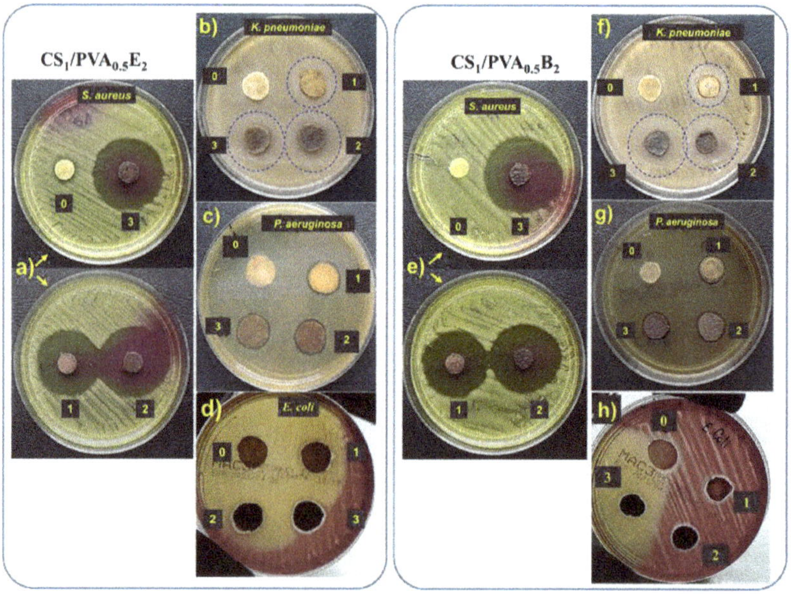

Figure 7. Antibacterial activity of $CS_1/PVA_{0.5}E_2$ and $CS_1/PVA_{0.5}B_2$ hydrogels against *S. aureus* (**a**,**e**), *K. pneumonia* (**b**,**f**), *P. aeruginosa* (**c**,**g**), and *E. coli* (**d**,**h**): zone 0, the hydrogel without AgNPs; zone 1, hydrogel with AgNPs obtained in 1 mM $AgNO_3$ solution; zone 2, hydrogel with AgNPs obtained in 5 mM $AgNO_3$ solution; zone 3, hydrogel with AgNPs obtained in 10 mM $AgNO_3$ solution.

Table 2. Antibacterial properties of the $CS_1/PVA_{0.5}E_2$ and $CS_1/PVA_{0.5}B_2$ hydrogels against *E. coli*, *P. aeruginosa*, *S. aureus*, and *K. pneumoniae*.

Hydrogel / Microorganism	$CS_1/PVA_{0.5}E_2$				$CS_1/PVA_{0.5}B_2$			
	Zone 0	Zone 1	Zone 2	Zone 3	Zone 0	Zone 1	Zone 2	Zone 3
E. coli	-	-	+	+	-	-	+	+
P. aeruginosa	-	+	+	+	-	+	+	+
S. aureus	-	+++	+++	+++	-	+++	+++	+++
K. pneumoniae	-	++	++	++	-	++	++	++

'-' no effect; '+'for d_{iz} < 15 mm; '++' for 15 mm < d_{iz} < 25 mm; '+++' for d_{iz} > 25 mm.

3.8. Viability Assay

Biocompatibility and the lack of cytotoxicity are imperative requirements when designing new wound or oral dressing materials. Silver nanoparticles and/or silver ions are known to be potentially toxic toward not only bacteria but also healthy, live cells [67].

Taking into account these aspects, the vitality of dermal fibroblasts in direct contact with hydrogels with and without AgNPs was investigated. The results showed a slight decrease of cell vitality, 83–85% after 24 h, for $CS_1/PVA_{0.5}$ hydrogels without AgNPs (Figure 8). However, as was expected, the cell viability decreased in the presence of AgNPs, even for the samples with a small number of AgNPs and low amount of Ag in hydrogel. Nevertheless, the cell viability for $CS_1/PVA_{0.5}B_2$ sample was 68%, which places the sample at the threshold of cytotoxicity and non-cytotoxicity. The hydrogel samples with higher content of AgNPs (>0.11 wt.%) and larger size (>100 nm) showed a cell viability smaller than 50% (data not shown) due to the higher concentration of released silver ions. Similar results were found by Li et al. [66] when they tested the growth of NIH3T3 cells in contact with AgCl NP-loaded hydrogels, a higher content of AgCl NPs (1.8 wt.%) in hydrogel causing a strong cytotoxic effect. In fact, the low toxicity reported by other researchers is due to the fact that they used another test approach, namely the extract method which does not involve direct contact between cells and nanoparticles [38,68]. We chose direct contact to test the skin's reaction to a potential applicability of the hydrogel in skin dressing.

Generally, it is a contradiction between antimicrobial activity and cell viability, and therefore the optimal concentration of nanoparticles must be chosen to satisfy both requirements. Taking into account these considerations, the $CS_1/PVA_{0.5}B_2$ sample loaded with 0.1 wt.% AgNPs smaller than 100 nm showed, at the same time, a good antimicrobial activity and an acceptable cytotoxicity.

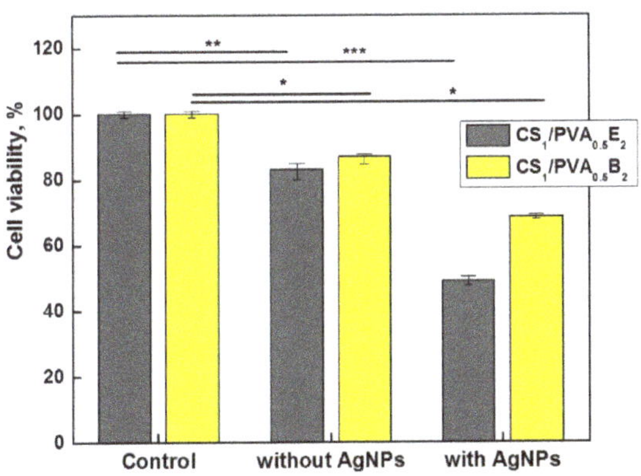

Figure 8. Cells viability of $CS_1/PVA_{0.5}E_2$ and $CS_1/PVA_{0.5}B_2$ hydrogels (t_T = 24 h) with and without AgNPs (1 mM $AgNO_3$), after 24 h (* $p < 0.001$ ** $p < 0.01$, *** $p < 0.05$).

4. Conclusions

New CS/PVA-based hydrogels were obtained using diluted PVA solutions by combining chemical and physical cross-linking methods. Two epoxy cross-linking agents with different chain lengths were used and the chemical reaction took place for 24 or 48 h. Then, the systems were subjected to F–T cycles in order to allow the formation of physical networks. The chemical structure was confirmed by FT-IR spectroscopy, and the morphology of the hydrogels was analyzed by SEM when well-defined pores having dimensions of 30–60 μm were observed. The gel fraction and the network morphology were influenced by the obtaining conditions. Hydrogels with the highest gel fractions were obtained at six F–T cycles, when the time at RT (when chemical cross-linking occurred) was 48 h, and the CS/PVA ratio was 1:0.5. The cross-linker agent with a longer chain, BDDE, confers a weaker cross-linking compared with EGDE, but at a lower cytotoxicity.

CS/PVA cryogels showed a fast swelling capacity when the equilibrium was reached in the first hour. The swelling degree in water and in PBS with pH = 6.8 was influenced by the chemical cross-linking: cross-linking agent, [epoxy]:[NH_2] ratio, and t_T. The mechanical tests showed an elastic behavior of hydrogels with low elastic modulus values and without cracking up to 70% compression.

The native antimicrobial activity of chitosan was enhanced by the in situ generations of silver nanoparticles under UV irradiation. The size of the AgNPs can be controlled by varying the concentration of $AgNO_3$ solutions used to immerse the CS/PVA hydrogels prior to UV irradiation. The increase in the size of Ag particles with the increase of the concentration of the $AgNO_3$ solution was proven by SEM and elemental analyses; AgNPs with dimensions of 88 nm AgNPs was obtained with a 1 mM $AgNO_3$ solution. The XRD analysis highlighted a face-centered cubic crystal of metallic Ag structure in hydrogel that confirms the successful synthesis of crystalline AgNPs.

The hydrogels with entrapped AgNPs present a high inhibitory activity against *S. aureus* (Gram-positive bacteria) and *K. pneumonia*, but low activity for *P. aeruginosa* and *E. coli*, the last three being Gram-negative bacteria. The cell vitality test showed that the hydrogels without and with AgNPs smaller than 100 nm may be considered non-cytotoxic.

The swelling behavior, the mechanical properties, and the antimicrobial activity of these hydrogels recommend their use in the manufacture of antibacterial devices with biomedical applications.

Author Contributions: Conceptualization, D.M.S.; Formal analysis, I.P., I.M.P., D.L.I. and O.M.D.; Investigation, D.M.S., I.P. and I.M.P.; Methodology, D.M.S.; Project administration, M.C. and G.F.; Supervision, M.C. and G.F.; Writing–original draft, D.M.S. All authors will be informed about each step of manuscript processing including submission, revision, revision reminder, etc. via emails from our system or assigned Assistant Editor. All authors have read and agreed to the published version of the manuscript.

Funding: This research was funded by Ministry of Research, Innovation and Digitization, CNCS/CCCDI–UEFISCDI, grant number PN–III–P2–2.1–PED–2019–1780 and PN–III–P4–ID–PCCF–2016–0050–P3, within PNCDI III.

Institutional Review Board Statement: Not applicable.

Informed Consent Statement: Not applicable.

Data Availability Statement: The data presented in this study are available on request from the corresponding author.

Conflicts of Interest: The authors declare no conflict of interest.

References

1. Doron, S.; Gorbach, S.L. Bacterial Infections: Overview. *Int. Encycl. Public Health* **2008**, 273–282. [CrossRef]
2. Ng, V.W.L.; Chan, J.M.W.; Sardon, H.; Ono, R.J.; García, J.M.; Yang, Y.Y.; Hedrick, J.L. Antimicrobial hydrogels: A new weapon in the arsenal against multidrug-resistant infections. *Adv. Drug Deliv. Rev.* **2014**, *78*, 46–62. [CrossRef] [PubMed]

3. Hemeg, H.A. Nanomaterials for alternative antibacterial therapy. *Int. J. Nanomed.* **2017**, *12*, 8211–8225. [CrossRef] [PubMed]
4. Veiga, A.S.; Schneider, J.P. Antimicrobial hydrogels for the treatment of infection. *Pept. Sci.* **2013**, *100*, 637–644. [CrossRef] [PubMed]
5. Chirani, N.; Yahia, L.H.; Gritsch, L.; Motta, F.L.; Chirani, S.; Faré, S. History and applications of hydrogels. *History and applications of hydrogels. J. Biomed. Sci.* **2015**, *4*, 2–23.
6. Guan, G.; Yu, C.; Xing, M.; Wu, Y.; Hu, X.; Wang, H.; Wang, L. Hydrogel small-diameter vascular graft reinforced with a braided fiber strut with improved mechanical properties. *Polymers* **2019**, *11*, 810. [CrossRef]
7. Wiraja, C.; Ning, X.; Cui, M.; Xu, C. Hydrogel-based technologies for the diagnosis of skin pathology. *Technologies* **2020**, *8*, 47. [CrossRef]
8. Calo, E.; Khutoryanskiy, V.V. Biomedical applications of hydrogels: A review of patents and commercial products. *Eur. Polym. J.* **2015**, *65*, 252–267. [CrossRef]
9. Hoffman, A.S. Hydrogels for biomedical applications. *Adv. Drug Deliv. Rev.* **2012**, *64*, 18–23. [CrossRef]
10. Hoare, T.R.; Kohane, D.S. Hydrogels in drug delivery: Progress and challenges. *Polymer* **2008**, *49*, 1993–2007. [CrossRef]
11. Hennink, W.E.; van Nostrum, C.F. Novel crosslinking methods to design hydrogels. *Adv. Drug Deliv. Rev.* **2012**, *64*, 223–236. [CrossRef]
12. Ullah, F.; Othman, B.M.H.; Javed, F.; Ahmad, Z.; Akil, H.M.d. Classification, processing and application of hydrogels: A review. *Mater. Sci. Eng. C* **2015**, *57*, 414–433. [CrossRef] [PubMed]
13. Bernkop-Schnürch, A.; Dünnhaupt, S. Chitosan-based drug delivery systems. *Eur. J. Pharm. Biopharm.* **2012**, *81*, 463–469. [CrossRef]
14. Madihally, S.V.; Matthew, H.W. Porous chitosan scaffolds for tissue engineering. *Biomaterials* **1999**, *20*, 1133–1142. [CrossRef]
15. Panjapheree, K.; Kamonmattayakul, S.; Meesane, J. Biphasic scaffolds of silk fibroin film affixed to silk fibroin/chitosan sponge based on surgical design for cartilage defect in osteoarthritis. *Mater. Des.* **2018**, *141*, 323–332. [CrossRef]
16. Pelin, I.M.; Suflet, D.M. Mucoadhesive buccal drug delivery systems containing polysaccharides. *Cellul. Chem. Technol.* **2020**, *54*, 889–902. [CrossRef]
17. Coviello, T.; Matricardi, P.; Marianecci, C.; Alhaique, F. Polysaccharide hydrogels for modified release formulations. *J. Control. Release* **2007**, *119*, 5–24. [CrossRef]
18. Dumitiu, S. *Polysaccharides in Biomedical Applications*; Publisher by Marcel Dekker Inc.: New York, NY, USA, 1996.
19. Felt, O.; Buri, P.; Gurny, R. Chitosan: A unique polysaccharide for drug delivery. *Drug Dev. Ind. Pharm.* **1998**, *24*, 979–993. [CrossRef]
20. Chopra, S.; Mahdi, S.; Kaur, J.; Iqbal, Z.; Talegaonkar, S.; Ahmad, F.J. Advances and potential applications of chitosan derivatives as mucoadhesive biomaterials in modern drug delivery. *J. Pharm. Pharmacol.* **2006**, *58*, 1021–1032. [CrossRef]
21. Varshosaz, J. The promise of chitosan microspheres in drug delivery systems. *Expert Opin. Drug Deliv.* **2007**, *4*, 263–273. [CrossRef]
22. Woodley, J. Bioadhesion. New possibilities for drug administration? *Clin. Pharmacokinet.* **2001**, *40*, 77–84. [CrossRef]
23. Thaya, R.; Vaseeharan, B.; Sivakamavalli, J.; Iswarya, A.; Govindarajan, M.; Alharbi, N.S.; Kadaikunnan, S.; Al-anbr, M.N.; Khaled, J.M.; Benelli, G. Synthesis of chitosan-alginate microspheres with high antimicrobial and antibiofilm activity against multi-drug resistant microbial pathogens. *Microb. Pathog.* **2018**, *114*, 17–24. [CrossRef]
24. Yang, K.; Han, Q.; Chen, B.; Zheng, Y.; Zhang, K.; Li, Q.; Wang, J. Antimicrobial hydrogels: Promising materials for medical application. *Int. J. Nanomed.* **2018**, *13*, 2217–2263. [CrossRef]
25. Meng, D.; Garba, B.; Ren, Y.; Yao, M.; Xia, X.; Li, M.; Wang, Y. Antifungal activity of chitosan against *Aspergillus ochraceus* and its possible mechanisms of action. *Int. J. Biol. Macromol.* **2020**, *158*, 1063–1070. [CrossRef]
26. Ngo, D.-H.; Kim, S.-K. Antioxidant effects of chitin, chitosan, and their derivatives. *Adv. Food Nutr. Res.* **2014**, *73*, 15–31.
27. Shikhi-Abadi, P.G.; Irani, M. A review on the applications of electrospun chitosan nanofibers for the cancer treatment. *Int. J. Biol. Macromol.* **2021**, *183*, 790–810. [CrossRef] [PubMed]
28. Liu, H.; Wang, C.; Li, C.; Qin, Y.; Wang, Z.; Yang, F.; Li, Z.; Wang, J. A functional chitosan-based hydrogel as a wound dressing and drug delivery system in the treatment of wound healing. *RSC Adv.* **2018**, *8*, 7533–7549. [CrossRef]
29. Kamoun, E.A.; Chen, X.; Mohy Eldin, M.S.; Kenawy, E.-R.S. Crosslinked poly(vinyl alcohol) hydrogels for wound dressing applications: A review of remarkably blended polymers. *Arab. J. Chem.* **2015**, *8*, 1–14. [CrossRef]
30. Maitz, M.F. Applications of synthetic polymers in clinical medicine. *Biosurf. Biotribol.* **2015**, *1*, 161–176. [CrossRef]
31. Kobayashi, M.; Oka, M. Characterization of a polyvinyl alcohol-hydrogel artificial articular cartilage prepared by injection molding. *J. Biomater. Sci. Polym. Edn.* **2004**, *5*, 741–751. [CrossRef]
32. Jensen, B.E.; Davila, I.; Zelikin, A.N. Poly(vinyl alcohol) physical hydrogels: Matrix-mediated drug delivery using spontaneously eroding substrate. *J. Phys. Chem. B.* **2016**, *120*, 5916–5926. [CrossRef]
33. Nakano, T.; Nakaoki, T. Coagulation size of freezable water in poly(vinyl alcohol) hydrogels formed by different freeze/thaw cycle periods. *Polym. J.* **2011**, *43*, 875–880. [CrossRef]
34. Sasaki, S.; Murakami, T.; Suzuki, A. Frictional properties of physically cross-linked PVA hydrogels as artificial cartilage. *Biosurf. Biotribol.* **2016**, *2*, 11–17. [CrossRef]
35. El-Salmawi, K.M. Gamma radiation-induced crosslinked PVA/chitosan blends for wound dressing. *J. Macromol. Sci. Part A Pure Appl. Chem.* **2007**, *44*, 541–545. [CrossRef]

36. Antunes, J.C.; Tavares, T.D.; Teixeira, M.A.; Teixeira, M.O.; Homem, N.C.; Amorim, M.T.P.; Felgueiras, H.P. Eugenol-containing essential oils loaded onto chitosan/polyvinyl alcohol blended films and their ability to eradicate Staphylococcus aureus or Pseudomonas aeruginosa from infected microenvironments. *Pharmaceutics* **2021**, *13*, 195. [CrossRef]
37. Lara, H.H.; Garza-Trevino, E.N.; Ixtepan-Turrent, L.; Singh, D.K. Silver nanoparticles are broad-spectrum bactericidal and virucidal compounds. *J. Nanobiotechnol.* **2011**, *9*, 30–38. [CrossRef]
38. Hiep, N.T.; Khon, H.C.; Niem, V.V.T.; Toi, V.V.; Quyen, T.N.; Hai, N.D.; Anh, M.N.T. Microwave-assisted synthesis of chitosan/polyvinyl alcohol silver nanoparticles gel for wound dressing applications. *Int. J. Polym. Sci.* **2016**, *2016*, 1584046. [CrossRef]
39. Krishnan, P.D.; Banas, D.; Durai, R.D.; Kabanov, D.; Hosnedlova, B.; Kepinska, M.; Fernandez, C.; Ruttkay-Nedecky, B.; Nguyen, H.V.; Farid, A.; et al. Silver nanomaterials for wound dressing applications. *Pharmaceutics* **2020**, *12*, 821. [CrossRef] [PubMed]
40. Gupta, S.; Goswami, S.; Sinha, A. A combined effect of freeze–Thaw cycles and polymer concentration on the structure and mechanical properties of transparent PVA gels. *Biomed. Mater.* **2012**, *7*, 015006. [CrossRef]
41. Stauffer, S.R.; Peppas, N.A. Poly(vinyl alcohol) hydrogels prepared by freezing-thawing cyclic processing. *Polymer* **1992**, *33*, 3932–3936. [CrossRef]
42. Hassan, C.M.; Peppas, N.A. Structure and morphology of freeze/thawed PVA hydrogels. *Macromolecules* **2000**, *33*, 2472–2479. [CrossRef]
43. Lin, S.-P.; Lo, K.-Y.; Tseng, T.-N.; Liu, J.-M.; Shih, T.-Y.; Cheng, K.-C. Evaluation of PVA/dextran/chitosan hydrogel for wound dressing. *Cell. Polym.* **2019**, *38*, 15–30. [CrossRef]
44. Gamzazade, A.I.; Slimak, V.M.; Skljar, A.M.; Stykova, E.V.; Pavlova, S.S.A.; Rogozin, S.V. Investigation of the hydrodynamic properties of chitosan solutions. *Acta Polym.* **1985**, *36*, 420–424. [CrossRef]
45. Yang, X.; Liu, Q.; Chen, X.; Yu, F.; Zhu, Z. Investigation of PVA/ws-chitosan hydrogels prepared by combined -irradiation and freeze-thawing. *Carbohydr. Polym.* **2008**, *73*, 401–408. [CrossRef]
46. Yeh, W.C.; Li, P.C.; Jeng, Y.M.; Hsu, H.C.; Kuo, P.L.; Li, M.L.; Yang, P.M.; Lee, P.H. Elastic modulus measurements of human liver and correlation with pathology. *Ultrasound Med. Biol.* **2002**, *28*, 467–474. [CrossRef]
47. Hudzicki, J. Kirby-Bauer Disk Diffusion Susceptibility Test Protocol. American Society for Microbiology. 2009. Available online: https://asm.org/Protocols/Kirby-Bauer-Disk-Diffusion-Susceptibility-Test-Pro (accessed on 5 February 2021).
48. Jeong, C.H.; Yune, J.H.; Kwon, H.C.; Shin, D.M.; Sohn, H.; Lee, K.H.; Choi, B.; Kim, E.S.; Kang, J.H.; Kim, E.K.; et al. In vitro toxicity assessment of crosslinking agents used in hyaluronic acid dermal filler. *Toxicol. Vitr.* **2021**, *70*, 105034. [CrossRef]
49. Bratskaya, S.; Privar, Y.; Nesterov, D.; Modin, E.; Kodess, M.; Slobodyuk, A.; Marinin, D.; Pestov, A. Chitosan gels and cryogels cross-linked with diglycidyl ethers of ethylene glycol and polyethylene glycol in acidic media. *Biomacromolecules* **2019**, *20*, 1635–1643. [CrossRef] [PubMed]
50. Abdel-Mohsen, A.M.; Aly, A.S.; Hrdina, R.; Montaser, A.S.; Hebeish, A. Eco-synthesis of PVA/chitosan hydrogels for biomedical application. *J. Polym. Environ.* **2011**, *19*, 1005–1012. [CrossRef]
51. Holloway, J.L.; Lowman, A.M.; Palmese, G.R. The role of crystallization and phase separation in the formation of physically cross-linked PVA hydrogels. *Soft Matter* **2013**, *9*, 826–833. [CrossRef]
52. Gupta, S.; Pramanik, A.K.; Kailath, A.; Mishra, T.; Guha, A.; Nayar, S.; Sinha, A. Composition dependent structural modulations in transparent poly(vinyl alcohol) hydrogels. *Colloids Surf. B* **2009**, *74*, 186–190. [CrossRef]
53. Agnihotri, S.; Mukherji, S.; Mukherji, S. Antimicrobial chitosan–PVA hydrogel as a nanoreactor and immobilizing matrix for silver nanoparticles. *Appl. Nanosci.* **2012**, *2*, 179–188. [CrossRef]
54. Hajji, S.; Slama-Ben Salema, R.B.; Hamdi, M.; Jellouli, K.; Ayadi, W.; Nasri, M.; Bouf, S. Nanocomposite films based on chitosan–poly(vinylalcohol) and silver nanoparticles with highantibacterial and antioxidant activities. *Process Saf. Environ. Prot.* **2017**, *111*, 112–121. [CrossRef]
55. Bajpai, A.K.; Saini, R. Preparation and characterization of biocompatible spongy cryogels of poly(vinyl alcohol)–Gelatin and study of water sorption behaviour. *Polym. Int.* **2005**, *54*, 1233–1242. [CrossRef]
56. Khalid, M.N.; Agnely, F.; Yagoubi, N.; Grossiord, J.L.; Couarraze, G. Water state characterization, swelling behavior, thermal and mechanical properties of chitosan based networks. *Eur. J. Pharm. Sci.* **2002**, *15*, 425–432. [CrossRef]
57. Tanaka, T.; Fillmore, D.J. Kinetics of swelling of gels. *J. Chem. Phys.* **1979**, *70*, 1214–1218. [CrossRef]
58. Sato-Matsuo, E.; Tanaka, T. Kinetics of discontinuous volume phase transition of gels. *J. Chem. Phys.* **1988**, *89*, 1695–1703. [CrossRef]
59. Zahouani, H.; Pailler-Mattei, C.; Sohm, B.; Vargiolu, R.; Cenizo, V.; Debret, R. Characterization of the mechanical properties of a dermal equivalent compared with human skin in vivo by indentation and static friction tests. *Skin Res. Technol.* **2009**, *15*, 68–76. [CrossRef]
60. Pailler-Mattei, C.; Bec, S.; Zahouani, H. In vivo measurements of the elastic mechanical properties of human skin by indentation tests. *Med. Eng. Phys.* **2008**, *30*, 599–606. [CrossRef] [PubMed]
61. Li, J.; Zhuang, S. Antibacterial activity of chitosan and its derivatives and their interaction mechanism with bacteria: Current state and perspectives. *Eur. Polym. J.* **2020**, *138*, 109984. [CrossRef]
62. Aziz, S.B.; Abdulwahid, R.T.; Rasheed, M.A.; Abdullah, O. Gh.; Ahmed, H.M. Polymer blending as a novel approach for tuning the SPR peaks of silver nanoparticles. *Polymers* **2017**, *9*, 486. [CrossRef] [PubMed]

63. Kumar, H.M.P.N.; Prabhakar, M.N.; Venkata, P.C.; Rao, K.M.; Reddy, T.V.A.K.; Rao, K.C.; Subha, M.C.S. Compatibility studies of chitosan/PVA blend in 2% aqueous acetic acidsolution at 30 °C. *Carbohydr. Polym.* **2010**, *82*, 251–255. [CrossRef]
64. Nguyen, N.-T.; Liu, J.-H. A green method for in situ synthesis of poly(vinyl alcohol)/chitosan hydrogel thin films with entrapped silver nanoparticles. *J. Taiwan Inst. Chem. Eng.* **2014**, *45*, 2827–2833. [CrossRef]
65. Trinh, N.D.; Nguyen, T.T.B.; Nguyen, T.H. Preparation and characterization of silver chloride nanoparticles as an antibacterial agent. *Adv. Nat. Sci. Nanosci. Nanotechnol.* **2015**, *6*, 045011. [CrossRef]
66. Li, G.; Zhang, D.; Qin, S. Preparation and performance of antibacterial polyvinylalcohol/polyurethane glycol/chitosan hydrogels containing silver chloride nanoparticles via one-step method. *Nanomaterials* **2019**, *9*, 972. [CrossRef]
67. Liao, C.; Li, Y.; Tjong, S.C. Bactericidal and cytotoxic properties of silver nanoparticles. *Int. J. Mol. Sci.* **2019**, *20*, 449. [CrossRef]
68. Alcântara, M.T.S.; Lincopan, N.; Santos, P.M.; Ramirez, P.A.; Brant, A.J.C.; Riella, H.G.; Lugão, A.B. Simultaneous hydrogel crosslinking and silver nanoparticle formulation by using ionizing radiation to obtain antimicrobial hydrogels. *Radiat. Phys. Chem.* **2019**, *165*, 108369. [CrossRef]

pharmaceutics

MDPI

Article

Kinetic Release Studies of Antibiotic Patches for Local Transdermal Delivery

Esra Altun [1], Esra Yuca [2], Nazmi Ekren [3], Deepak M. Kalaskar [4,*], Denisa Ficai [5,6], Georgiana Dolete [6,7], Anton Ficai [6,7,8,*] and Oguzhan Gunduz [1,*]

1 Centre for Nanotechnology & Biomaterials Research, Department of Metallurgical and Materials Engineering, Faculty of Technology, Goztepe Campus, Marmara University, Istanbul 34722, Turkey; esra.altun@ymail.com
2 Department of Molecular Biology and Genetics, Faculty of Arts and Sciences, Davutpasa Campus, Yildiz Technical University, Istanbul 34220, Turkey; eyuca@yildiz.edu.tr
3 Centre for Nanotechnology & Biomaterials Research, Department of Electrical-Electronics Engineering, Faculty of Technology, Goztepe Campus, Marmara University, Istanbul 34722, Turkey; nazmiekren@marmara.edu.tr
4 UCL Division of Surgery and Interventional Science, Royal Free Hospital Campus, University College London, Rowland Hill Street, London NW3 2PF, UK
5 Department of Inorganic Chemistry, Physical Chemistry and Electrochemistry, Faculty of Applied Chemistry and Materials Science, University POLITEHNICA of Bucharest, 060042 Bucharest, Romania; denisaficai@yahoo.ro
6 National Centre for Micro- and Nanomaterials, University POLITEHNICA of Bucharest, 060042 Bucharest, Romania; dolete.georgiana@gmail.com
7 Department of Science and Engineering of Oxide Materials and Nanomaterials, Faculty of Applied Chemistry and Materials Science, University POLITEHNICA of Bucharest, 060042 Bucharest, Romania
8 Academy of Romanian Scientists, 060042 Bucharest, Romania
* Correspondence: d.kalaskar@ucl.ac.uk (D.M.K.); anton.ficai@upb.ro (A.F.); ucemogu@ucl.ac.uk (O.G.)

Abstract: This study investigates the usage of electrohydrodynamic (EHD)-3D printing for the fabrication of bacterial cellulose (BC)/polycaprolactone (PCL) patches loaded with different antibiotics (amoxicillin (AMX), ampicillin (AMP), and kanamycin (KAN)) for transdermal delivery. The composite patches demonstrated facilitated drug loading and encapsulation efficiency of drugs along with extended drug release profiles. Release curves were also subjected to model fitting, and it was found that drug release was optimally adapted to the Higuchi square root model for each drug. They performed a time-dependent and diffusion-controlled release from the patches and followed Fick's diffusion law by the Korsmeyer–Peppas energy law equation. Moreover, produced patches demonstrated excellent antimicrobial activity against Gram-positive (*Staphylococcus aureus*) and Gram-negative (*Escherichia coli*) strains, so they could be helpful in the treatment of chronic infectious lesions during wound closures. As different tests have confirmed, various types of antibiotics could be loaded and successfully released regardless of their types from produced BC/PCL patches. This study could breathe life into the production of antibiotic patches for local transdermal applications in wound dressing studies and improve the quality of life of patients.

Keywords: electrohydrodynamic printing; antibiotic patches; drug release; polymer; bacterial cellulose

1. Introduction

Proper wound management is essential to guard wounded areas from pathogens responsible for secondary infections. For example, if germs rapidly multiply in a wound, and the infection spreads, it could cause a serious internal infection [1]. Moreover, an internal infection can delay wound healing by extending the inflammatory stage while the immune system responds and clears the infection [2]. In this respect, antibiotic patches are an innovative approach, i.e., to deliver improved healthcare, reduce overall healthcare costs, increase drug effectiveness, and balance the toxicity of drugs involved in local wound management [3]. They are essential in managing local infections where high concentrations of antibiotics are required locally [4]. On the other hand, their concentrations in the patches should be kept under control as high amounts of antibiotics can lead to systemic toxicity [5]. Beta-lactam antibiotics (amoxicillin, ampicillin) and aminoglycosides (kanamycin) are well-known drugs applied to inhibit bacterial spread in wound dressing applications [6]. Their loadings into transdermal patches for local wound dressing applications can provide beneficial antibacterial activity while being absorbed into the body. Furthermore, local administration can limit the potential for systemic absorption of antibiotics, diminishing antibiotic resistance [6].

Drug delivery systems have been extensively explored in recent decades to provide more effective and side effect-free treatments for patients suffering from diseases. The most common forms of drug delivery systems include oral tablets, liquid injections, microspheres, hydrogels, liposomes, micelles, and nanoparticles [7]. However, such delivery systems have challenges in maintaining precise control of the rate and duration of drug release, require repeated administration, and have long-term release problems [8–13]. To overcome these difficulties, drug embedded thin polymer patches have been proposed as a versatile approach to prolong drug release, and they are currently being widely used [14]. Thin polymer patches are flexible and can be produced in various shapes and sizes [15]. Three-dimensional (3D) printers, such as inkjet printers and thermal inject printers, can create small-sized patches (with high placement accuracy) using thermal or mechanical actuating mechanisms [16]. However, inkjet printing is limited in writing operations that involve small micrometers, and cannot be utilized for drug reservoirs requiring several micrometers [17]. The electrohydrodynamic (EHD)-3-dimensional (3D) printing method is capable of achieving micrometer-scale, complex, high-resolution 3D structures, in ambient temperature operations, with single-step productions [18]. The EHD uses an electric field on the polymer solution using micro/nano-scaled nozzle tips [19]. The EHD-3D printing process is the same as the conventional EHD process, yet the sample is collected close to the needle, which is quite small in diameter. Therefore, it is possible to achieve higher print resolution without replacing the nozzle tip with a smaller diameter nozzle, using highly viscous colloidal solutions, or viscous polymeric fluids [20]. Moreover, the EHD-3D printing method can provide essential leads for cell alignment and cell contact guidance, for better cell proliferation and tissue formation, with optimal porosity, inter-fibrous pore size, and pore distribution [21–23]. The dosage of the antibiotics can be adjusted according to the patient's individual needs by altering the size of the EHD-3D printed scaffolds and the content of the solid materials. In addition, through EHD-3D printing, the formulation can be achieved during the production and purification of the drugs, improving efficiency, and decreasing the costs for pharmaceuticals industries.

Polymeric biomaterials and composites are widely used in the EHD-3D printing method due to their polymerizing capabilities, by various methods, to make them "3D printable". They provide physical support for cell attachment, but they also offer transport of therapeutic agents, such as drugs, proteins, growth factors, and chemical agents [24]. PCL is one of the biomaterials that has semi-crystalline, biocompatible, and biodegradable features, with ideal mechanical properties and biological performance [25]. The usage of PCL in drug dosage forms is quite common due to its ability to be easily absorbed into the porous structure of the drug active substance, its capability of returning the active substance to the body at high rates, its complete resorption from the body, and its extended bioavailability on the tissue [26]. The high biocompatibility and slow(er) biodegradation rate, as well as the required flexibility of PCL, make it applicable for use in local transdermal applications. Furthermore, PCL patches show great potential for the delivery of therapeutics, and they can be combined with other biomaterials, such as bacterial cellulose (BC). BC is a biomaterial that shares common features with extracellular matrix (ECM) components [27]. Its porosity can provide the release of antimicrobial agents, drugs, and other bio-functional materials to the targeted tissue in a controlled manner [28,29]. Blending this remarkable polymer with PCL can offer improved cell viability rates while an antibiotic patch releases its therapeutic agent [30]. Moreover, thanks to the hydrophilic nature of BC, moisture can be maintained, and dehydration of the wound and crust formation can be prevented.

This research aims to produce layer-on-layer stacked antibiotic patches from BC/PCL blend polymers carrying various broad-spectrum antibiotics using the EHD-3D printing method, and compare their releasing properties with drug release studies and drug release kinetic models for local transdermal applications in wound dressing applications. The use of these antibiotics will exhibit the versatility of the antibiotic patches in terms of different antibiotic loadability.

2. Materials and Methods

2.1. Materials

Polycaprolactone (PCL, Mw 80,000 g/mol), dimethylformamide (DMF), and dichloromethane (DCM) were purchased from Sigma-Aldrich (St. Louis, MO, USA). Bacterial cellulose (BC) pellicle was provided by the Department of Medical Microbiology, Medipol University (Istanbul, Turkey). Amoxicillin (AMX, Mw 365.4 g/mol), ampicillin (AMP, Mw 349.4 g/mol), kanamycin (KAN, Mw 600.6 g/mol), and phosphate buffered saline (PBS, pH 7.4) were also obtained from Sigma-Aldrich (Steinheim, Germany).

2.2. Preparation of Solutions

BC pellicle was sonicated in DMF for 1 h using a sonicator (Branson Sonifier 250, BRANSON, Danbury, CT, USA) at a power output of 100% and then was centrifuged at 6000 rpm for 600 s. After each centrifugation step, the supernatant was removed and, finally, fragmented BC was obtained with almost no DMF residue. After that, 10 wt.% PCL solution was prepared in DCM with continuous magnetic stirring (WiseStir®, MSH-20A, Germany) for 24 h at 23 °C. As the last step, fragmented BC was blended with 10 wt.% PCL solution (5:95 wt. ratio), and the final blend was magnetically stirred for 1 h at an ambient temperature (23 °C).

The direct blending of the drug with a polymeric solution is the simplest method for fabricating drug-containing patches through the processing of a polymeric [31]. Therefore, three different types of antibiotics (AMX, AMP, and KAN) were individually blended with BC/PCL solution under constant stirring at an ambient temperature (23 °C) for 12 h prior to EHD-3D printing. The initial loading of each drug was 50 mg/mL. After blending, final solutions were stirred at an ambient temperature (23 °C) for another 12 h before EHD-3D printing. Prepared solutions were measured for their viscosity, density, electric conductivity, and surface tension using a Brookfield DV-E programmable viscometer, a standard density bottle (5 mL), a KRÜSS K100SF tensiometer, and a WTW Cond-3110-SET1 conductivity meter, respectively. The measurements were carried out in triplicate at the ambient temperature of 23 °C and relative humidity of 50 ± 3%.

2.3. EHD-3D Printing

The EHD-3D printing was accomplished by using an altered lab system, which consisted of an X–Y stage motion control head, a glass substrate holder, a nozzle holder with Z-axis control, a high voltage power supply (output DC voltage from 0 to 40 kV), and a solution feeding system (Figure 1). The solution feeding system consisted of a programmable syringe pump (NE-1000, New Era Pump System Inc., Farmingdale, NY, USA) and a stainless steel nozzle (with 21G orifice). The syringe was filled with a prepared solution, which gave a constant solution source through the EHD-3D printing process.

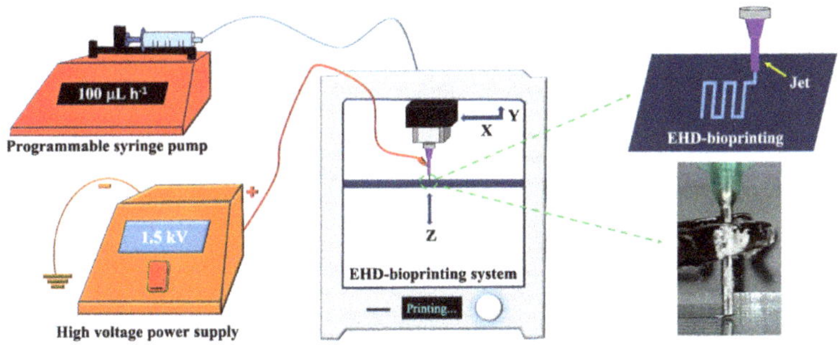

Figure 1. Experimental EHD-3D printing setup with a closer view of the EHD-3D patterning.

For EHD-3D printing, SolidWorks software was used to design the square patches as 10 x 10 mm^2, and layers were piled in 5 min with a 90°angle as pre-designed. The distance between the nozzle and substrate was stabilized to 1 mm for experimental setup after the preliminary investigations [23,30]. According to the study, as mentioned earlier, the flow rate was optimized for reproducible and uniform fiber formation to 100 mLh^{-1} for all samples. EHD-3D printed microfibers were deposited using 1.5 kV and stacked with 0°/90° angles as two layers. Upon completing the EHD-3D printing process, drug patches were removed from the substrate. The vacuum oven was used to evaporate residue solvents that may have remained in the samples before characterization tests. Finally, yield percentages of samples were calculated using Equation (1) and all calculations were repeated three times:

$$\text{Yield\%} = 100 \times (\text{weight of collected product}/\text{weight of solid materials applied}) \quad (1)$$

2.4. Characterization of Drug Patches

Morphological characterization of the neat and the drug-containing patches was investigated by a ZEISS MA/EVO10 scanning electron microscope (SEM). Samples were placed on microscopic glass slides and mounted on a specimen stub with double-sided carbon tape. Then, they were sputter-coated with Au for 60 s using a Quorum SC7620 Mini Sputter Coater before SEM imaging. After SEM imaging at an acceleration voltage of 10 kV, diameters of the individual fibers, inter-fibrous pore sizes, and porous sizes and porosity on the fibers within the EHD-3D printed patches were measured using ImageJ.

A JASCO FT/IR-4000 Fourier transform infrared spectrometer (FTIR) was used to investigate the chemical characterization of the EHD-3D printed patches. Each FTIR spectra was taken at 4 cm^{-1} resolution over the range of 4000–400 cm^{-1}.

Uniaxial tensile tests were performed using an INSTRON 4411 tensile device (Instron, Norwood, MA, USA) with a 50 N load cell under a crosshead speed of 5 mm/min at the ambient temperature (23 °C). All patches were subjected to tensile testing. Each measurement was taken in triplicate, and the ultimate tensile strength and Young's modulus values were calculated from the stress–strain curves of the tested samples.

The swelling characteristics of the EHD-3D printed patches were measured according to Equation (2) after the samples were submerged in PBS at 37 °C for 24 h:

$$\text{Degree of swelling (\%)} = 100 \times (M - Md)/Md \quad (2)$$

where M is the weight of each sample after submersion in the PBS solution for 24 h and Md is the weight of the sample after submersion in the buffer solution for 24 h in its dry state.

Prior to AMX, AMP, and KAN drug loading (DL) and encapsulation efficiency (EE) studies, all drug-containing patches were utterly dissolved in DCM. Thereafter, AMX, AMP, and KAN concentrations in DCM were determined using UV Spectrophotometer

(UV1280, Shimadzu, Kyoto, Japan) at 229, 230 and 207 nm. The DL and EE percentages of AMX, AMP, and KAN in patches were calculated using Equations (3) and (4) [31]:

$$\text{DL (\%)} = 100 \times (\text{Weight of antibiotic entrapped in the sample/weight of the patch}) \quad (3)$$

$$\text{EE (\%)} = 100 \times (\text{Determined drug content/theoretical drug content}) \quad (4)$$

2.5. *Drug Release Studies*

The in vitro release kinetics of antibiotics from EHD-3D printed patches in PBS medium were determined using a UV Spectroscopy by recording the wavelengths at 229, 230 and 207 nm for AMX, AMP, and KAN, respectively. A linear calibration curve was established based on standard solutions with concentrations ranging from 5–200 μg/mL. The amount of the drug present in a patch was calculated from the obtained calibration curve for each drug. All drug-release experiments were carried out using the total immersion method with a known quantity of sample immersed in a sealed vial bottle containing 20 mL of PBS and incubated in a shaker at a constant incubation temperature 37 °C. At designated time intervals ranging between 0 and 14 days (336 h), a 1 mL release medium was taken out and replenished with an equal volume of fresh PBS. Supernatants were filtered using 0.45 μm Millipore. The amount of the drugs in the withdrawn release mediums was determined at the same wavelengths previously mentioned against the pre-determined calibration curves for each AMX, AMP, and KAN drugs using the UV spectrophotometer. The obtained data were used for assessing cumulative release and mathematical modeling of release mechanisms.

2.6. *Application of Drug Release Data to Mathematical Models*

The application of drug release data to mathematical models is usually a few mathematical equations that define the dissolution profile. Once a suitable function has been selected, an assessment of the dissolution profile can be performed, and the drug release profile can be correlated with drug release kinetic models.

2.6.1. Zero Order Kinetics

Zero order kinetics describe the process of continuous drug release from a drug delivery system, and the level of the drug in the blood remains constant throughout the drug release. Hence, the drug release kinetics from an in vitro release study are plotted against time to study the data. In this kinetic model, the drug release rate is independent of its concentration. Used zero order equation (Equation (5)) is [32]:

$$C_t = C_0 + K_0 t \quad (5)$$

In the equation, C_t shows the amount of drug released in a given time, C_0 is the initial concentration of the drug, and K_0 is the zero order rate constant.

2.6.2. First Order Kinetics

First order kinetics describe a concentration-dependent drug release rate from the delivery system. The used first order equation (Equation (6)) is [32]:

$$dC/dt = K_1(C_t - C_0) \quad (6)$$

In the Equation (6), dC/dt shows the rate of change in concentration with respect to time, and K_1 represents the first order rate constant expressed in units of time (t).

2.6.3. Higuchi Model

The Higuchi model describes the release of drugs from the insoluble matrix as a square root of a time-dependent process based on Fickian diffusion. The used Higuchi equation (Equation (7)) is [32]:

$$Q = K_H t^{1/2} \quad (7)$$

In the Equation (7), Q represents the cumulative amount of drug released at the time (t per unit area), K_H shows the Higuchi dissolution constant.

2.6.4. Korsmeyer–Peppas Model

In order to understand the mechanism of drug diffusion from patches, release data were detected using the well-known empirical equation proposed by Korsmeyer and Peppas. The Korsmeyer–Peppas model describes the drug release mechanism from a polymeric system. The used Korsmeyer–Peppas equation (Equation (8)) is [33]:

$$M_t/M\infty = K_{kp} t^{n} \quad (8)$$

In the Equation (8), M_t represents the amount of drug released at time t, $M\infty$ shows the amount of drug released after ∞ period, n symbolize the release exponent, and K_{kp} depicts the Korsmeyer–Peppas release rate constant.

For a cylindrical geometry with a non-swellable drug release system, when n = 0.45, the drug release mechanism is Fickian diffusion, when $n = 1$, Case II transport occurs leading to the zero-order release and when the value of n is between 0.45 and 1, Non-Fickian diffusion occurs [33].

2.7. Antibacterial Activity Assay

The inhibition method zone was used to investigate the antibacterial activity of the drug patches against Gram-positive (*S. aureus*) and Gram-negative (*E. coli*) bacterial strains as the model organisms. The bacterial cultures from different individual colonies were cultivated in Luria-Bertani (LB) broth (10 g/L bacto-tryptone, 5 g/L yeast extract, 5 g/L NaCl) overnight at 37 °C. The bacterial strains (107 colony-forming unit (cfu)/mL) were inoculated on LB-agar (LB broth with 15 g/L agar) plate using sterile cotton swabs. The equal weight of the patches was gently placed on the bacteria inoculated agar plates and incubated at 37 °C for 24 h. Zones of inhibition of the patches were evaluated by measuring the clear area that formed around each patch. Experiments were repeated in triplicate, and the average diameter of zones of inhibition was measured.

3. Results and Discussion

3.1. Characterization of Fabricated Samples

The EHD systems are governed by processing parameters and the properties of the working polymer solution [34]. EHD-3D printing is a single-step production process that provides both nano and micro-scale writing and combines the electrohydrodynamic atomization and 3D writing methods. All of the properties of the working polymer solution, including viscosity, electrical conductivity, density, and surface tension, are given in Table 1. The presence of the drugs in the resulting patches did not cause any significant influence on the physical properties of the BC/PCL solutions.

Table 1. Physical properties of EHD-3D printed solutions. (Mean ± SD, $n = 3$).

Sample Name (wt. ratio)	Viscosity (Pa.S)	Electrical Conductivity (μ.S cm^{-1})	Density (kg m^{-3})	Surface Tension (mN m^{-1})
BC/PCL (5:95)	363 ± 0.5	1.20 ± 0.03	1445 ± 2	52.8 ± 0.1
BC/PCL (5:95) +AMX	368 ± 0.3	1.20 ± 0.02	1448 ± 1	52.9 ± 0.1
BC/PCL (5:95) +AMP	371 ± 0.1	1.20 ± 0.02	1452 ± 1	53.2 ± 0.5
BC/PCL (5:95) +KAN	376 ± 0.2	1.20 ± 0.01	1459 ± 1	53.4 ± 0.3

As stated previously [23], the EHD-3D printing process could be influenced by the concentration of the printed solution and the electric field strength. In this study, the diameters of resultant fibers changed with variant PCL solution concentration values under 1.5 kV, and 10 wt.% PCL found to be the best sample with 100 ± 5 μm mean fiber diameter (Figure 2A). Additionally, different voltage values from 1 to 6 kV were applied for detecting the optimal fiber structure of 10 wt.% PCL (Figure 2B). Consequently, 1.5 kV was found to be the optimal value for EHD-3D printing of 10 wt.% PCL solution. In another preliminary study [30], different ratios of BC (3, 5, 7, 10, and 12 wt.%) were mixed with 10 wt.% PCL to produce layer-on-layer stacked patches. According to optical microscopy images, the 5:95 (wt. ratio) BC/PCL solution sample showed the best fiber structure and inter-fibrous pore size with 100 × 100 μm^2 (Figure 2C). Therefore, BC/PCL 5:95 (wt. ratio) blend solution was selected to carry AMX, AMP, and KAN antibiotics for further research made in this study.

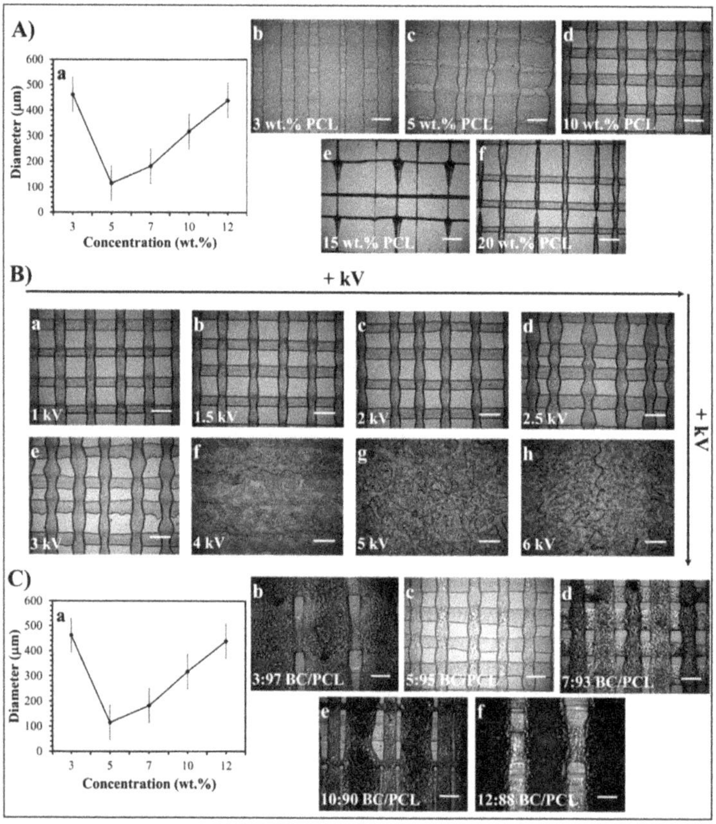

Figure 2. (**A**) EHD-3D printed fiber diameter graph (a) using different concentrations of PCL solution and optical images of the patches (b–f). (**B**) Optical images of the various voltage value (1–6 kV) applied EHD-3D printed deposited fibers of 10 wt.% PCL with different alignments and structures. (**C**) EHD-3D printed fiber diameter graph (a) using different concentrations of BC/PCL composition and optic images of the samples (b–f). (All scale bars indicate 100 μm).

Microfibers in the antibiotic patches were pictured with SEM after being left in the vacuum oven and represented along with digital photographs of relevant samples in Figure 3. According to morphological observations on SEM and measurements in the ImageJ program, inter-fibrous pore sizes were 100 × 100 μm^2 for BC/PCL (5:95 wt. ratio) and BC/PCL+AMX samples and sub-100 × 100 μm^2 for BC/PCL+AMP and BC/PCL+KAN samples, with mean fiber diameters of 50.1 ± 0.7

μm for BC/PCL, 77.1 ± 4.8 μm for BC/PCL+AMX, 86.3 ± 45.8 μm for BC/PCL+AMP, and 140.2 ± 9.1 μm for BC/PCL+KAN, respectively.

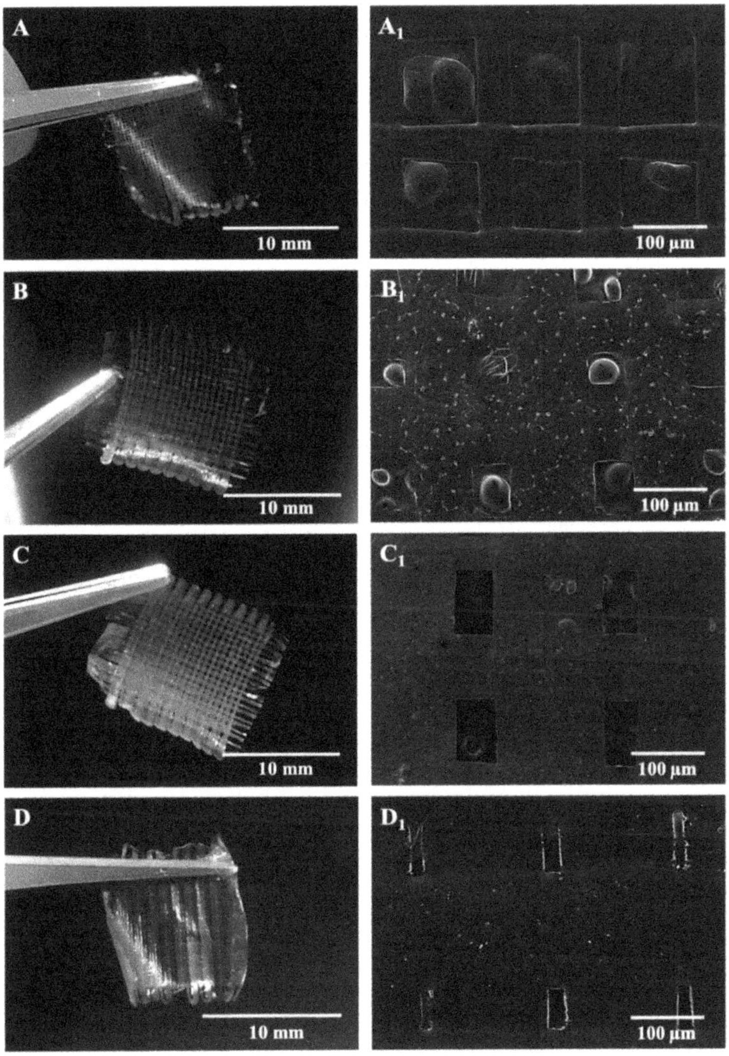

Figure 3. SEM images of EHD-3D printed patches with digital photographs of BC/PCL 5:95 (wt. ratio) patch with inter-fibrous pore size 100 × 100 μm^2 (**A**,**A$_1$**), AMX-loaded patch with inter-fibrous pore size 100 × 100 μm^2 (**B**,**B$_1$**), AMP-loaded patch with inter-fibrous pore size sub-100 × 100 μm^2 (**C**,**C$_1$**) and KAN-loaded patch with inter-fibrous pore size sub-100 × 100 μm^2 (**D**,**D$_1$**).

The variation in fiber diameter can be explained due to the varied molecular weight difference caused by different drugs since high molecular weight promotes obtaining fibers with a larger diameter [35]. Thus, the EHD-3D printing yields thinner fibers when the molecular weight decreases. However, the AMP drug had more molecular weight than AMX and BC/PCL+AMX fibers were thinner in diameter than BC/PCL+AMP fibers. In this case, this phenomenon can be explained based on the chemical structure of the monomer unit of BC/PCL 5:95 (wt. ratio) and its chemical interaction with AMP antibiotic.

Therefore, FTIR tests were performed to examine the presence and possible chemical interaction between polymers and antibiotics.

SEM images of patches revealed the interconnected pores on the microfibers. The rapid evaporation of DCM may have created a rise in local phase separation, and the solvent-rich regions may have transformed into pores during the process [36]. In this study, pore sizes were in the micro-range, and porosity varied with different types of antibiotics, as shown in Table 2. As a general rule, the pore size and porosity increase as a consequence of the addition of the antibiotics; the highest increase being recorded for the patches with AMX (highest pore size—10.20 ± 0.35 μm^2) and AMP (highest porosity—53.21 ± 1.76%) during the patch with KAN exhibit the lowest increase.

Table 2. Pore size and porosity of EHD-3D printed patches. (Mean ± SD, n = 3).

Sample Name (wt. ratio)	Pore Size (μm^2)	Porosity (%)
BC/PCL (5:95)	2.22 ± 0.85	11.52 ± 1.14
BC/PCL (5:95) +AMX	10.20 ± 0.35	32.43 ± 1.36
BC/PCL (5:95) +AMP	6.18 ± 0.12	53.21 ± 1.76
BC/PCL (5:95) +KAN	3.06 ± 1.09	19.11 ± 1.44

Biomedical applications require highly porous patches that should mimic the biological environment they are intended to interact with to improve biocompatibility and allow increased drug loading capacity [37]. Biological environments usually have porous structures ranging from nano- to macroscale [38].

FTIR spectroscopy was used to confirm the possible interaction between antibiotics and polymers in produced patches. Figure 4 shows the FTIR absorbance spectra of BC/PCL 5:95 (wt. ratio), neat AMX, BC/PCL+AMX, neat AMP, BC/PCL+AMP, neat KAN, and BC/PCL+KAN samples. Typical BC and PCL absorption bands were observed in all samples [30,39]. When the functional groups of the drug and the polymer interact chemically, FTIR spectra is expected to show band shifts and dilatations in the spectrum compared to the pure drug and polymer spectra [40]. No chemical binding between AMX and BC or PCL was observed when examining the AMX containing BC/PCL (5:95 weight ratio) patch. In addition, no chemical interaction between KAN and BC/PCL (5:95 weight ratio) was observed in the BC/PCL+KAN patch. AMX or KAN drugs were not chemically bound to the hydrogen bonds of BC or the carbonyl groups of PCL. They were physically bounded to the patches of the BC/PCL (5:95 weight ratio) blend. A physically bounded drug can be efficiently released from the patch. It is expected to have a faster release. Thus, a burst, such as a release, and high initial antimicrobial activity is expected [41].

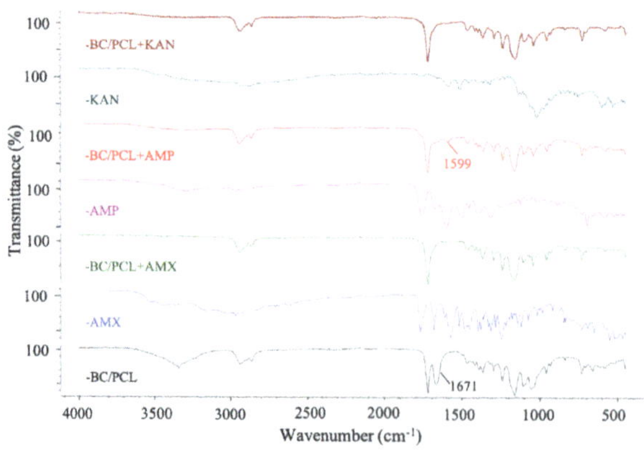

Figure 4. FTIR spectra of EHD-3D printed patches.

On the other hand, the band of BC/PCL at 1671 cm^{-1} shifted to 1599 cm^{-1} in BC/PCL+AMP patch, and this chemical interaction can be the reason for the increased fiber diameter of the BC/PCL+AMP patch. Moreover, solution characteristics can also be the reason for this, even though the AMP has a higher molecular weight than AMX. This difference can be mainly associated with the chemical structure, AMP being a very soluble molecule, oligosaccharide (Mw = 484.499 g/mol) with six free hydroxyl and four free amino groups, which can strongly interact via hydrogen bonds with the polymeric patch.

Mechanical properties of EHD-3D printed patches (BC/PCL, BC/PCL+AMX, BC/PCL+AMP, and BC/PCL+KAN) were investigated using a tensile testing device, and relevant stress–strain curves for patches are represented in Figure 5. Ultimate tensile strength and Young's modulus of the BC/PCL patch were 5.64 and 9.72 MPa, respectively. AMX, AMP, and KAN loaded BC/PCL patches exhibited lower ultimate tensile strength (3.87, 3.38, and 5.35 MPa, respectively) than the BC/PCL patch. As stated previously [42], poor interface interaction and unstable phase dispersion can compromise the mechanical properties of polymer composite systems, which can be the reason behind the reduction in tensile strength with the addition of antibiotics to the BC/PCL blend. AMX, AMP, and KAN contain patches that also exhibited poor Young's modulus values (7.58, 5.61, and 8.63 MPa, respectively) compared to the BC/PCL patch. As it is known, a decrease in Young's modulus results from an increase in the porosity of that material [43]. In this study, porosity values of AMX, AMP, and KAN loaded patches are found higher than the BC/PCL patch and this increase in porosity reduced Young's modulus values. The Young's modulus of the human soft tissues ranges from 0.6 to 1.7 MPa [44]. Hence, it is thought that the produced drug patches can meet the mechanical requirements needed at various anatomical locations, such as the skin, muscle, or tissues where the drug needs to be delivered.

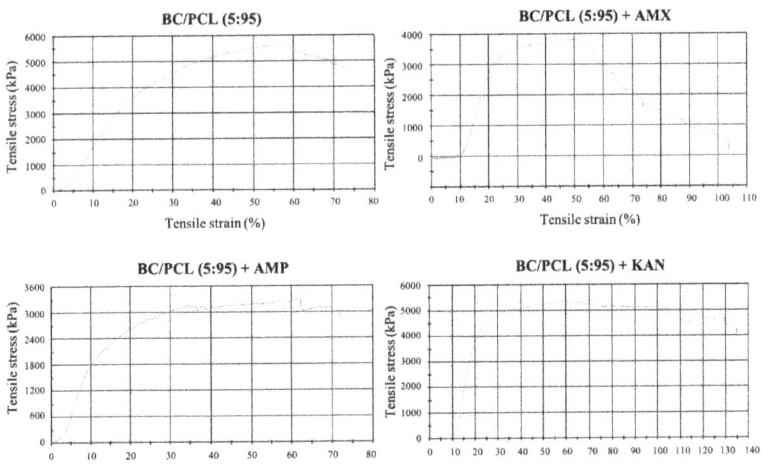

Sample Name	Ultimate Tensile Strength (MPa)	Young's modulus (MPa)
BC/PCL (5:95)	5.64 ± 0.04	9.72 ± 0.05
BC/PCL (5:95) + AMX	3.87 ± 0.03	7.58 ± 0.03
BC/PCL (5:95) + AMP	3.38 ± 0.02	5.61 ± 0.03
BC/PCL (5:95) + KAN	5.35 ± 0.03	8.63 ± 0.04

Figure 5. Stress-strain curves of EHD-3D printed patches with ultimate tensile strength and Young's modulus values.

Since swelling properties of a scaffold are vital because of its role in enabling liquid absorption of physiological secretions and allow the efficient exchange of nutrients and waste of regenerated tissue [34], the neat and the drug-containing patches were further characterized to determine their swelling behavior after submersion in the phosphate buffer saline (PBS, pH = 7.4) at 37 °C for 24 h. At the end of 24 h of immersions, the degree of swelling of the BC/PCL patch (168%) was comparable with those of the patches loaded with antibiotics: AMX loaded patch ≈ 162%, AMP loaded patch ≈ 159%, and KAN loaded patch ≈ 164% (Figure 6). Considering a release rate of about 30–35% of the antibiotics, it can assume that the presence of antibiotics increases the swelling ratio because of the additional porosity assured after the antibiotic release. The swelling and water uptake depends on the hydrophobicity of the polymer [45], as it is well-known that the PCL is a hydrophobic polyester and can prevent water uptake [46]. Besides, since BC is a hydrophilic polymer, it can pick the water that diffuses into the PCL matrix while increasing the equilibrium swelling of the water [47].

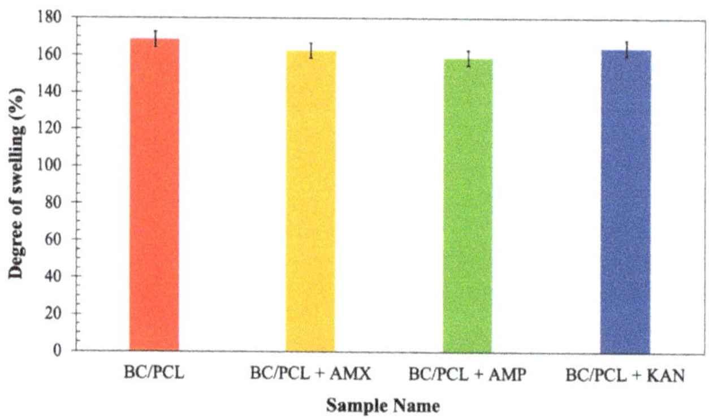

Figure 6. The degree of swelling (%) of EHD-3D printed patches.

Antibiotic addition to the BC/PCL blend in the selected quantities did not interfere with patch water uptake capability. Consequently, produced patches can prevent loss of body fluid and nutrients due to the enhanced swelling ability while releasing their loaded drugs.

The yield percentage, drug loading, and antibiotic encapsulation efficiency of the patches are presented in Table 3. The drug loading was similar for the three different antibiotics (21%), while the encapsulation efficiency of the drugs was similar for the three different antibiotics and between 96% and 98%.

Table 3. Yield percentage, drug loading, and encapsulation efficiency of the patches. (Mean ± SD, n = 3).

Sample Name (wt. ratio)	Yield (%)	Drug Loading (%)	Encapsulation Efficiency (%)
BC/PCL (5:95)	98.4 ± 0.6	-	-
BC/PCL (5:95) +AMX	98.2 ± 1.3	21.9 ± 0.3	97.4 ± 0.9
BC/PCL (5:95) +AMP	96.3 ± 2.1	21.7 ± 0.7	96.8 ± 1.6
BC/PCL (5:95) +KAN	97.9 ± 1.9	21.9 ± 0.5	98.6 ± 1.4

3.2. Drug Release and Mathematical Modeling of Release Mechanisms

Measurement of drug release from the patches was performed using the UV spectroscopy method. For this purpose, AMX, AMP, and KAN calibration curves were prepared to determine release kinetics (Figure 7).

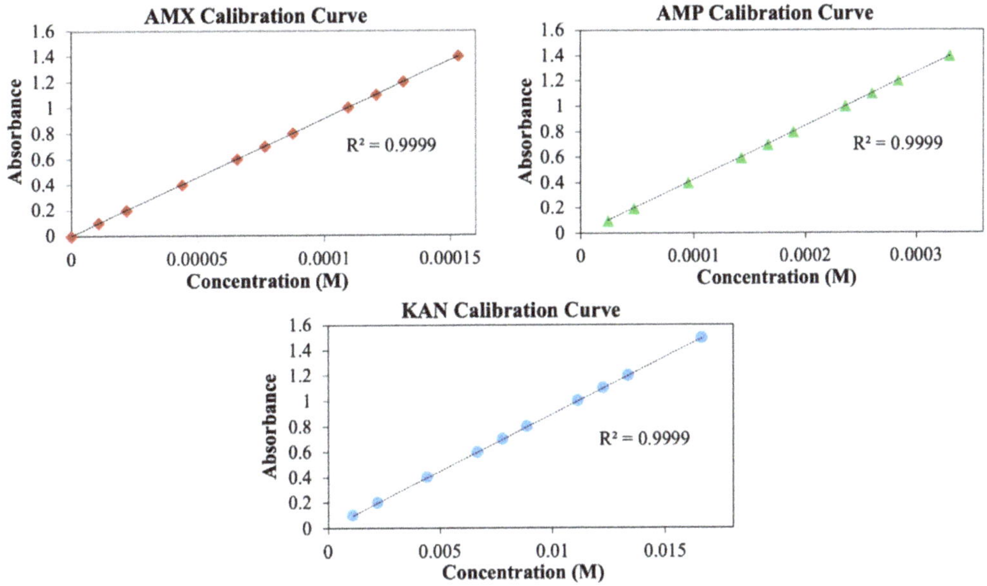

Figure 7. Calibration curves for AMX, AMP, and KAN drugs.

It is crucial to investigate drug release profiles for several days due to the advantages of increasing medication and drug efficacy, reducing side effects, and reducing the frequency of administration of these drugs [48]. In this study, the release of AMX, AMP, and KAN drugs from EHD-3D printed patches were examined for 14 days (336 h) in PBS, pH 7.4 (Figure 8).

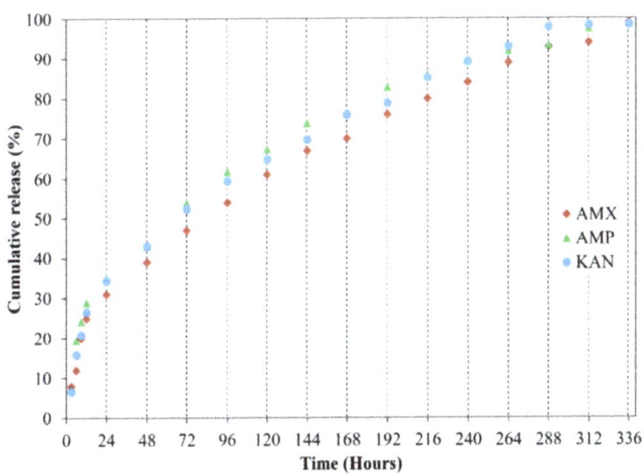

Figure 8. Cumulative release (%) of loaded antibiotic patches.

All release curves mainly consisted of two phases as the first phase, a "burst release", and then "slow-release", so that the second phase took place. The initial burst release is beneficial for releasing antibiotic drugs since it is essential to remove the spreading bacteria before they start to multiply [49]. Hence, to prevent the remaining population for several organisms that survive the first immediate release, followed by a sustained release of antibiotics is required [50]. After the burst release phase, increases were observed in the release of drugs due to the concentration of AMX, AMP, and KAN in the patches, and then these increases were found to be linear and prolonged over many days. The AMX loaded patch showed 99.1 ± 0.6% cumulative release after 336 h, whereas the cumulative release of AMP was about 98.1 ± 0.3% after 336 h. The cumulative release of KAN was 98.6 ± 0.6% after 336 h of release studies. The increased surface area for the release of the drugs can be responsible for improving the release of the drugs when formulated as microfibers [51].

Wang et al. [52] produced antibiotically loaded (tetracycline hydrochloride) patches using polycaprolactone (PCL), polyvinyl pyrrolidone (PVP), and their composite system (PVP-PCL) by the EHD-3D printing method. The researchers examined the release of antibiotic drug from PCL-PVP dosage forms for 5 days. They witnessed a slower antibiotic release compared to neat PCL or PVP. It was observed that the size of gaps between the fibers in the printed patches influenced the release of the antibiotic material. As a result of this study, it was found that the EHD-3D printing method is a promising method for adapting the dosage forms with minimum adjuvants and large-scale procedures.

The release curves were also subjected to model fitting consisting of various models such as zero order, first order, Higuchi, and Korsmeyer–Peppas models (Figures 9–11). Graphical representations of the cumulative percentage of drug release against time showed that the release of AMX, AMP, and KAN from the patches correctly followed the Higuchi model. Drug release profiles were very close to the trend line or regression line and had the highest coefficient value (R^2) for AMX, AMP, and KAN loaded patches (0.9961, 0.9761, and 0.9888, respectively).

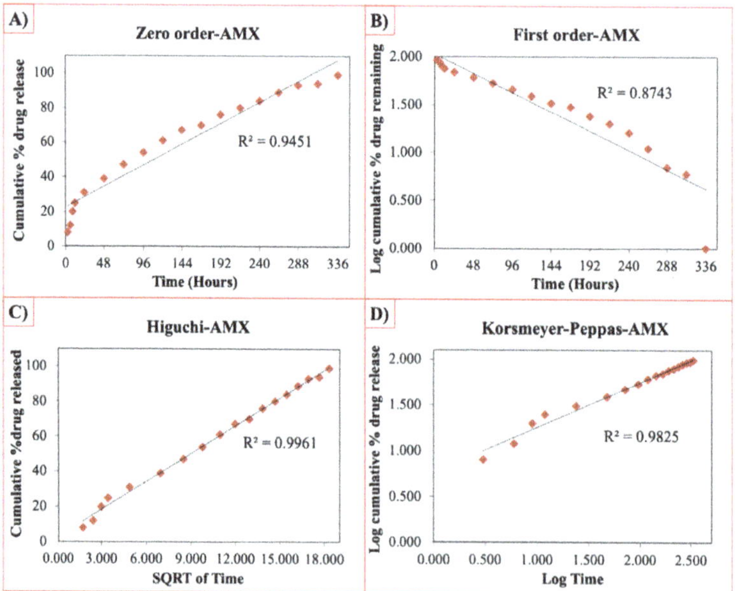

Figure 9. Zero order (**A**), first order (**B**), Higuchi (**C**), Korsmeyer–Peppas (**D**) kinetic releases of BC/PCL+AMX patch.

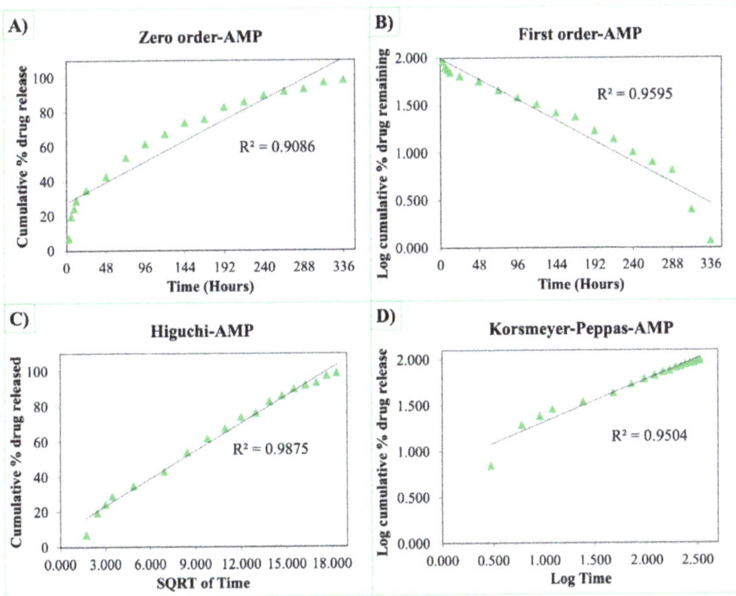

Figure 10. Zero order (**A**), first order (**B**), Higuchi (**C**), Korsmeyer–Peppas (**D**) kinetic releases of BC/PCL+AMP patch.

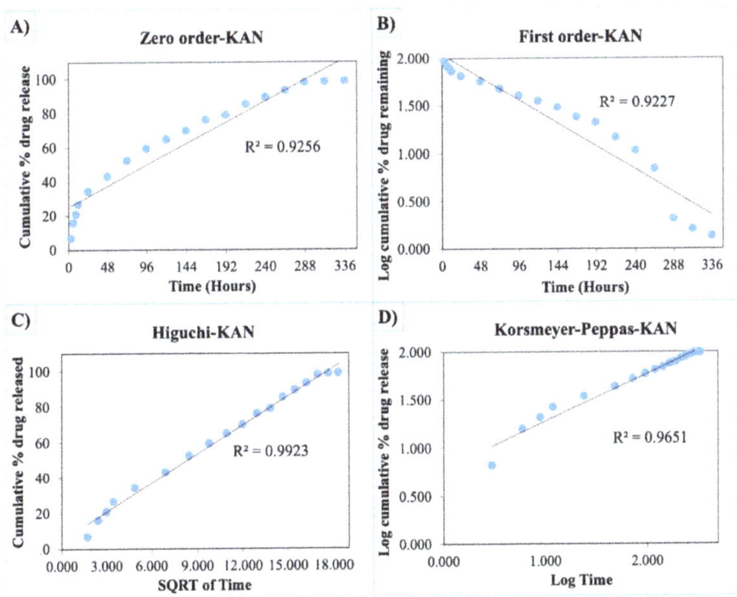

Figure 11. Zero order (**A**), first order (**B**), Higuchi (**C**), Korsmeyer–Peppas (**D**) kinetic releases of BC/PCL+KAN patch.

The Korsmeyer–Peppas model was applied to release profiles to determine the diffusion types. After the Korsmeyer–Peppas model application, the diffusional exponent, which is indicative of the release mechanism (n), was found for all samples. The n = 0.45

value indicates Fickian diffusion, while values between 0.45 and 1.00 suggest non-Fickian transport and values higher than 1.00 suggest zero order/case II transport of a drug [53].

The diffusion exponent 'n' was 0.49, 0.46, and 0.48 for AMX, AMP, and KAN loaded patches, respectively. The 'n' values for AMX, AMP, and KAN loaded patches indicated that the releases of the drugs followed Fick's law of diffusion. As a general conclusion, regardless of the solubility of the antibiotic, the release of the drugs from patches was likely to be controlled by a diffusion mechanism [54].

3.3. Antibacterial Assessment of Fabricated Patches

The antibacterial potentials of the patches were evaluated as the diameter of the zone around the 3D printed scaffolds. The experiment was performed in triplicate, and the average of the zones of inhibition used to determine the antibacterial activity.

Table 4 shows the inhibition zones after 24 h of incubation. The inhibition zones were observed as a clear circular area surrounding the samples specifying the zone of bacteria that was destroyed or prevented from growing. BC/PCL was used as a negative control and did not have any antibacterial properties, as they showed no activity against the two bacterial strains (*S. aureus* and *E. coli*). According to the results in Table 4, BC/PCL+AMX, BC/PCL+AMP, and BC/PCL+KAN drug patches showed 43, 39, and 30 mm diameter inhibition zones against *S. aureus*, respectively. On the other hand, the same drug patches showed 35, 37, and 35 mm inhibition zones against *E. coli*, respectively. Results suggest that all three different antibiotics (AMX, AMP, and KAN) containing patches can be used as an effective antibacterial patch on both Gram-positive and Gram-negative bacteria.

Table 4. Antibacterial properties of the EHD-3D printed patches against *S. aureus* and *E. coli* bacteria. All scale bars indicate 10 mm.

Sample / Bacterial Strain	BC/PCL	BC/PCL+AMX	BC/PCL+AMP	BC/PCL+KAN
S. aureus	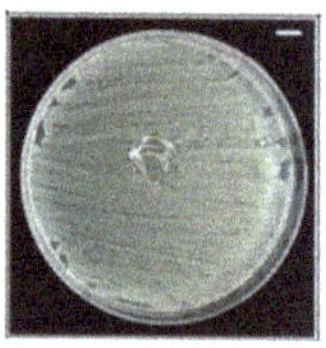	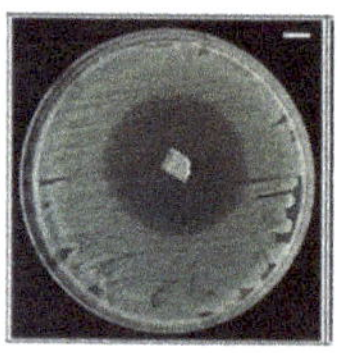	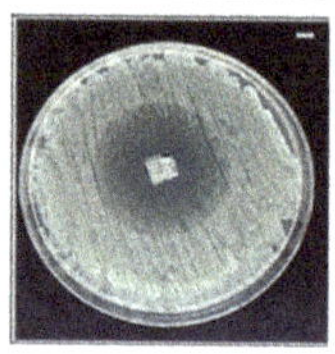	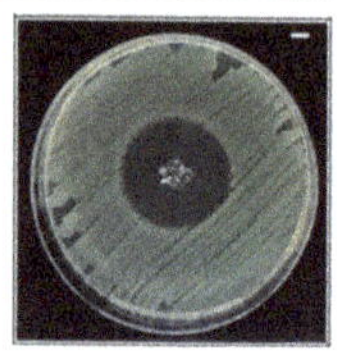
E. coli	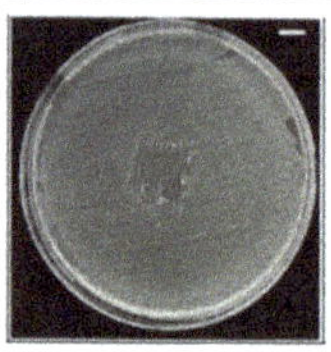	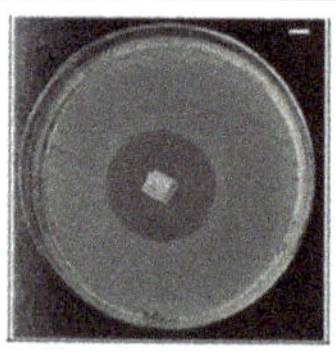	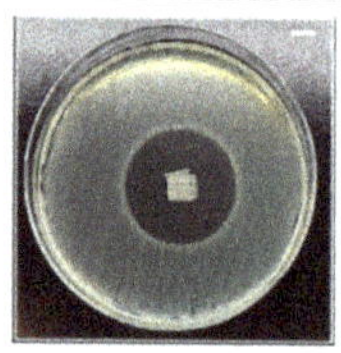	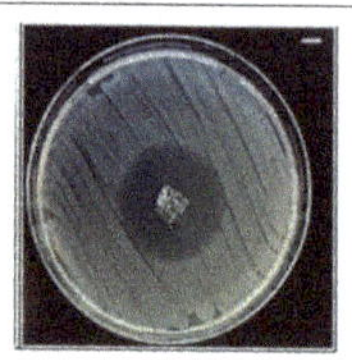

4. Conclusions

Antibiotic patches composed of BC/PCL, AMX, AMP, and KAN, were prepared by the EHD-3D printing method, resulting in aligned layer-on-layer stacked fibers. The drug release properties and drug release kinetic models of different antibiotics loaded patches were evaluated. Drug-containing patches displayed micron-scaled fibers and inter-fibrous pore size. Moreover, microfibers exhibited a porous structure, which provided improved biocompatibility and increased drug loading capacity. All patches displayed enhanced swelling ability. They were found able to meet mechanical requirements needed at various anatomical locations according to tensile test results. The release of AMX, AMP, and KAN antibiotics from BC/PCL dosage forms was shown at 14 days (336 h). The kinetic models (zero order, first order, Higuchi, and Korsmeyer–Peppas) on drug release were profiled for each antibiotic patches. It was found that the release of the drugs from the patches followed Fick's law of diffusion. The antibacterial potential was evaluated, and

AMX, AMP, and KAN were found effective on both Gram-positive and Gram-negative bacteria to provide antimicrobial activity to BC/PCL patches. It is suggested that the single-step EHD-3D printing method offers rapidly tailored dosage forms with minimal excipients. Moreover, various types of antibiotics could be loaded and successfully released, regardless of their types, from produced BC/PCL patches. Produced antibiotic patches can meet the characteristic requirements needed in local transdermal applications with accelerated release characteristics, adjustability, and versatility for wound dressing studies, and decrease the pharmaceuticals industries' costs.

Author Contributions: Conceptualization, D.M.K. and O.G.; Data curation, E.A., E.Y., N.E., D.F. and G.D.; Formal analysis, E.A., E.Y., N.E. and G.D.; Investigation, G.D.; Methodology, E.A., E.Y. and N.E.; Project administration, O.G.; Supervision, D.M.K., D.F. and O.G.; Visualization, D.M.K., D.F. and A.F.; Writing—original draft, E.A., E.Y., N.E., D.F., G.D., A.F. and O.G.; Writing—review & editing, D.M.K., A.F. and O.G. All authors have read and agreed to the published version of the manuscript.

Funding: The APC was funded by UPB.

Institutional Review Board Statement: Not applicable.

Informed Consent Statement: Not applicable.

Data Availability Statement: The raw/processed data required to reproduce these findings cannot be shared at this time due to technical or time limitations. Data will be made available on request.

Acknowledgments: This study was supported by BAPKO, Marmara University; project no: FEN-C-YLP-110117-0021. D.M.K. would like to acknowledge UKIERI funding, project IND/CONT/G/17-18/46 for collaborative work.

Conflicts of Interest: The authors declare no conflict of interest.

References

1. Negut, I.; Grumezescu, V.; Grumezescu, A.M. Treatment Strategies for Infected Wounds. *Molecules* **2018**, *23*, 2392. [CrossRef] [PubMed]
2. Misic, A.M.; Gardner, S.E.; Grice, E.A. The Wound Microbiome: Modern Approaches to Examining the Role of Microorganisms in Impaired Chronic Wound Healing. *Adv. Wound Care* **2014**, *3*, 502–510. [CrossRef] [PubMed]
3. Mathur, S.; Sutton, J. Personalized medicine could transform healthcare. *Biomed. Rep.* **2017**, *7*, 3–5. [CrossRef] [PubMed]
4. Ramasubbu, D.A.; Smith, V.; Hayden, F.; Cronin, P. Systemic antibiotics for treating malignant wounds. *Cochrane Database Syst. Rev.* **2017**, *8*, CD011609. [CrossRef] [PubMed]
5. Everts, R. How to Treat Wound Infection. Prevention and Treatment. 2016. Available online: https://www.acc.co.nz/assets/provider/treating-wound-infections.pdf (accessed on 7 April 2020).
6. Smith, R.; Russo, J.; Fiegel, J.; Brogden, N. Antibiotic Delivery Strategies to Treat Skin Infections When Innate Antimicrobial Defense Fails. *Antibiotics* **2020**, *9*, 56. [CrossRef] [PubMed]
7. Langer, R. New methods of drug delivery. *Science* **1990**, *249*, 1527–1533. [CrossRef] [PubMed]
8. Saltzman, W.M. *Drug Delivery–Engineering Principles for Drug Therapy*; Oxford University Press: Oxford, UK, 2001.
9. Berchane, N.; Jebrail, F.; Andrews, M. Optimization of PLG microspheres for tailored drug release. *Int. J. Pharm.* **2010**, *383*, 81–88. [CrossRef]
10. Wei, L.; Cai, C.; Lin, J.; Chen, T. Dual-drug delivery system based on hydrogel/micelle composites. *Biomaterials* **2009**, *30*, 2606–2613. [CrossRef]
11. Ranganath, S.H.; Kee, I.; Krantz, W.B.; Chow, P.K.-H.; Wang, C.-H. Hydrogel Matrix Entrapping PLGA-Paclitaxel Microspheres: Drug Delivery with Near Zero-Order Release and Implantability Advantages for Malignant Brain Tumour Chemotherapy. *Pharm. Res.* **2009**, *26*, 2101–2114. [CrossRef]
12. Hoare, T.R.; Kohane, D.S. Hydrogels in drug delivery: Progress and challenges. *Polymer* **2008**, *49*, 1993–2007. [CrossRef]
13. Bhattacharyya, S.; Wang, H.; Ducheyne, P. Polymer-coated mesoporous silica nanoparticles for the controlled release of macromolecules. *Acta Biomater.* **2012**, *8*, 3429–3435. [CrossRef]
14. Akhtar, N.; Singh, V.; Yusuf, M.; Khan, R.A. Non-invasive drug delivery technology: Development and current status of transdermal drug delivery devices, techniques and biomedical applications. *Biomed. Tech. Eng.* **2020**, *65*, 243–272. [CrossRef]
15. Pastore, M.N.; Kalia, Y.N.; Horstmann, M.; Roberts, M.S. Transdermal patches: History, development and pharmacology. *Br. J. Pharmacol.* **2015**, *172*, 2179–2209. [CrossRef]
16. Jamróz, W.; Szafraniec, J.; Kurek, M.; Jachowicz, R. 3D Printing in Pharmaceutical and Medical Applications—Recent Achievements and Challenges. *Pharm. Res.* **2018**, *35*, 1–22. [CrossRef]

17. Park, J.-U.; Lee, J.H.; Paik, U.; Lu, Y.; Rogers, J.A. Nanoscale Patterns of Oligonucleotides Formed by Electrohydrodynamic Jet Printing with Applications in Biosensing and Nanomaterials Assembly. *Nano Lett.* **2008**, *8*, 4210–4216. [CrossRef]
18. Huang, Y.; Bu, N.; Duan, Y.; Pan, Y.; Liu, H.; Yin, Z.; Xiong, Y. Electrohydrodynamic direct-writing. *Nanoscale* **2013**, *5*, 12007–12017. [CrossRef]
19. Meier, W. Polymer nanocapsules. *Chem. Soc. Rev.* **2000**, *29*, 295–303. [CrossRef]
20. Park, J.U. High-Resolution Electrohydrodynamic Jet Printing Methods for Applications in Electronics and Biotechnology. Ph.D. Thesis, University of Illinois at Urbana-Champaign, Champaign, IL, USA, 2009.
21. Wang, K.; Cai, L.; Zhang, L.; Dong, J.; Wang, S. Biodegradable Photo-Crosslinked Polymer Substrates with Concentric Microgrooves for Regulating MC3T3-E1 Cell Behavior. *Adv. Heal. Mater.* **2012**, *1*, 292–301. [CrossRef]
22. Ahn, S.H.; Lee, H.J.; Kim, G.H. Polycaprolactone Scaffolds Fabricated with an Advanced Electrohydrodynamic Direct-Printing Method for Bone Tissue Regeneration. *Biomacromolecules* **2011**, *12*, 4256–4263. [CrossRef]
23. Altun, E.; Akyol, S.; Ekren, N.; Kilic, O.; Gündüz, O.; Kılıç, O. Electrohydrodynamic (EHD) Bioprinting of Polycaprolactone Scaffolds. *Mater. Sci. Forum* **2018**, *923*, 93–97. [CrossRef]
24. Béduneau, A.; Saulnier, P.; Anton, N.; Hindré, F.; Passirani, C.; Rajerison, H.; Noiret, N.; Benoit, J.-P. Pegylated Nanocapsules Produced by an Organic Solvent-Free Method: Evaluation of their Stealth Properties. *Pharm. Res.* **2006**, *23*, 2190–2199. [CrossRef]
25. Albrecht, L.D.; Sawyer, S.W.; Soman, P. Developing 3D Scaffolds in the Field of Tissue Engineering to Treat Complex Bone Defects. *3D Print. Addit. Manuf.* **2016**, *3*, 106–112. [CrossRef]
26. Bernards, D.A.; Lance, K.D.; Ciaccio, N.A.; Desai, T.A. Nanostructured Thin Film Polymer Devices for Constant-Rate Protein Delivery. *Nano Lett.* **2012**, *12*, 5355–5361. [CrossRef]
27. Altun, E.; Aydogdu, M.O.; Koc, F.; Crabbe-Mann, M.; Brako, F.; Kaur-Matharu, R.; Ozen, G.; Kuruca, S.E.; Edirisinghe, U.; Gunduz, O.; et al. Novel Making of Bacterial Cellulose Blended Polymeric Fiber Bandages. *Macromol. Mater. Eng.* **2018**, *303*, 1700607. [CrossRef]
28. Shah, N.; Ul-Islam, M.; Khattak, W.A.; Park, J.K. Overview of bacterial cellulose composites: A multipurpose advanced material. *Carbohydr. Polym.* **2013**, *98*, 1585–1598. [CrossRef]
29. Qiu, Y.; Qiu, L.; Cui, J.; Wei, Q. Bacterial cellulose and bacterial cellulose-vaccarin membranes for wound healing. *Mater. Sci. Eng. C* **2016**, *59*, 303–309. [CrossRef]
30. Altun, E.; Ekren, N.; Kuruca, S.E.; Gunduz, O. Cell studies on Electrohydrodynamic (EHD)-3D-bioprinted Bacterial Cellulose\Polycaprolactone scaffolds for tissue engineering. *Mater. Lett.* **2019**, *234*, 163–167. [CrossRef]
31. Puppi, D.; Piras, A.M.; Detta, N.; Dinucci, D.; Chiellini, F. Poly(lactic-co-glycolic acid) electrospun fibrous meshes for the controlled release of retinoic acid. *Acta Biomater.* **2010**, *6*, 1258–1268. [CrossRef] [PubMed]
32. Samuelov, Y.; Donbrow, M.; Friedman, M. Sustained Release of Drugs from Ethylcellulose-Polyethylene Glycol Films and Kinetics of Drug Release. *J. Pharm. Sci.* **1979**, *68*, 325–329. [CrossRef] [PubMed]
33. Ritger, P.L.; Peppas, N.A. A simple equation for description of solute release I. Fickian and non-fickian release from non-swellable devices in the form of slabs, spheres, cylinders or discs. *J. Control. Release* **1987**, *5*, 23–36. [CrossRef]
34. Aydogdu, M.O.; Altun, E.; Crabbe-Mann, M.; Brako, F.; Koç, F.; Ozen, G.; Kuruca, S.E.; Edirisinghe, U.; Luo, C.; Gündüz, O.; et al. Cellular interactions with bacterial cellulose: Polycaprolactone nanofibrous scaffolds produced by a portable electrohydrodynamic gun for point-of-need wound dressing. *Int. Wound J.* **2018**, *15*, 789–797. [CrossRef]
35. Roldán, G.J.C.; Bolivariana, U.P.; Martínez, Y.Q.; Gómez, L.M.A.; Vinasco, L.F.R.; Palacio, L.M.H. Influence of the molecular weight of polymer, solvents and operational condition in the electrospinning of polycaprolactone. *Revista Facultad de Ingeniería Universidad de Antioquia* **2017**, *84*, 35–45. [CrossRef]
36. Celebioglu, A.; Uyar, T. Electrospun porous cellulose acetate fibers from volatile solvent mixture. *Mater. Lett.* **2011**, *65*, 2291–2294. [CrossRef]
37. Wendorff, J.H.; Agarwal, S.; Greiner, A. *Medicinal Applications for Electrospun Nanofibers. Electrospinning*; Wiley-VCH Verlag GmbH & Co. KGaA: Weinheim, Germany, 2012; pp. 217–236.
38. Currey, J.D. Biomechanics of Mineralized Skeletons. In *Skeletal Biomineralization: Patterns, Processes and Evolutionary Trends*; Carter, J.G., Ed.; Wiley: Hoboken, NJ, US, 2013.
39. Elzein, T.; Nasser-Eddine, M.; Delaite, C.; Bistac, S.; Dumas, P. FTIR study of polycaprolactone chain organization at interfaces. *J. Colloid Interface Sci.* **2004**, *273*, 381–387. [CrossRef]
40. Silverstein, R.M.; Bassler, G.C.; Morril, T.C. *Spectrometric Identification of Organic Compounds*; Wiley: New York, NY, USA, 1991.
41. Altun, E.; Aydogdu, M.O.; Koc, F.; Kutlu, O.; Gozuacik, D.; Yucel, S.; Gunduz, O. Amoxicillin Loaded Hollow Microparticles in the Treatment of Osteomyelitis Disease Using Single-Nozzle Electrospinning. *BioNanoScience* **2018**, *8*, 790–801. [CrossRef]
42. Wang, J.-C.; Chang, M.-W.; Ahmad, Z.; Li, J.-S. Fabrication of patterned polymer-antibiotic composite fibers via electrohydrodynamic (EHD) printing. *J. Drug Deliv. Sci. Technol.* **2016**, *35*, 114–123. [CrossRef]
43. Niinomi, M.; Hattori, T.; Niwa, S.; Yaszemski, M.; Trantolo, D.; Lewandrowski, K.-U.; Hasirci, V.; Altobelli, D.; Wise, D. Material Characteristics and Biocompatibility of Low Rigidity Titanium Alloys for Biomedical Applications. In *Biomaterials in Orthopedics*; Yaszemski, M., Trantolo, D., Lewandrowski, K., Hasirci, V., Altobelli, D., Wise, D., Eds.; Marcel Dekker, INC.: Basel, Switzerland, 2004; pp. 41–62.
44. Griffin, M.; Premakumar, Y.; Seifalian, A.; Butler, P.E.; Szarko, M. Biomechanical Characterization of Human Soft Tissues Using Indentation and Tensile Testing. *J. Vis. Exp.* **2016**, *118*, e54872. [CrossRef]

45. Rodríguez-Galán, A.; Franco, L.; Puiggalí, J.; García, M.L.F. Biodegradable Polyurethanes and Poly(ester amide)s. In *Handbook of Biodegradable Polymers: Isolation, Synthesis, Characterization and Applications*; Lendlein, A., Sisson, A., Eds.; Wiley: Hoboken, NJ, USA, 2011; Chapter 6; pp. 133–154.
46. Dash, T.K.; Konkimalla, V.B. Poly-ϵ-caprolactone based formulations for drug delivery and tissue engineering: A review. *J. Control. Release* **2012**, *158*, 15–33. [CrossRef]
47. Altun, E.; Aydogdu, M.O.; Togay, S.O.; Sengil, A.Z.; Ekren, N.; Haskoylu, M.E.; Oner, E.T.; Altuncu, N.A.; Ozturk, G.; Crabbe-Mann, M.; et al. Bioinspired scaffold induced regeneration of neural tissue. *Eur. Polym. J.* **2019**, *114*, 98–108. [CrossRef]
48. Valarezo, E.; Tammaro, L.; González, S.; Malagón, O.; Vittoria, V. Fabrication and sustained release properties of poly(ε-caprolactone) electrospun fibers loaded with layered double hydroxide nanoparticles intercalated with amoxicillin. *Appl. Clay Sci.* **2013**, *72*, 104–109. [CrossRef]
49. Gimeno, M.; Pinczowski, P.; Pérez, M.; Giorello, A.; Martínez, M.Á.; Santamaría, J.; Arruebo, M.; Luján, L. A controlled antibiotic release system to prevent orthopedic-implant associated infections: An in vitro study. *Eur. J. Pharm. Biopharm.* **2015**, *96*, 264–271. [CrossRef]
50. Shin, Y.M.; Hohman, M.M.; Brenner, M.P.; Rutledge, G.C. Electrospinning: A whipping fluid jet generates submicron polymer fibers. *Appl. Phys. Lett.* **2001**, *78*, 1149–1151. [CrossRef]
51. Chou, S.-F.; Carson, D.; Woodrow, K.A. Current strategies for sustaining drug release from electrospun nanofibers. *J. Control. Release* **2015**, *220*, 584–591. [CrossRef]
52. Wang, J.-C.; Zheng, H.; Chang, M.-W.; Ahmad, Z.; Li, J.-S. Preparation of active 3D film patches via aligned fiber electrohydrodynamic (EHD) printing. *Sci. Rep.* **2017**, *7*. [CrossRef]
53. Peppas, N.A.; Duncan, R.; Wnek, G.E.; Hoffman, A.S.; Gao, G.H.; Kim, S.W.; Lee, D.S.; Hadjiargyrou, M.; Touitou, E.; Ainbinder, D.; et al. Highly cited research articles in Journal of Controlled Release: Commentaries and perspectives by authors. *J. Control. Release* **2014**, *190*, 29–74. [CrossRef]
54. Fu, Y.; Kao, W.J. Drug release kinetics and transport mechanisms of non-degradable and degradable polymeric delivery systems. *Expert Opin. Drug Deliv.* **2010**, *7*, 429–444. [CrossRef]

pharmaceutics

MDPI

Article

PLGA/PEG Nanoparticles Loaded with Cyclodextrin-*Peganum harmala* Alkaloid Complex and Ascorbic Acid with Promising Antimicrobial Activities

Sherif Ashraf Fahmy [1,†], Noha Khalil Mahdy [1,†], Hadeer Al Mulla [1], Aliaa Nabil ElMeshad [2,3], Marwa Y. Issa [4] and Hassan Mohamed El-Said Azzazy [1,*]

1 Department of Chemistry, School of Sciences & Engineering, The American University in Cairo, AUC Avenue, P.O. Box 74, New Cairo 11835, Egypt; sheriffahmy@aucegypt.edu (S.A.F.); noha.khalil@aucegypt.edu (N.K.M.); hadeer.almulla@aucegypt.edu (H.A.M.)
2 Department of Pharmaceutics and Industrial Pharmacy, Faculty of Pharmacy, Cairo University, Kasr El-Aini Street, Cairo 11562, Egypt; aliaa.elmeshad@pharma.cu.edu.eg
3 Department of Pharmaceutics, Faculty of Pharmacy and Drug Technology, The Egyptian Chinese University, Gesr El Suez Street, Cairo 11786, Egypt
4 Department of Pharmacognosy, Faculty of Pharmacy, Cairo University, Kasr El-Aini Street, Cairo 11562, Egypt; marwa.issa@pharma.cu.edu.eg
* Correspondence: hazzazy@aucegypt.edu; Tel.: +20-226-152-559; Fax: +20-227-957-565
† These authors contributed equally to this work.

Abstract: Antimicrobial drugs face numerous challenges, including drug resistance, systemic toxic effects, and poor bioavailability. To date, treatment choices are limited, which warrants the search for novel potent antivirals, including those extracted from natural products. The seeds of *Peganum harmala* L. (*Zygophyllaceae* family) have been reported to have antimicrobial, antifungal, and anticancer activities. In the present study, a 2-hydroxy propyl-β-cyclodextrin (HPβCD)/harmala alkaloid-rich fraction (HARF) host–guest complex was prepared using a thin-film hydration method to improve the water solubility and bioavailability of HARF. The designed complex was then co-encapsulated with ascorbic acid into PLGA nanoparticles coated with polyethylene glycol (HARF–HPßCD/AA@PLGA-PEG NPs) using the W/O/W multiple emulsion-solvent evaporation method. The average particle size, PDI, and zeta potential were 207.90 ± 2.60 nm, 0.17 ± 0.01, and 31.6 ± 0.20 mV, respectively. The entrapment efficiency for HARF was 81.60 ± 1.20% and for ascorbic acid was 88 ± 2.20%. HARF–HPßCD/AA@PLGA-PEG NPs had the highest antibacterial activity against *Staphylococcus aureus* and *Escherichia coli* (MIC of 0.025 mg/mL). They also exhibited high selective antiviral activity against the H1N1 influenza virus (IC50 2.7 μg/mL) without affecting the host (MDCK cells). In conclusion, the co-encapsulation of HPCD–HARF complex and ascorbic acid into PLGA-PEG nanoparticles significantly increased the selective H1N1 killing activity with minimum host toxic effects.

Keywords: *Peganum harmala*; ascorbic acid (Vitamin C); 2-hydroxy propyl-β-cyclodextrin; PLGA; PEG; antiviral; antibacterial

1. Introduction

Microbial infections, including bacterial and viral infections, represent a significant health burden responsible for many deaths worldwide. Influenza A virus causes the hospitalization of about 4 million cases and the death of 600,000 patients each year. The pervasive use of antimicrobials is the leading cause of resistance to the existing medications [1]. Moreover, most antimicrobial agents suffer from many drawbacks, including systemic adverse reactions, allergy, poor bioavailability, and narrow spectrum [2,3].

Nature provides an enormous library of novel chemicals that can be explored to develop antimicrobials. *Peganum harmala* L. (*Zygophyllaceae* family) is a member of the *Zygophyllaceae* family native to the area from the eastern Mediterranean region to India [3,4].

The seeds of *P. harmala* are rich in harmala alkaloids such as quinazoline and β-carboline alkaloids. The major alkaloid, harmine, is responsible for the various pharmacological activities of the harmala seeds [5–7]. Several studies have reported the promising antimicrobial activities of *P. harmala* extracts [8,9].

Ascorbic acid is an essential dietary nutrient required for various biological functions [10]. Several in vivo studies have reported the ability of ascorbic acid to prevent and alleviate many types of viral infections [10,11]. Ascorbic acid was found to improve the immune response to viral infections by stimulating the function and proliferation of T-lymphocyte and NK-lymphocyte and the production of interferon [11].

Despite the global trend to replace synthetic antimicrobials with natural ones, the use of natural products to treat microbial infections may face challenges similar to those of synthetic drugs, such as poor water solubility, low bioavailability, and non-selective targeting the infected organ. Thus, numerous nanoplatforms, including liposomes, supramolecular systems, and polymeric nanoparticles, were designed, being reported to remarkably ameliorate the therapeutic activities of different types of biologically active compounds [12–18].

Supramolecular host molecules, such as cyclodextrins (CDs), are involved in overcoming the poor water solubility of the natural compounds via the formation of host–guest complexes. CDs are amphiphilic cyclic oligosaccharides obtained from the enzymatic hydrolysis of starch. They comprise three types, namely, α, β, and γ-CDs, consisting of six, seven, and eight glucopyranose units, respectively [19]. In this context, 2-hydroxy propyl ß cyclodextrin (HPßCD) is a promising derivative of CDs due to its safety, high water solubility, and ability to selectively form inclusion complexes with various drug molecules [20].

Polymeric nanoparticles (NPs) have been extensively involved in drug delivery, owing to their ability to encapsulate different therapeutically active compounds into their matrix and the ability of their surfaces to be decorated with different functional groups [3,21,22]. Consequently, this could improve the stability of their cargos and prolong their half-lives. Poly (lactic-co-glycolic acid) (PLGA) is commonly employed in the fabrication of polymeric NPs since it is a biocompatible and biodegradable polymer with favorable physicochemical properties [3,21,22]. The surface decoration of PLGA NPs with polyethylene glycol (PEG) has been reported to have many advantages, such as (i) improving the water solubility of the PLGA NPs and their hydrophobic cargos, (ii) increasing the surface charge and the steric hindrance on the PLGA NPs thus extending the stability of the NPs and minimizing their aggregation, (iii) causing a stealth effect that prolongs the systemic circulation duration of drugs via inhibiting the opsonization effect, and (iv) enhancing the release of the drugs out of the polymeric matrix [23,24]. Many studies have reported using PEGylated PLGA NPs as promising nanocarriers for various natural bioactive compounds, biotherapeutics (as peptides, proteins, or vaccines), and synthetic drugs [23–27].

In the present study, a 2-hydroxy propyl-β- cyclodextrin (HPβCD)/harmala alkaloid-rich fraction (HARF) host–guest complex was fabricated using the thin-film hydration method to alleviate the water solubility of HARF. The complex formation was investigated by ^{1}H NMR spectroscopy and phase solubility study. The prepared complex was then co-encapsulated with ascorbic acid into PEGylated PLGA (HARF–HPßCD/AA@PLGA-PEG NPs) utilizing the W/O/W multiple emulsion-solvent evaporation method. The designed NPs were physically characterized in terms of average size, surface charge, polydispersity index (PDI), entrapment efficiency, and morphology. Additionally, the release of HARF–HPβCD complex and ascorbic acid (AA) from HARF–HPßCD/AA@PLGA-PEG NPs was studied, and the release profile was acquired. Finally, the antibacterial and antiviral activities of HARF–HPßCD/AA@PLGA-PEG NPs were evaluated.

2. Materials and Methods

2.1. Materials

2-Hydroxy propyl ß cyclodextrin was purchased from BLD Pharmatech Co., Limited (Cincinnati, OH, USA). Poly (D,L-lactide-co-glycolide) (PLGA) and polyethylene glycol

6000 (PEG) were obtained from Sigma-Aldrich (St. Louis, MO, USA). Polyvinyl alcohol (PVA; 98% hydrolyzed, MW $\approx$ 13,000) was obtained from Acros Organics (Geel, Belgium). Dimethylsulfoxide was purchased from Fisher Chemicals (Fair Lawn, NJ, USA). Tween 80 was purchased from El-Nasr Pharmaceutical Chemicals Co. (Cairo, Egypt). Tryptic soy agar was purchased from Millipore (Bedford, MA, USA). Phosphate-buffered saline, streptomycin, penicillin, fetal bovine serum, trichloroacetic acid, Dulbecco's modified Eagle's medium (DMEM), SRB, and tris (hydroxymethyl)aminomethane were purchased from Lonza (Basel, Switzerland).

2.2. Extraction and Isolation of Major *P. harmala* Alkaloids

Dried mature seeds of *P. harmala* L. were purchased from the local Egyptian market. The major alkaloids of *P. harmala* seeds were extracted, isolated, and characterized according to our previously reported method [5,6].

2.3. Preparation of Harmala Alkaloid-Rich Fraction/2-Hydroxy Propyl ß Cyclodextrin Complex (HARF–HPßCD)

HARF–HPßCD inclusion complex was designed via the thin-film hydration method, as described elsewhere, with some modifications [28]. Briefly, a thin film was formed by dissolving HARF in methanol followed by drying in a round bottom flask using a rotary evaporator. Separately, 5% (*w*/*v*) of HPßCD solution was prepared and subsequently added to the film, followed by 30 min sonication in a SONOREX DIGITAL 10 P sonicator (BANDELIN electronic GmbH & Co. KG, Darmstadt, Germany). Afterward, the complex solution was stirred for another 30 min and purified through filtration using a 0.22 µm nylon filter. The filtrate was then dried in a lyophilizer (TOPTION TOPT-10C Freeze dryer, Toption Group Co., Xi'an, China).

2.3.1. ^{1}H NMR Spectroscopy of HARF–HPßCD Complex

^{1}H NMR spectra of HPßCD and the inclusion complex were measured on a 400 MHz FT-NMR spectrometer (ECA-500, JEOL, Tokyo, Japan) in D_2O solutions. The scanning range of the hydrogen spectrum is 1–13 ppm.

2.3.2. Phase Solubility Study

The phase solubility study was performed, as described elsewhere with some modifications [29], to confirm the formation of the HARF–HPßCD inclusion complex using UV spectroscopy (FLUOstar Omega microplate reader, BMG Labtech, Offenburg, Germany) at λ = 373 nm [30]. Briefly, an excess amount of HARF was added to 5 mL of several mixtures containing successively increasing concentrations of HPßCD (ranging from 0 to 200 mm) all in deionized water. These mixtures were then sealed and stirred at room temperature until equilibrium had been reached. The mixtures were then centrifuged at 4000 rpm for 10 min (Hermle Z326K, Wehingen, Germany), and the absorbance was measured at 373 nm using UV spectrophotometer.

The stability constant ($K_{1:1}$) was estimated utilizing the slope of the linear correlation and the intrinsic solubility (S_0) using Equation (1) [31,32].

$$K_{1:1} = \frac{Slope}{S_0(1 - Slope)} \quad (1)$$

Then, the complexation efficiency (*CE*) was computed employing the intrinsic solubility and the stability constant using Equation (2) [31,32].

$$CE = S_0 * K_{1:1} \quad (2)$$

where S_0 is the intrinsic solubility, and $K_{1:1}$ is the stability constant.

2.4. Preparation of PEG-Coated PLGA Nanoparticles Dual-Loaded with Ascorbic Acid (AA) and HARF–HPβCD Inclusion Complex (HARF–HPβCD/AA@PLGA-PEG NPs)

The designed HARF–HPßCD complex was co-loaded with AA into PLGA NPs coated with PEG (HARF–HPßCD/AA@PLGA-PEG NPs) using the W/O/W multiple emulsion-solvent evaporation method, as previously reported with some modifications [3,7,33,34].

Briefly, 300 mg of PLGA and 100 mg PEG 6000 were dissolved in 8 mL dichloromethane (DCM), forming the emulsion's organic phase (O). Then, HARF–HPßCD complex and AA were dissolved in water (the W/W ratio of each drug/PLGA was 0.2:1) to obtain the inner aqueous phase (W1). Into the previously prepared organic phase, we emulsified 800 μL of the internal aqueous phase using a homogenizer at 11,000 rpm for 8 min in an ice bath to form a W1/O emulsion (first emulsion). Then, 2.5% polyvinyl alcohol was dissolved in 100 mL deionized water to form the external aqueous phase (W2). The first emulsion was added to the external aqueous phase (W2) and emulsified by homogenization at 11,000 rpm for 8 min in an ice bath. The resulting W1/O/W2 emulsion (second emulsion) was stirred overnight using a magnetic stirrer to allow solvent evaporation and particle hardening. Afterward, the nanoparticle colloid was lyophilized (TOPTION TOPT-10C Freeze dryer, Toption Group Co., Limited).

2.5. Characterization of the Designed HARF–HPβCD/AA@PLGA-PEG NPs

The average particle size and polydispersity index (PDI) were determined using a Zetasizer Nano ZS equipped with a 10 mW HeNe laser, allowing for measurements at 633 nm and a detection angle of 173° backscatter (Malvern Instruments, Worcestershire, UK). The ζ-potential was measured using laser Doppler velocimetry [35–37].

The morphological features of the prepared nanoparticles were investigated using scanning electron microscopy (SEM) via high-resolution field emission scanning electron microscopy (FESEM) (LEO SUPRA 55, Carl Zeiss, Oberkochen, Germany) with an acceleration voltage of 1.00 kV. Before conducting the SEM study, we coated freeze-dried PLGA NPs via gold sputtering (current: 10 mA) for 2 min under a nitrogen atmosphere. The measurements were performed using the image processing program ImageJ (NIH, Bethesda, MD, USA).

2.6. Entrapment Efficiency

The entrapment efficiency (*EE*) of HARF–HPßCD/AA@PLGA-PEG NPs was measured indirectly by centrifugation (15,000 rpm, 1 h, 4 °C), followed by ultrafiltration of the supernatant. The quantities of the HARF and AA unencapsulated in the ultrafiltrate were determined by a FLUOstar Omega microplate reader (BMG Labtech, Offenburg, Germany) spectrophotometer at 373 and 287 nm, respectively, using Equation (3) [37–39].

$$EE\ (\%) = \frac{Initial\ feeded\ drug - Final\ loaded\ drug}{Initial\ feeded\ drug} \times 100 \tag{3}$$

2.7. In Vitro Release Study of HARF–HPβCD Complex and Ascorbic Acid from HARF–HPβCD/AA@PLGA-PEG NPs

The cumulative release % of HARF and AA from HARF–HPßCD/AA@PLGA-PEG NPs was investigated utilizing the dialysis membrane method. Briefly, 1 mL of the NPs was loaded to a dialysis bag (cutoff molecular weight of 12,000–14,000 Da). The dialysis bag was inserted into 50 mL of PBS at pH 7.4 with 2% Tween in a proper jar. The whole system was left on a stirrer at 37 °C. A 1 mL aliquot of the sample was withdrawn for analysis and immediately replaced with another equal volume of warmed buffer at specific time intervals. The concentrations of HARF and AA were determined by a FLUOstar Omega microplate reader (BMG Labtech, Offenburg, Germany) spectrophotometer at 373 and 287 nm wavelengths, respectively.

The cumulative release (%) was calculated according to Equation (4) [40].

$$Cumulative\ release\ (\%) = \frac{\mathrm{Vs}\sum_{1}^{n-1} c(n-1) + VoCn}{\mathrm{mo}} \times 100 \quad (4)$$

where Vs is the volume of each sample removed, $c(n-1)$ is the bulk concentration before sampling, Cn is the concentration of the sample, Vo is the bulk volume of the release medium, and mo is the original amount of the drug in the nanoparticles tested.

2.8. Antibacterial Assay for the Designed HARF–HPßCD/AA@PLGA-PEG NPs

2.8.1. Preparation of the Inoculum Using Colony Suspension Approach

Staphylococcus aureus ATCC® 6538 (lot no. 4600502) and *Escherichia coli* ATCC® 8739 (lot no. 380063) were obtained from American Type Culture Collection (Manassas, VA, USA). Tryptic soy broth (100 mL) was used as inoculation medium for either the Gram-positive or Gram-negative bacteria, then incubated at 37.0 ± 1.0 °C for 24 ± 2 h. A loopful of the broth was streaked onto tryptic soy agar medium and incubated at 37.0 °C for 21 ± 3 h. Three or four colonies (from each plate) were inoculated in broth, and the suspension was incubated till turbidity reached 0.5 McFarland standard. The inoculum density was standardized utilizing a 0.5 McFarland standard and DensiCHEK© optical device (BioMérieux, Marcy l'Etoile, France). The obtained suspensions contained about 1.0×10^8 CFU/mL of *Staphylococcus aureus* or *Escherichia coli*.

2.8.2. Broth Macrodilution Method

The antibacterial activities of plain PLGA-PEG NPs, HARF, and HARF–HPßCD/AA@PLGA-PEG NPs were evaluated by adding and mixing 5.0 mL from each sample to 5.0 mL broth (1:2 dilution) according to a previously reported method [41]. Next, 5.0 mL of the 1:2 dilution was pipetted using a new tip and mixed with 5.0 mL broth (1:4). Ten dilutions for each bacterial strain were prepared and inoculated in a 24-well plate. Then, 100 μL of prepared inoculum was added to each well, resulting in a final concentration of 5.0×10^5 CFU/mL. An additional 100 μL from each bacterial strain suspension was diluted and cultured to confirm the inoculum density. All plates were incubated at 37.0 °C for 24 h and then kept in the dark to monitor growth. All control wells yielded a turbid solution. The inoculum density was 4–6 × 10^5 CFU/mL for both evaluated bacterial strains, with reference to the Clinical and Laboratory Standards Institute (CLSI) procedures (document M07, A09) [41].

2.9. Cytotoxicity and Antiviral Assays

The human influenza H1N1 and Madin–Darby Canine Kidney (MDCK) cells were obtained from American Type Culture Collection (University Boulevard, Manassas, VA, USA). Vero E6 cells were grown in DMEM medium supplemented with 10% fetal bovine serum and 0.1% antibiotic/antimycotic solution. Gibco BRL provided the antibiotic and antimycotic solution, trypsin-EDTA, fetal bovine serum, and DMEM medium (Grand Island, NY, USA).

Cytopathic effect (CPE) reduction assay was used to evaluate the antiviral activities of plain PLGA-PEG NPs, HARF, and HARF–HPßCD/AA@PLGA-PEG NPs in cell culture systems, as described elsewhere with some modifications [42,43]. Briefly, MDCK cells seeded into a 96-well culture plate at a density of 2×10^4 cells per well were infected with 0.1 mL (CCID50) influenza (H1N1) of the diluted viral suspension of Influenza H1N1 and then incubated at 37 °C for 60 min to facilitate the adsorption of the virus. The antiviral activity for each sample was determined using a twofold diluted concentration of 1–100 μg/mL. Culture plates were incubated at 37 °C in 5% CO_2 for 72 h. The development of the cytopathic effect was monitored by light microscopy. The cell monolayers were stained (0.03% crystal violet in 2% ethanol and 10% formalin). After washing and drying, the optical density was measured at 540/630 nm. The antiviral activities were calculated according to Pauwels et al. [44], employing Equation (5).

$$Antiviral\ activity\ \% = \frac{(\text{mean optical density of cell controls} - \text{mean optical density of virus controls})}{(\text{Optical density of the test} - \text{mean optical density of virus controls})} \times 100 \quad (5)$$

The cytotoxicity was evaluated against MDCK cells before conducting the aforementioned assay. Cells were seeded at a density of 2×10^4 cells per well in a 96-well culture plate. After 24 h, the culture medium containing serially diluted samples was added to the cells and incubated for 72 h before being removed, and the cells were washed with PBS. The next steps were carried out in the same manner as described above for the antiviral activity assay. The 50% cytotoxic concentrations (CC50) and the 50% inhibitory concentration (IC50) were computed using GraphPad PRISM Software (Graph-Pad Software, San Diego, CA, USA). Selectivity index (SI) was estimated as the ratio of CC50 to IC50.

3. Results and Discussion

3.1. Characterization of the Prepared HARF–HPßCD Complex

3.1.1. ^{1}H NMR Spectroscopy of the HARF–HPßCD Complex

^{1}H NMR analyses performed in D_2O could shed more light on the host–guest complexation between HPßCD and other guest molecules and could provide clues on the inclusion of guest molecules into the hydrophobic cavity of the host molecule. Upon the inclusion of the guest molecule into the HPßCD cavity, an upfield shift of the H-3 and H-5 (hydrogen atoms placed in the interior cavity of HPßCD) occurred due to their shielding by the guest molecule (the chemical structure of HPßCD is presented in Figure S1A). On the other hand, H-1, H-2, H-4, and H-6 (hydrogen atoms localized on the outer surface of HPßCD) will experience milder chemical shifts. In the present work, the suggestion of the HARF–HPßCD complex foundation was grounded on modifying the ^{1}H NMR spectra of the pure HPßCD upon its complexation with HARF (Figure S1). The ^{1}H NMR chemical shifts (in ppm) of HPßCD protons in the presence and absence of HARF are presented in Table 1. The induced chemical shift, δ (ppm), is the difference in chemical shifts of HPßCD protons before and after the formation of the HARF–HPßCD complex. In this context, a positive sign of δ (ppm) indicates a downfield shift, while a negative sign indicates an upfield one. It was shown that the H-3 and H-5 protons experienced more remarkable HARF-induced chemical shifts than the outer surface protons (H-1, H-2, H-4, and H-6), indicating that HARF was only included within the HPßCD cavity. Furthermore, the clear upfield shift of H-3 and H-5 protons signals was due to the inclusion of π-electron-rich aromatic groups of HARF inside the host cavity. These findings agree very well with the previous studies, which reported that the upfield proton shifts of H-3 and H-5 indicate inclusion complexation [45–48].

Table 1. The chemical shifts (ppm) of HPβCD before and after forming the HARF–HPßCD complex.

	Chemical Shifts (δ, ppm)					
	H-1	**H-2**	**H-3**	**H-4**	**H-5**	**H-6**
HPßCD	4.899	3.310	3.768	3.329	3.552	3.696
HARF-HPßCD complex	4.877	3.307	3.741	3.307	3.526	3.675
Δδ	−0.022	−0.003	−0.027	−0.022	−0.026	−0.021

3.1.2. Phase Solubility Study

The solubility of HARF was determined by plotting the change with HPβCD concentration, as presented in Figure 1. The R^2 value of the extrapolated curve was 0.9807, suggesting that a remarkable linear correlation between HARF solubility and HPβCD concentration where the solubility of HARF increased with increasing the concentration of HPβCD. As previously described by Higuchi and Connors, this phase solubility diagram displays the A_L-type pattern, indicating a complex stoichiometry of 1:1 (HARF/HPβCD) [29,31]. Moreover, the slope was >0.0 and <1.0, while the K value was >0.0, confirming the formation of a 1:1 complex [31].

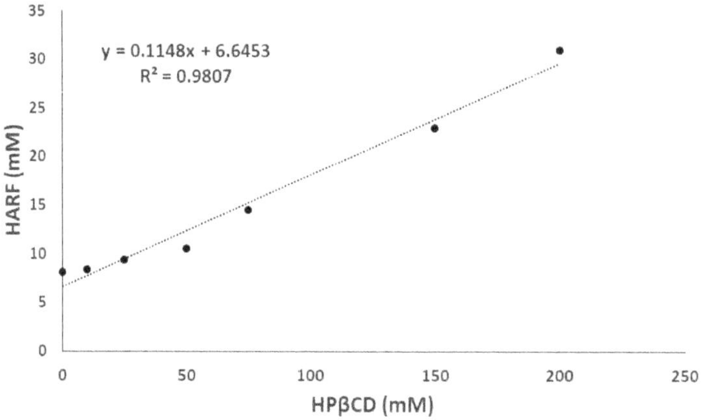

Figure 1. Phase solubility diagram of HARF in the presence of several mixtures containing successively increasing concentrations of HPβCD.

The stability constant and the complexation efficiency (CE) were 81.4 M^{-1}, and 0.66, respectively, similar to values reported previously [31,32,49]. This showed that the HPβCD had improved the water solubility of HARF via the formation of the host–guest complex.

3.2. *Characterization of the HARF–HPßCD/AA@PLGA-PEG NPs*

3.2.1. Particle Size, Polydispersity Index (PDI), ζ-Potential, Entrapment Efficiency (EE), and Morphological Features

The average particle size and PDI of the designed HARF–HPßCD/AA@PLGA-PEG NPs were 207 ± 2.60 nm and 0.17 ± 0.01, respectively (Table 2). These values lie within the range of the previously reported nanoparticles that exhibited improved biomedical applications [50,51]. The particle size distribution of the nanoparticles is presented in Figure 2. The ζ-potential of the designed nanoparticles showed a high negative surface charge of −31.60 ± 0.20 mV. This is attributed to the presence of the highly negatively charged PEG coating the PLGA NPs. Using PEG as a coat for the NPs improves the stability of the PLGA NPs in aqueous media, enabling their long shelf life. Moreover, PEG inhibits the aggregation of NPs through its steric hindrance effect [49].

Table 2. Average particle size, PDI, ζ-potential, and entrapment efficiency of HARF–HPßCD/AA@PLGA-PEG NPs.

Formula	Average Size (nm)	PDI	ζ-Potential (mV) ± SD	Encapsulation Efficiency (%)	
				HARF	AA
HARF–HPßCD/AA@ PLGA-PEG NPs	207 ± 2.60	0.17 ± 0.01	−31.60 ± 0.20	81.60 ± 1.20	87 ± 2.20

The entrapment efficiencies of the HARF and AA are presented in Table 2.

The high entrapment efficiencies of HARF and AA are attributed to the PEG coating the PLGA NPs founding a barrier that surrounds and protects the loaded drugs, preventing their loss during the homogenization step and retaining their stability [52]. The capacity to entrap high concentrations of drugs supports sustained drug release [53].

As demonstrated in Figure 3A, SEM analysis showed a spherical shape of the prepared NPs. The mean average size of the NPs was determined by utilizing the image processing program ImageJ (NIH, Bethesda, MD, USA) and was found to be 204.5 ± 70.6 nm (Figure 3B, which is close to that obtained from the dynamic light scattering (207 ± 2.6).

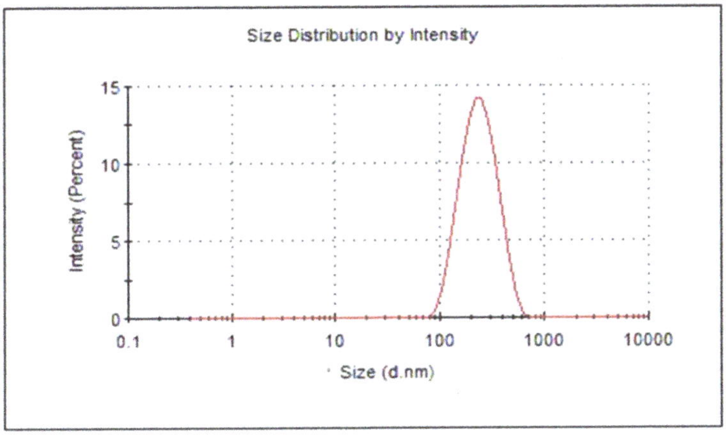

Figure 2. Particle size distribution of the prepared HARF–HPßCD/AA@PLGA-PEG NPs using DLS analysis.

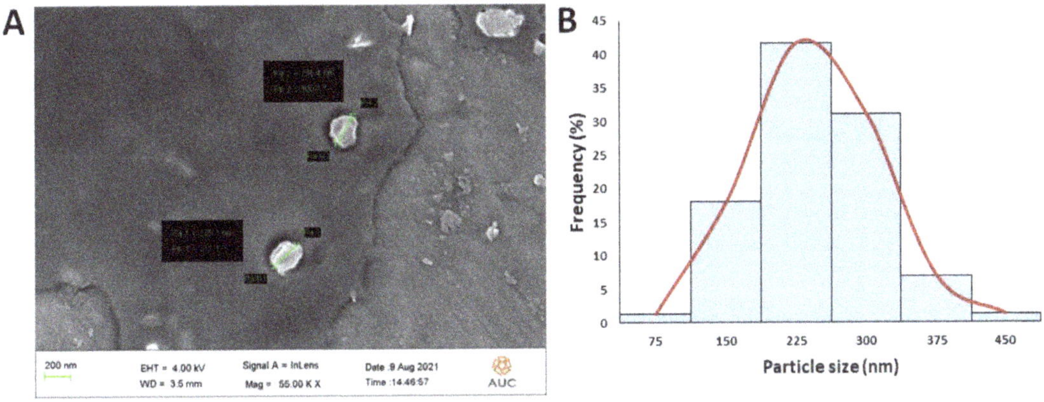

Figure 3. (**A**) Scanning electron microscopy (SEM) images of the HARF–HPßCD/AA@PLGA-PEG NPs. (**B**) Particle size (nm) histogram of the designed HARF–HPßCD/AA@PLGA-PEG NPs generated through employing the image processing program ImageJ (NIH, Bethesda, MD, USA).

3.2.2. In Vitro Release Study

Figure 4 portrays the release profile of HARF–HPßCD complex and AA from the HARF–HPßCD/AA@PLGA-PEG NPs. The release profile showed an initial release of 29.8 ± 4.9% and 66.7 ± 6.8% for HARF–HPßCD complex and AA, respectively, in the first 2 h, which increased to 64.6 ± 3% and 76.6 ± 1.3%, respectively, over 48 h. These release profiles are attributed to the hydrolytic cleavage of the ester bonds of the PLGA copolymer in the presence of water [54]. Moreover, coating NPs with PEG was reported to enhance the permeability and release of the loaded drugs [55]. The faster release rates of AA compared to HARF complex could be due to the higher molecular weight of the complex, making its diffusion out of the dialysis membrane much slower than that of AA.

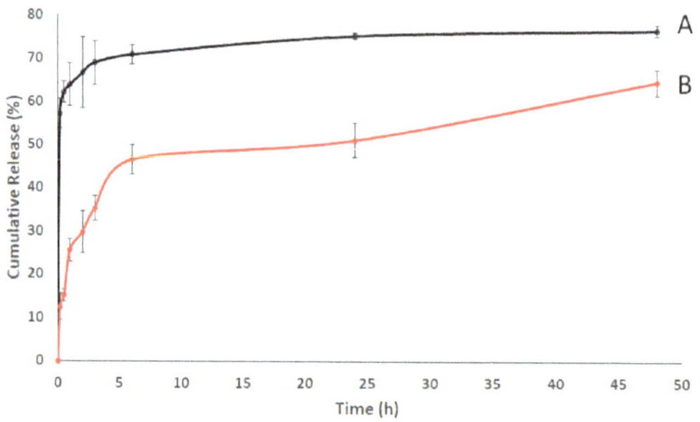

Figure 4. Time-dependent release profiles of (**A**) AA and (**B**) HARF–HPßCD complex from HARF–HPßCD/AA@PLGA-PEG NPs at 37 °C into phosphate-buffered saline.

3.2.3. Antibacterial Assay for the Designed HARF–HPßCD/AA@PLGA-PEG NPs

The bacterial activities of the HARF, plain PLGA-PEG NPs, and HARF–HPßCD/AA@PLGA-PEG NPs were assessed against *S. aureus* and *E. coli*. HARF–HPßCD/AA@PLGA-PEG NPs exhibited high antibacterial activity against *S. aureus* and *E. coli* (MIC of 0.025 mg/mL), as compared to free HARF (MIC of 0.5 mg/mL), as shown in Table 3. On the other hand, PLGA-PEG NPs showed no antibacterial activities. The HARF–HPßCD/AA@PLGA-PEG NPs showed higher bactericidal activity as compared to free HARF. This could possibly be attributed to the improved hydrophilicity of HARF upon its inclusion inside the HPßCD cavities. Furthermore, the NPs can adhere to the bacterial cell wall due to their effective surface areas, releasing their cargo across the cell wall [34,56–58]. The bactericidal activity of HARF is attributed to the presence of high concentrations of β-carboline and quinazoline alkaloids, which have been reported to intercalate the bacterial DNA [6].

Table 3. Bactericidal activity of HARF and HARF–HPßCD/AA@PLGA-PEG NPs against *Staphylococcus aureus* and *Escherichia coli*.

Bacterial Strain	Minimum Inhibitory Concentration (MIC in mg/mL)		
	PLGA-PEG NPs	HARF	HARF–HPßCD/AA@PLGA-PEG NPs
Staphylococcus aureus	0	0.5	0.025
Escherichia coli	0	0.5	0.025

3.2.4. Cytotoxicity and Antiviral Activity Assays

Treatment choices for viral infections are limited and face many challenges, such as drug resistance, systemic toxic effects, and poor bioavailability. This warrants the search for effective antiviral agents that can overcome the shortcomings of their current counterparts. The antiviral activities of the HARF, PLGA-PEG NPs, and HARF–HPßCD/AA@PLGA-PEG NPs against influenza A (H1N) were investigated through employing the cytopathic effect (CPE) reduction assay. The PLGA-PEG showed no antiviral activity. The developed HARF–HPCD/PLGA-PEG NPs exhibited high selective antiviral activity against the H1N1 influenza virus (Table 4) without affecting the host, MDCK cells, (IC50 of 2.7 µg/mL; CC50 of 110.4 µg/mL; and selective index, CC50/IC50, of 41.2) compared to free HARF (IC50 of 30.2 µg/mL, CC50 of 238.8 µg/mL, and selective index of 7.9). The NPs decompose, releasing their cargos that kill the virus selectively without harming the host cells. HARF is reported to exert its antiviral activity by inhibiting the viral polymerase activity and consequently viral RNA replication [59]. On the other hand, the co-encapsulation of ascorbic acid causes a synergistic antiviral effect against the influenza virus by increasing the generation of interferon (IFN)-α/β [60,61].

Table 4. Antiviral activity of HARF and HARF–HPßCD/AA@PLGA-PEG NPs against influenza A (H1N1) virus.

Sample	CC50 (µg /mL)	IC50 (µg /mL)	SI
Plain PLGA-PEG	0	0	0
HARF	238.8	30.2	7.9
HARF–HPßCD/AA@PLGA-PEG NPs	110.4	2.7	41.2

4. Conclusions

In this work, HARF was first mixed with HPßCD as a host molecule to form an inclusion complex. This was done to improve the solubility of HARF, which is highly hydrophobic. Finally, AA and HARF–HPßCD inclusion complex were co-loaded into the PLGA-PEG nanoparticles to form the final biodegradable delivery system with enhanced release and bioavailability. The fabricated HARF–HPßCD/AA@PLGA-PEG NPs demonstrated acceptable cytotoxicity and enhanced antibacterial and antiviral activities compared to free HARF. They showed a higher bactericidal effect against *Staphylococcus aureus* and *Escherichia coli*, with a MIC value of 0.025 mg/mL, as compared to free HARF (MIC of 0.50 mg/mL). Moreover, the designed NPs exhibited high selective antiviral activity against the H1N1 influenza virus without affecting the host, MDCK cells. Biocompatible nanocarriers loaded with natural extracts with antimicrobial activity could be explored as future effective treatments of infections.

Supplementary Materials: The following are available online at https://www.mdpi.com/article/10.3390/pharmaceutics14010142/s1, Figure S1. (A) Structure of HPßCD with appropriate protons numbered. ^{1}H NMR spectra of (B) HPßCD, (C) HARF, and (D) HARF–HPßCD complex.

Author Contributions: Conceptualization, S.A.F. and H.M.E.-S.A.; methodology, S.A.F. and H.A.M.; data curation, S.A.F. and N.K.M.; software, S.A.F. and H.A.M.; Resources, A.N.E., M.Y.I. and H.M.E.-S.A.; writing—original draft preparation, S.A.F. and H.M.E.-S.A.; writing—review and editing, S.A.F., H.A.M., A.N.E., M.Y.I. and H.M.E.-S.A.; funding acquisition, H.M.E.-S.A.; supervision, H.M.E.-S.A.; project administration, H.M.E.-S.A. All authors have read and agreed to the published version of the manuscript.

Funding: This project was funded by a grant from the American University in Cairo to H.M.E.-S.A. Funding number: SSE-CHEM-H.A.-FY2I-FY2 2-RG (2-20)-2020-Fcb-28-21-32-34.

Institutional Review Board Statement: Not applicable.

Informed Consent Statement: Not applicable.

Conflicts of Interest: The authors declare no conflict of interest.

References

1. Influenza (Seasonal). Available online: https://www.who.int/en/news-room/fact-sheets/detail/influenza-(seasonal) (accessed on 22 November 2021).
2. Mogana, R.; Adhikari, A.; Tzar, M.N.; Ramliza, R.; Wiart, C. Antibacterial activities of the extracts, fractions and isolated compounds from canarium patentinervium miq. Against bacterial clinical isolates. *BMC Complement. Med. Ther.* **2020**, *20*, 228. [CrossRef] [PubMed]
3. Fahmy, S.A.; Mamdouh, W. Garlic oil–loaded PLGA nanoparticles with controllable size and shape and enhanced antibacterial activities. *J. Appl. Polym. Sci.* **2018**, *135*, 46133. [CrossRef]
4. Omrani, M.; Keshavarz, M.; Nejad Ebrahimi, S.; Mehrabi, M.; McGaw, L.J.; Ali Abdalla, M.; Mehrbod, P. Potential Natural Products Against Respiratory Viruses: A Perspective to Develop Anti-COVID-19 Medicines. *Front. Pharmacol.* **2021**, *11*, 2115. [CrossRef]
5. Fahmy, S.A.; Fawzy, I.M.; Saleh, B.M.; Issa, M.Y.; Bakowsky, U.; Azzazy, H.M.E.-S. Green Synthesis of Platinum and Palladium Nanoparticles Using *Peganum harmala* L. Seed Alkaloids: Biological and Computational Studies. *Nanomaterials* **2021**, *11*, 965. [CrossRef] [PubMed]
6. Fahmy, S.A.; Issa, M.Y.; Saleh, B.M.; Meselhy, M.R.; Azzazy, H.M.E.A. *Peganum harmala* alkaloids self-assembled supramolecular nanocapsules with enhanced antioxidant and cytotoxic activities. *ACS Omega* **2021**, *6*, 11954–11963. [CrossRef]

7. Azzazy, H.M.E.-S.; Fahmy, S.A.; Mahdy, N.K.; Meselhy, M.R.; Bakowsky, U. Chitosan-Coated PLGA Nanoparticles Loaded with *Peganum harmala* Alkaloids with Promising Antibacterial and Wound Healing Activities. *Nanomaterials* **2021**, *11*, 2438. [CrossRef] [PubMed]
8. Pathan Aslam, R.; Vadnere Gautam, P.; Singhai Abhay, K.; Kulkarni Bharti, U. *Peganum harmala*: A Phyto-pharmacological Review. *Inventi Rapid: Planta Activa* **2012**, *2012*, 2–4.
9. Jinous, A.; Ramezanloo, F. Chemistry, pharmacology and medicinal properties of *Peganum harmala* L. *Afr. J. Pharm. Pharmacol.* **2012**, *6*, 1573–1580. [CrossRef]
10. Grosso, G.; Bei, R.; Mistretta, A.; Marventano, S.; Calabrese, G.; Masuelli, L.; Giganti, M.G.; Modesti, A.; Galvano, F.; Gazzolo, D. Effects of vitamin C on health: A review of evidence. *Front. Biosci.* **2013**, *18*, 1017–1029. [CrossRef]
11. Hemilä, H. Vitamin C and infections. *Nutrients* **2017**, *9*, 339. [CrossRef] [PubMed]
12. Fahmy, S.A.; Ponte, F.; Fawzy, I.M.; Sicilia, E.; Bakowsky, U.; Azzazy, H.M.E.-S. Betaine host–guest complexation with a calixarene receptor: Enhanced in vitro anticancer effect. *RSC Adv.* **2021**, *11*, 24673–24680. [CrossRef]
13. Fahmy, S.A.; Alawak, M.; Brüßler, J.; Bakowsky, U.; El Sayed, M.M.H. Nanoenabled Bioseparations: Current Developments and Future Prospects. *BioMed Res. Int. J.* **2019**, *2019*, 4983291. [CrossRef] [PubMed]
14. Fahmy, S.A.; Ponte, F.; Abd El-Rahman, M.K.; Russo, N.; Sicilia, E.; Shoeib, T. Investigation of the host-guest complexation between 4-sulfocalix[4]arene and nedaplatin for potential use in drug delivery. *Spectrochim. Acta-Part A Mol. Biomol. Spectrosc.* **2018**, *193*, 528–536. [CrossRef] [PubMed]
15. Fahmy, S.A.; Ponte, F.; Sicilia, E.; El-Said Azzazy, H.M. Experimental and Computational Investigations of Carboplatin Supramolecular Complexes. *ACS Omega* **2020**, *5*, 31456–31466. [CrossRef]
16. Fahmy, S.A.; Ponte, F.; Fawzy, I.M.; Sicilia, E.; Bakowsky, U.; Azzazy, H.M.E.A. Host-Guest Complexation of Oxaliplatin and Para-Sulfonatocalix[n]Arenes for Potential Use in Cancer Therapy. *Molecules* **2020**, *25*, 5926. [CrossRef]
17. Fahmy, S.A.; Brüßler, J.; Ponte, F.; Abd El-Rahman, M.K.; Russo, N.; Sicilia, E.; Bakowsky, U.; Shoeib, T. A study on the physicochemical properties and cytotoxic activity of *p*-sulfocalix[4]arene-nedaplatin complex. *J. Phys. Conf. Ser.* **2019**, *1310*, 012011. [CrossRef]
18. Fahmy, S.A.; Brüßler, J.; Alawak, M.; El-Sayed, M.M.H.; Bakowsky, U.; Shoeib, T. Chemotherapy based on supramolecular chemistry: A promising strategy in cancer therapy. *Pharmaceutics* **2019**, *11*, 292. [CrossRef]
19. Fahmy, S.A.; Ramzy, A.; Saleh, B.M.; Azzazy, H.M.E.-S. Stimuli-Responsive Amphiphilic Pillar[n]arene Nanovesicles for Targeted Delivery of Cancer Drugs. *ACS Omega* **2021**, *6*, 25876–25883. [CrossRef]
20. Chen, J.; Qin, X.; Zhong, S.; Chen, S.; Su, W.; Liu, Y. Characterization of curcumin/cyclodextrin polymer inclusion complex and investigation on its antioxidant and antiproliferative activities. *Molecules* **2018**, *23*, 1179. [CrossRef]
21. Mohan, L.J.; McDonald, L.; Daly, J.S.; Ramtoola, Z. Optimising PLGA-PEG nanoparticle size and distribution for enhanced drug targeting to the inflamed intestinal barrier. *Pharmaceutics* **2020**, *12*, 1114. [CrossRef]
22. Tammam, S.N.; Azzazy, H.M.E.A.; Lamprecht, A. Biodegradable Particulate Carrier Formulation and Tuning for Targeted Drug Delivery. *J. Biomed Nanotechnol.* **2015**, *11*, 555–577. [CrossRef]
23. Perinelli, D.R.; Cespi, M.; Bonacucina, G.; Palmieri, G.F. PEGylated polylactide (PLA) and poly (lactic-co-glycolic acid) (PLGA) copolymers for the design of drug delivery systems. *J. Pharm. Investig.* **2019**, *49*, 443–458. [CrossRef]
24. Devasvaran, K.; Jairaman, S.; Yahaya, N.A.; Jaganath, I.B.S.; Khung, Y.L.; Lim, V.; Ngalim, S.H. PEG-b-PLGA nanoparticles loaded with Geraniin from Phyllanthus watsonii extract as a phytochemical delivery model. *Appl. Sci.* **2020**, *10*, 4891. [CrossRef]
25. Alhakamy, N.A. Development and Evaluation of Icariin-Loaded PLGA-PEG Nanoparticles for Potentiation the Proapoptotic Activity in Pancreatic Cancer Cells. *AAPS PharmSciTech* **2021**, *22*, 1–11. [CrossRef] [PubMed]
26. Del Amo, L.; Cano, A.; Ettcheto, M.; Souto, E.B.; Espina, M.; Camins, A.; García, M.L.; Sánchez-López, E. Surface functionalization of plga nanoparticles to increase transport across the bbb for alzheimer's disease. *Appl. Sci.* **2021**, *11*, 4305. [CrossRef]
27. Rafiei, P.; Haddadi, A. Docetaxel-loaded PLGA and PLGA-PEG nanoparticles for intravenous application: Pharmacokinetics and biodistribution profile. *Int. J. Nanomed.* **2017**, *12*, 935–947. [CrossRef]
28. Plenagl, N.; Duse, L.; Seitz, B.S.; Goergen, N.; Pinnapireddy, S.R.; Jedelska, J.; Brüßler, J.; Bakowsky, U. Photodynamic therapy–hypericin tetraether liposome conjugates and their antitumor and antiangiogenic activity. *Drug Deliv.* **2019**, *26*, 23–33. [CrossRef]
29. Higuchi, T.; Connors, K.A. Phase Solubility Techniques. *Adv. Anal. Chem. Instrum.* **1965**, *4*, 117–212.
30. Liu, L.; Zhao, T.; Cheng, X.; Wang, C.; Wang, Z. Characterization and determination of trace alkaloids in seeds extracts from *Peganum harmala* linn. Using LC-ESI-MS and HPLC. *Acta Chromatogr.* **2013**, *25*, 221–240. [CrossRef]
31. Kurkov, S.V.; Ukhatskaya, E.V.; Loftsson, T. Drug/cyclodextrin: Beyond inclusion complexation. *J. Incl. Phenom. Macrocycl. Chem.* **2011**, *69*, 297–301. [CrossRef]
32. Loftsson, T.; Hreinsdóttir, D.; Másson, M. The complexation efficiency. *J. Incl. Phenom. Macrocycl. Chem.* **2007**, *57*, 545–552. [CrossRef]
33. Tewes, F.; Munnier, E.; Antoon, B.; Okassa, L.N.; Cohen-Jonathan, S.; Marchais, H.; Douziech-Eyrolles, L.; Soucé, M.; Dubois, P.; Chourpa, I. Comparative study of doxorubicin-loaded poly(lactide-co-glycolide) nanoparticles prepared by single and double emulsion methods. *Eur. J. Pharm. Biopharm.* **2007**, *66*, 488–492. [CrossRef]
34. Wang, Y.; Li, P.; Kong, L. Chitosan-modified PLGA nanoparticles with versatile surface for improved drug delivery. *AAPS PharmSciTech* **2013**, *14*, 585–592. [CrossRef] [PubMed]

35. Pinnapireddy, S.R.; Duse, L.; Strehlow, B.; Schäfer, J.; Bakowsky, U. Composite liposome-PEI/nucleic acid lipopolyplexes for safe and efficient gene delivery and gene knockdown. *Colloids Surf. B Biointerfaces* **2017**, *158*, 93–101. [CrossRef]
36. Preis, E.; Baghdan, E.; Agel, M.R.; Anders, T.; Pourasghar, M.; Schneider, M.; Bakowsky, U. Spray dried curcumin loaded nanoparticles for antimicrobial photodynamic therapy. *Eur. J. Pharm. Biopharm.* **2019**, *142*, 531–539. [CrossRef]
37. Duse, L.; Baghdan, E.; Pinnapireddy, S.R.; Engelhardt, K.H.; Jedelská, J.; Schaefer, J.; Quendt, P.; Bakowsky, U. Preparation and Characterization of Curcumin Loaded Chitosan Nanoparticles for Photodynamic Therapy. *Phys. Status Solidi Appl. Mater. Sci.* **2018**, *215*, 1–5. [CrossRef]
38. Chen, M.X.; Li, T.; Peng, S.; Tao, D. Supramolecular nanocapsules from the self-assembly of amphiphilic calixarene as a carrier for paclitaxel. *N. J. Chem.* **2016**, *40*, 9923–9929. [CrossRef]
39. El-Shafie, S.; Fahmy, S.A.; Ziko, L.; Elzahed, N.; Shoeib, T.; Kakarougkas, A. Encapsulation of nedaplatin in novel pegylated liposomes increases its cytotoxicity and genotoxicity against a549 and u2os human cancer cells. *Pharmaceutics* **2020**, *12*, 863. [CrossRef] [PubMed]
40. Li, B.; Shan, M.; Di, X.; Gong, C.; Zhang, L.; Wang, Y.; Wu, G. A dual pH- and reduction-responsive anticancer drug delivery system based on PEG-SS-poly(amino acid) block copolymer. *RSC Adv.* **2017**, *7*, 30242–30249. [CrossRef]
41. Weinstein, M.P.; Patel, J.B.; Burnhman, C.-A.; ZImmer, B.L. *M07-Methods for Dilution Antimicrobial Susceptibility Tests for Bacteria That Grow Aerobically*, 11th ed.; Approved Standard—2018.
42. Khiralla, A.; Spina, R.; Varbanov, M.; Philippot, S.; Lemiere, P.; Slezack-Deschaumes, S.; André, P.; Mohamed, I.; Yagi, S.M.; Laurain-Mattar, D. Evaluation of antiviral, antibacterial and antiproliferative activities of the endophytic fungus curvularia papendorfii, and isolation of a new polyhydroxyacid. *Microorganisms* **2020**, *8*, 1353. [CrossRef]
43. Jang, Y.; Shin, J.S.; Lee, J.Y.; Shin, H.; Kim, S.J.; Kim, M. In vitro and in vivo antiviral activity of nylidrin by targeting the hemagglutinin 2-mediated membrane fusion of influenza A virus. *Viruses* **2020**, *12*, 581. [CrossRef]
44. Pauwels, R.; Balzarini, J.; Baba, M.; Snoeck, R.; Schols, D.; Herdewijn, P.; Desmyter, J.; De Clercq, E. Rapid and automated tetrazolium-based colorimetric assay for the detection of anti-HIV compounds. *J. Virol. Methods* **1988**, *20*, 309–321. [CrossRef]
45. Gao, S.; Liu, Y.; Jiang, J.; Li, X.; Zhao, L.; Fu, Y.; Ye, F. Encapsulation of thiabendazole in hydroxypropyl-β-cyclodextrin nanofibers via polymer-free electrospinning and its characterization. *Pest Manag. Sci.* **2020**, *76*, 3264–3272. [CrossRef] [PubMed]
46. Veiga, F.J.B.; Fernandes, C.M.; Carvalho, R.A.; Geraldes, C.F.G.C. Molecular modelling and 1H-NMR: Ultimate tools for the investigation of tolbutamide: β-cyclodextrin and tolbutamide: Hydroxypropyl-β-cyclodextrin complexes. *Chem. Pharm. Bull.* **2001**, *49*, 1251–1256. [CrossRef] [PubMed]
47. Yuan, C.; Jin, Z.; Xu, X. Inclusion complex of astaxanthin with hydroxypropyl-β-cyclodextrin: UV, FTIR, ^{1}H NMR and molecular modeling studies. *Carbohydr. Polym.* **2012**, *89*, 492–496. [CrossRef]
48. Ge, X.; He, J.; Qi, F.; Yang, Y.; Huang, Z.; Lu, R.; Huang, L. Inclusion complexation of chloropropham with β-cyclodextrin: Preparation, characterization and molecular modeling. *Spectrochim. Acta-Part A Mol. Biomol. Spectrosc.* **2011**, *81*, 397–403. [CrossRef] [PubMed]
49. Meinguet, C.; Masereel, B.; Wouters, J. Preparation and characterization of a new harmine-based antiproliferative compound in complex with cyclodextrin: Increasing solubility while maintaining biological activity. *Eur. J. Pharm. Sci.* **2015**, *77*, 135–140. [CrossRef]
50. Imran, M.; Shah, M.R.; Ullah, F.; Ullah, S.; Sadiq, A.; Ali, I.; Ahmed, F.; Nawaz, W. Double-tailed acyl glycoside niosomal nanocarrier for enhanced oral bioavailability of Cefixime. *Artif. Cells Nanomed. Biotechnol.* **2017**, *45*, 1440–1451. [CrossRef] [PubMed]
51. Cho, K.; Wang, X.; Nie, S.; Chen, Z.; Shin, D.M. Therapeutic nanoparticles for drug delivery in cancer. *Clin. Cancer Res.* **2008**, *14*, 1310–1316. [CrossRef]
52. Li, Y.P.; Pei, Y.Y.; Zhang, X.Y.; Gu, Z.H.; Zhou, Z.H.; Yuan, W.F.; Zhou, J.J.; Zhu, J.H.; Gao, X.J. PEGylated PLGA nanoparticles as protein carriers: Synthesis, preparation and biodistribution in rats. *J. Control. Release* **2001**, *71*, 203–211. [CrossRef]
53. Ullah, S.; Shah, M.R.; Shoaib, M.; Imran, M.; Elhissi, A.M.A.; Ahmad, F.; Ali, I.; Shah, S.W.A. Development of a biocompatible creatinine-based niosomal delivery system for enhanced oral bioavailability of clarithromycin. *Drug Deliv.* **2016**, *23*, 3480–3491. [CrossRef] [PubMed]
54. Kumari, A.; Yadav, S.K.; Yadav, S.C. Biodegradable polymeric nanoparticles based drug delivery systems. *Colloids Surf. B Biointerfaces* **2010**, *75*, 1–18. [CrossRef]
55. Locatelli, E.; Franchini, M.C. Biodegradable PLGA-b-PEG polymeric nanoparticles: Synthesis, properties, and nanomedical applications as drug delivery system. *J. Nanoparticle Res.* **2012**, *14*, 1–17. [CrossRef]
56. Zhang, L.; Pornpattananangkul, D.; Hu, C.-M.; Huang, C.-M. Development of Nanoparticles for Antimicrobial Drug Delivery. *Curr. Med. Chem.* **2010**, *17*, 585–594. [CrossRef]
57. Esmaeili, F.; Hosseini-Nasr, M.; Rad-Malekshahi, M.; Samadi, N.; Atyabi, F.; Dinarvand, R. Preparation and antibacterial activity evaluation of rifampicin-loaded poly lactide-co-glycolide nanoparticles. *Nanomed. Nanotechnol. Biol. Med.* **2007**, *3*, 161–167. [CrossRef] [PubMed]
58. Koopaei, M.N.; Maghazei, M.S.; Mostafavi, S.H.; Jamalifar, H.; Samadi, N.; Amini, M.; Malek, S.J.; Darvishi, B.; Atyabi, F.; Dinarvand, R. Enhanced antibacterial activity of roxithromycin loaded pegylated poly lactide-co-glycolide nanoparticles. *DARU J. Pharm. Sci.* **2012**, *20*, 1–8. [CrossRef]

59. Moradi, M.T.; Karimi, A.; Rafieian-Kopaei, M.; Fotouhi, F. In vitro antiviral effects of *Peganum harmala* seed extract and its total alkaloids against Influenza virus. *Microb. Pathog.* **2017**, *110*, 42–49. [CrossRef] [PubMed]
60. Kim, Y.; Kim, H.; Bae, S.; Choi, J.; Lim, S.Y.; Lee, N.; Kong, J.M.; Hwang, Y.; Kang, J.S.; Lee, W.J. Vitamin C is an Essential Factor on the Anti-viral Immune Responses through the Production of Interferon-α/β at the Initial Stage of Influenza A Virus (H3N2) Infection. *Immune Netw.* **2013**, *13*, 70. [CrossRef]
61. Mahmoud, D.B.; Ismail, W.M.; Moatasim, Y.; Kutkat, O.; ElMeshad, A.N.; Ezzat, S.M.; El Deeb, K.S.; El-Fishawy, A.M.; Gomaa, M.R.; Kandeil, A.; et al. Delineating a potent antiviral activity of Cuphea ignea extract loaded nano-formulation against SARS-CoV-2: In silico and in vitro studies. *J. Drug Deliv. Sci. Technol.* **2021**, *66*, 102845. [CrossRef] [PubMed]

pharmaceutics

MDPI

Article

Formulation and Optimization of Butenafine-Loaded Topical Nano Lipid Carrier-Based Gel: Characterization, Irritation Study, and Anti-Fungal Activity

Wael A. Mahdi [1,†], Sarah I. Bukhari [1,†], Syed Sarim Imam [1], Sultan Alshehri [1], Ameeduzzafar Zafar [2] and Mohd Yasir [3,*]

1 Department of Pharmaceutics, College of Pharmacy, King Saud University, Riyadh 11451, Saudi Arabia; wmahdi@ksu.edu.sa (W.A.M.); sbukhari@ksu.edu.sa (S.I.B.); simam@ksu.edu.sa (S.S.I.); salshehri1@ksu.edu.sa (S.A.)

2 Department of Pharmaceutics, College of Pharmacy, Jouf University, Sakaka 72341, Saudi Arabia; zzafarpharmacian@gmail.com

3 Department of Pharmacy, College of Health Sciences, Arsi University, Asella P.O. Box 396, Ethiopia

* Correspondence: mohdyasir31@gmail.com

† These authors contributed equally to this paper.

Abstract: The present study aims to prepare and optimize butenafine hydrochloride NLCs formulation using solid and liquid lipid. The optimized selected BF-NLCopt was further converted into Carbopol-based gel for topical application for the treatment of fungal infection. Box Behnken design was employed to optimize the nanostructure lipids carriers (NLCs) using the lipid content (A), Tween 80 (B), and homogenization cycle (C) as formulation factors at three levels. Their effects were observed on the particle size (Y_1) and entrapment efficiency (Y_2). The selected formulation was converted into gel and further assessed for gel characterization, drug release, anti-fungal study, irritation study, and stability study. The solid lipid (Compritol 888 ATO), liquid lipid (Labrasol), and surfactant (tween 80) were selected based on maximum solubility. The optimization result showed a particle size of 111 nm with high entrapment efficiency of 86.35% for BF-NLCopt. The optimized BF-NLCopt converted to gel (1% w/v, Carbopol 934) and showed ideal gel evaluation results (drug content 99.45 ± 2.11, pH 6.5 ± 0.2, viscosity 519 ± 1.43 CPs). The drug release study result depicted a prolonged drug release (65.09 ± 4.37%) with high drug permeation 641.37 ± 46.59 µg (32.07 ± 2.32%) than BF conventional gel. The low value of irritation score (0.17) exhibited negligible irritation on the skin after application. The anti-fungal result showed greater efficacy than the BF gel at both time points. The overall conclusion of the results revealed NLCs-based gel of BF as an ideal delivery system to treat the fungal infection.

Keywords: butenafine; NLCs; topical gel; anti-fungal; optimization; irritation study

1. Introduction

Butenafine hydrochloride (BF) is a novel synthetic antifungal drug used in the treatment of different antifungal diseases. Chemically it is a benzylamine and naphthalene having the molecular formula $C_{23}H_{27}N{\cdot}HCl$ and molecular weight of 353.93 gm/mol [1]. It comes under a broad-spectrum antifungal agent and acts by inhibiting fungal squalene epoxidase synthesis of ergosterol, an important intermediate of fungal cell membrane synthesis [2,3]. BF has shown better therapeutic efficacy than terbinafine [4]. It is not used by the oral route as highly metabolized by the liver (1.5–3% oral bioavailability) and excreted in the urine.

Many different types of fungal germs live in the soil and other places in the environment. They can grow and multiply on the body surface and cause infection to the mouth, nails, and skin and produce fungal infection. The topical application of nanocarriers has shown a potential delivery system for fungal infection treatment. These nanocarriers have also been used in the topical treatment and possess superior therapeutic efficacy as compared to conventional cream and gel. These delivery systems improve the therapeutic efficacy by a different mechanism including enhancing permeation through subcutaneous, enhanced solubility, sustained drug release, and targeting potential to macrophage, and deeper skin layer [5,6]. Nano lipid carrier (NLC) is a second-generation lipid formulation prepared with a blend of solid lipid and liquid lipid [7]. They are prepared with physiological solid and liquid lipid, which is completely biodegradable. The used lipids for the NLCs preparation resemble the skin and sebum making them biocompatible with limited risk [8]. It possesses a lower melting point due to the presence of liquid lipid, maintains the particulate character, and remains solid at room temperature [9]. The presence of solid lipid in the NLCs helps to provide a prolonged drug-release pattern and also reduces systemic absorption. After application to the skin, it gives occlusive property [10,11]. The drug loading and entrapment are higher than the solid lipid nanoparticles due to the greater drug solubility in both solid and liquid lipid [12]. The nanoparticle size gives close contact to the stratum corneum and enhances the amount of drug permeation into the skin. Recently, NLCs are gaining popularity in the field of cosmetic, pharmaceutical, and food-delivery systems with a great potential for commercialization in the future [13].

The gel formulation is a semisolid dispersion system made up of inorganic and organic molecules enclosing and interpenetrated by a liquid. The inorganic particles give a three-dimensional structure [1,14]. The gel system should be washable, non-staining, stable at room temperature, and should not affect the biological nature of the drug [15]. There are numerous NLCs-based topical systems that have been reported to enhance the therapeutic efficacy of poorly soluble drugs like valdecoxib [11], luliconazole [16,17], oxiconazole [18], amphotericin B [7], itraconazole [19].

The design of the present study was divided into two steps. In the first step, BF-NLCs were prepared by the homogenization method. The formulation was optimized by the Box Behnken design (Design of expert, version 8.0.06, State-Ease Inc., Minneapolis, MN, USA) approach using independent variables as total lipid (A), tween 80 (B), and homogenization cycle (C) as independent variables. Their effects were observed on the particle size (Y_1) and entrapment efficiency (Y_2). In the second step, the optimized selected formulation (BF-NLCopt) was converted into a topical gel system using Carbopol as gelling agent to enhance the skin adhesion (BF-NLCopt gel). Carbopol offers excellent biocompatibility with the nano-drug delivery carriers, good gelling properties (easily swells at a small concentration of 1–2% *w*/*v*) with the ease of spreadability, thermal stability, and optimum rheological properties. The concentrations between 1% and 2% gave good consistency for the gel and it is generally recognized as a safe (GRAS) property. It also helps in controlling the release of the drug from the gel matrix. The prepared gel system was further assessed for gel characterization, drug release, permeation study, irritation study, and antifungal study.

2. Materials and Methods

2.1. Materials

BF was procured from Sigma Aldrich (St. Louis, MO, USA). The used lipids Preciol ATO 5, Compritol 888 ATO, and Labrasol were procured from Gattefosse Ltd. (Mumbai, India). Glyceryl monostearate, beeswax, stearic acid were obtained from SD fine Ltd. (Mumbai, India). Castor oil, almond oil, sunflower oil, and olive oil were obtained from a local market, Colaba (South Mumbai India). Cremophore EL, tween 20, tween 80, span 80, poloxamer 188, poloxamer F127 and span 20, methanol, HPLC grade water were procured from Sigma Aldrich (USA). Buffering reagents were obtained from the Central Drug House (Delhi, India).

2.2. Methods

2.2.1. Screening of Solid and Liquid Lipid

The screening of lipid was performed to select the best lipid (solid and liquid) to formulate NLCs. The selection was performed on the criteria of the highest solubility of a drug into a lipid. The appropriate quantity (2 mg) of each solid lipid (Precirol ATO 5, Compritol 888 ATO, glyceryl monostearate, stearic acid, beeswax) was taken in glass vials and melted above 5 °C of the melting point. An excess amount of BF was added to the melted lipid. Similarly, the liquid lipids (castor oil, sunflower oil, labrasol, cremophore EL, almond oil, olive oil) were taken in glass vials. An excess of BF was added to the liquid lipid (2 mL). Both the lipids solid and liquid were vortexed (5 min) for complete solubilization and was allowed to stand for 72 h in an orbital shaker (Thermo-fisher Scientific, Mumbai, India). The mixture was centrifuged at 6000 rpm for 30 min and the supernatant separated. The supernatant was appropriately diluted and the concentration of BF dissolved in lipids was measured by a UV-spectrophotometer (Shimadzu-1800, Kyoto, Japan).

2.2.2. Selection of Surfactant

The selection of surfactant for NLCs formulation was determined by the criteria of maximum solubility of BF in tested surfactants (Tween 20, Tween 80, Span 20, Span 80, Poloxamer, F127, Poloxamer F188). The appropriate volume of surfactant (2 mL) was taken in a glass vial and an excess amount of BF was added. The samples were and vortexed for 5 min and kept in an orbital shaker for 72 h. After that, the supernatant was centrifuged at 6000 rpm for 30 min. The supernatant was appropriately diluted and BF content in each surfactant was measured by a UV-spectrophotometer after dilution.

2.2.3. Solid and Liquid Lipid Miscibility

The melted solid and liquid lipid in different compositions (9:1, 8:2, 7:3, 6:4, 5:5, 4:6) were mixed in a glass vial and vortexed. The mixture was kept aside at room temperature. One drop of the mixture was added to the filter paper, and the presence of an oil droplet was checked. The ratio of solid and liquid lipids at which they did not separate and exhibiting better miscibility (showing a single-phase system) was considered for further study.

2.2.4. Optimization

From the preliminary screening study, parameters like lipid concentration, surfactant concentration, and homogenization cycle were chosen as pre-optimized parameters [19, 20]. The optimization was done by design expert software (version 8.0.06, State-Ease Inc., Minneapolis, MN, USA) using the Box–Behnken design (3 factors at 3 levels). The three independent variables, lipid concentration (%, A), surfactant concentration (%, B), and homogenization cycle (number, C) were set at three levels i.e., low (−), medium (0), and high (+), as shown in Table 1. The effect of these three independent parameters was observed on two dependent variables, particle size (nm, Y_1) and entrapment efficiency (%, Y_2). The software showed a total of 17 compositions (5 commons) after adding the lower and higher values of lipid, surfactant, and homogenization cycle in the design expert software as shown in Table 2. The design depicted the effect of the independent parameters individually (A or B or C), their interaction effect (AB, AC, BC), and quadratic effect (A^2, B^2, and C^2) on the particle size and entrapment efficiency. The outcome of the design was obtained in the form of a polynomial equation, and three-dimensional response surface graphs (3D and contour) [21,22].

Table 1. Formulation variables range used to optimize BF-loaded NLCs.

Parameters	Constraints		
	Lower Value	Medium Value	Upper Value
	Independent variables		
Lipid concentration (A, %)	2	3	4
Surfactant concentration (B, %)	1	2	3
Homogenization cycle (C, n)	2	4	6
Dependent variables		Goals	
Particle size (Y_1, nm)		Minimize	
Entrapment efficiency (Y_2, %)		Maximize	

Table 2. Composition of BF-loaded NLCs with their actual and predicted particle size and encapsulation efficiency.

Batch No.	Lipid (A, %)	Surfactant (B, %)	Homogenization Cycle (C, n)	Particle Size (nm) (Y_1)		Entrapment Efficiency (%) (Y_2)	
				Actual	Predicted	Actual	Predicted
BF1	2	3	4	182.55	179.48	66.74	69.31
BF2	3	1	2	302.77	299.76	69.82	71.65
BF3	4	3	4	276.36	276.57	76.44	76.60
BF4	3	2	4	118.14	117.81	83.91	84.85
BF5	2	2	6	146.23	143.44	77.76	78.02
BF6	3	2	4	118.37	117.81	83.38	84.85
BF7	4	2	6	124.25	127.23	82.51	84.15
BF8	4	1	4	311.75	311.97	77.81	75.24
BF9	3	2	4	117.85	117.81	84.42	83.85
BF10	3	3	6	190.44	193.45	75.65	73.82
BF11	3	2	4	119.47	117.81	85.65	84.85
BF12	3	3	2	268.28	271.18	76.59	75.66
BF13	2	2	2	200.69	197.71	78.70	76.06
BF14	3	2	4	117.21	117.81	83.89	84.85
BF15	3	1	6	274.38	276.14	76.61	77.54
BF16	2	1	4	240.78	246.76	70.59	70.39
BF17	4	2	2	176.78	179.57	81.33	82.07

2.2.5. Formulation of BF-NLCs

BF-NLCs were prepared according to the reported method by Muller et al., 2000 with slight modification [23]. Lipid was melted around 5 °C above its melting point and the weighed amount of BF was added to the melted lipid. Separately, an aqueous surfactant solution was prepared and the same temperature was maintained as lipid. The hot surfactant solution was poured into the melted lipid phase with continuous stirring at 12,000 rpm using a magnetic stirrer for 30 min. The primary emulsion was formed and converted to the NLCs dispersion using a high-pressure homogenizer (Stansted Fluid Ltd., Harlo, UK) as shown in Table 2. The prepared NLCs dispersion was cooled down at room temperature and then lyophilized using mannitol (5% *w*/*v*) as a cryoprotectant.

2.2.6. BF-NLCs Characterization

The particle size (PS), polydispersity index (PDI), zeta potential (ZP)of the prepared BF-NLCs were evaluated by a particle size analyzer (Nano-sizer, Malvern, UK). The sample was diluted 100-fold with Mili-Q water and the particle size, PDI, and ZP were measured. The morphology of the selected sample was assessed by transmission electron microscopy (TEM-TECNAI-G2, FEI, Eindhoven, The Netherlands). One drop of diluted sample was placed over a copper grid, and phosphotungstic acid dye was added to stain the sample. The grid was placed in an instrument and the image was captured.

2.2.7. Entrapment and Loading Efficiency

EE and DL of BF in BF-NLCs were analyzed to check the amount of BF loaded and entrapped in the prepared NLCs. BF-NLCs were filled in a centrifugation tube and centrifuged at 15,000 rpm for 30 min (cooling centrifuge, Remi Lab, Mumbai, India). The supernatant was separated and diluted further to evaluate the unentrapped amount of drug in NLCs by measuring the absorbance with the help of a UV-spectrophotometer. The resulted absorbance was mathematically converted into a drug concentration. %EE and %DL were calculated by the given below equation:

$$\%\ \text{EE} = \frac{\text{Total BF} - \text{Unentrapped BF}}{\text{Total BF}} \times 100$$

$$\%\ \text{DL} = \frac{\text{Total BF} - \text{Unentrapped BF}}{\text{Weight of NLCs}} \times 100$$

2.2.8. Formulation of BF-NLCs Gel

The point prediction-based optimized BF-NLCopt was converted into a gel by using Carbopol 934 as a gelling agent. Carbopol (0.75, 1, 1.25% *w*/*v*) was dispersed in distilled water overnight, and then the prepared BF-NLCopt dispersion was added with continuous stirring to form the uniformed dispersion. Finally, the triethanolamine was added to the gel system for neutralization of pH.

2.2.9. Gel Evaluation

The prepared BF-NLCopt gel (10 mg) was taken and dissolved in 10 mL of methanol. The sample was centrifuged at 15,000 rpm for 30 min (Cooling centrifuge, Remi Lab, Mumbai, India). The supernatant was collected, filtered, diluted, and the absorbance was measured by UV-spectrophotometer. The absorbance was mathematically converted into the amount of drug present in the formulation. The viscosity of the gel sample was evaluated by Brookfield viscometer at room temperature. The viscosity was measured by using spindle number 6 at 15 rpm [24]. The pH was measured by dispersing 100 mg of gel into distilled water (10 mL) using a digital pH meter (AS218, Eutech Instrument, Singapore) [25]. The measurement of gel spreadability is an important parameter to check the flow of gel. The measured quantity of BF-NLCopt gel (0.5 g) was placed on the glass slide within the pre-marked diameter of 1 cm and covered with another slide. The weight (500 g) was placed over the slide for 10 min, and the increase in diameter due to the spread of gel was noted [26]. The extrudability test was carried out to assess the extrusion of gel from a collapsible tube on the application of constant weight. The collapsible tube containing BF-NLCopt gel (20 g) was pressed by applying a constant load of 1000 gm at the crimp end. The extruded gel was collected after opening the cap and weighed.

2.2.10. In Vitro Release Study

The in vitro release study of BF-NLCopt gel and BF-NLCopt was determined by dialysis membrane (MW12,000 kDa, Sigma Aldrich, USA). The prepared formulations (BF-NLC gel and BF-NLCopt) containing ~5 mg BF were filled into a pretreated dialysis bag (1.0 cm diameter) and both ends were tightly tied. The dialysis bag was dipped into a beaker containing release media (500 mL, phosphate buffer, pH 6.8). The temperature was maintained at 37 ± 0.5 °C throughout the study with regular stirring (50 rpm). The released content (5 mL) from each sample was withdrawn at a specified time and replaced with the same volume. The sample was filtered, diluted with the same medium and the drug concentration at each time point was measured by UV-spectrophotometer [1]. The drug release was calculated and the graph between time vs.% drug release was plotted to calculate the maximum BF release. The release study was performed in triplicate and the release data of BF-NLCopt gel was applied into various release kinetic models to identify the release mechanism. The data were fitted into zero-order, first-order, Higuchi,

Korsmeyer Peppas, and Hixson-Crowell model, and the best fit was selected based on the maximum regression (R^2) value.

2.2.11. In Vitro Permeation Study

The eggshell membrane was used to evaluate the prepared formulations for the permeation study because it is similar to the stratum corneum of human skin [27,28]. The eggs were procured from a poultry house and dipped into concentrated 0.1*N* HCl. The shell was removed, the membrane was collected and cleaned with distilled water. The membrane was mounted between the donor and acceptor compartment of the diffusion cell (area 1.5 cm^2). The diffusion media (10 mL) phosphate buffer (pH-6.8) were filled in the receptor compartment. The temperature was maintained at 37 $\pm$ 0.5 °C throughout the study. The formulations (BF-NLCopt, BF-NLCopt gel, BF-gel containing equivalent to 2 mg of BF) were filled into the donor compartment. At a predetermined time interval, 2 mL of sample was withdrawn and replaced with the same volume to maintain the uniformity in volume. The collected sample was filtered through a membrane filter (0.45 μm) and absorbance was measured by the reported HPLC method to calculate the amount of BF permeated [29].

2.2.12. Antifungal Activity

The antifungal sensitivity activity was analyzed by cup plate diffusion method using Sabouraud's dextrose agar medium following a previously reported study [30]. The fungal broth culture (*Candida albicans* and *Aspergillus fumigatus*) was standardized in the growth medium. The required quantity medium was prepared and transferred in a clean sterilized Petri plate at 121 °C (15 PSI). The plates were kept aside for 15 min to solidify the media. The wells (6 mm) were prepared using a sterilized stainless-steel borer in the sterilized condition. The samples BF-NLCs-gel and BF-gel were placed into the wells. The plates were kept for 2 h at room temperature to diffuse the sample into the medium and then transferred into the incubator. The zone of inhibition (mm) (ZOI) was measured at 12 h and 24 h for each test sample and compared with the standard.

2.2.13. Irritation Study

The irritation study of BF-NLCs-opt-gel was done by using chick chorioallantoic membrane (CAM) [31]. The irritancy of a developed gel was measured by measuring its ability to induce toxicity in the CAM. The fertilized hen eggs were collected from a poultry farm and incubated into a humidity incubator (Komeg, Tech industrial Ltd., Shenzhen, Guangdong, China) for ten days at 37 $\pm$ 0.5 °C and 55 $\pm$ 2% RH. The eggs were manually rotated every 24 h. On the 10th day, the eggs were removed from the incubator and the outer shell of the eggs was removed from the air chamber side with the help of forceps. Sterilized normal saline was added for clear visibility of CAM. The formulations (BF-NLCs-opt-gel), normal saline (negative control), and sodium lauryl sulphate (1% *w*/*v*, positive control) were added over CAM for evaluation of irritation. The irritation score was noted at a definite time interval (0.5, 2, and 5 min) and compared to the standard irritation score. No visible hemorrhage score 0–0.9 (non-irritant), membrane decolorization score 1–4.9 (mild irritant), hemorrhage and structured partially covered score 5–8.9 (moderate irritant), hemorrhage and structure covered completely score 9–20 (severe irritant) were seen [32].

2.2.14. Statistical Analysis

All experimental study data were expressed in mean $\pm$ SD (n = 3). The GraphPad Prism software (San Diego, CA, USA) was applied for statistics. One-way ANOVA was employed followed by the Tukey–Kramer multiple comparison test to analyze statistically significant different samples and control and $p < 0.05$ was taken for statistical significance.

3. Result and Discussion

3.1. Screening of Lipids and Surfactant

The solubility profile of the drug in solid, liquid lipid, and surfactant is shown in Figure 1A,B. The criterion of each component selection was the maximum solubility of the drug. The order of solubility of BF in various lipids is compritol 888 ATO > precirol ATO 5 > glyceryl monostearate > stearic acid > beeswax. The highest solubility of BF was found in compritol 888 ATO (65.24 ± 7.48 mg/mL). The order of solubility of BF in liquid lipid is in the order of labrasol > cremophore EL > almond oil > sunflower oil > olive oil > castor oil (Figure 1A). The highest solubility of BF was found in labrasol (43.12 ± 6.43 mg/mL). The tested surfactant depicted the BF solubility in order of tween 80 > tween 20 > poloxamer 188 > poloxamer F127 > span 20 > span 80 (Figure 1B). Tween 80 showed the maximum solubility of BF (72.45 ± 6.45 mg/mL). So, from the preliminary solubility study, compritol 888 ATO, labrasol, and tween 80 were selected as solid lipid, liquid lipid, and surfactant for the preparation of NLCs.

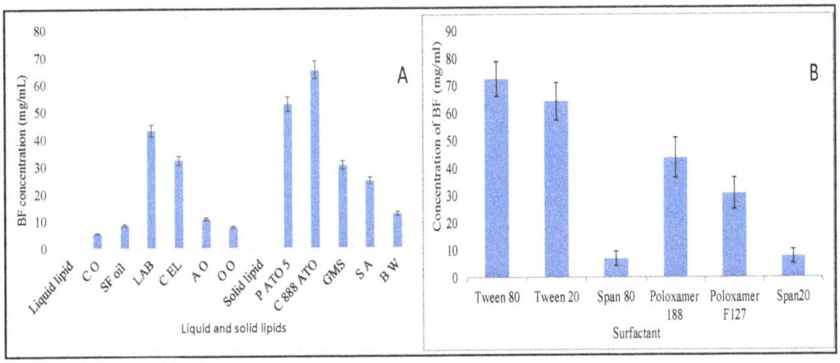

Figure 1. (**A**): Solubility study result of BF in different liquid and solid lipids. Each study performed in triplicate and data shown as mean ± SD. (**B**): Solubility study result of BF in different surfactants. Each study performed in triplicate and data shown as mean ± SD.

3.2. Miscibility of Solid and Liquid Lipid

The selected solid and liquid lipid were mixed at different proportions to select the optimized blend to use in the preparation. Among the tested ratios, 6:4 was found to be the best composition. This ratio showed no phase separation, as well as no oil droplet was found after smear formation in the filter paper, so the ratio 6:4 was used for further study in the preparation of BF-NLCs.

3.3. Optimization

BF-NLCs were optimized using a three-factor three-level Box Behnken formulation design approach to select the optimum composition with minimum particle size and high entrapment efficiency. The design showed 17 formulation compositions with five common compositions to check the error in the results [21,22]. The result of particle size (Y_1) and entrapment efficiency (Y_2) was added to the software to interpret the results. The software generated predicted values, polynomial equations, contour, and 3D response surface graphs to evaluate the interaction of independent factors on dependent factors. The actual particle size and entrapment efficiency were compared to the predicted value generated by the software and the result was found to a closer value (Table 2). The summary of regression analysis for the particle size and entrapment efficiency is depicted in Table 3. For all responses, the best fit model was found to be quadratic with the highest value of correlation coefficient (R^2). The quadratic response is ideal for the optimization because the used independent variable showed individual as well as a combined effect on the dependent variables. The predicted R^2 was found to be in good agreement with the adjusted R^2 for both responses. ANOVA value was found to be significant and the F value showed a value greater than 4. Lack of fit was found to be non-significant and residual values results support the optimization (Table 4). The calculation of percentage prediction was performed to determine the accuracy of the software. The negative sign in the polynomial equation corresponds to an opposite relationship and the positive sign favors the interaction of independent variables with the dependent variables. The overall combined desirability was found to be closer to unity, which confirms that the tested independent variable was found suitable for the optimization.

Table 3. Statistical model summary of regression analysis results for response Y_1, and Y_2.

Model	R^2	Adjusted R^2	Predicted R^2	SD	% CV	Remark
			Particle size (Y_1)			
Linear	0.9147	0.8155	0.8465	7.14		
2F1	0.9315	0.9295	0.9012	8.02		
Quadratic	0.9979	0.9951	0.9718	5.04	2.61	Suggested
			Entrapment efficiency (Y_2)			
Linear	0.9031	0.8423	0.8391	5.90		
2F1	0.9183	0.8305	0.8026	6.60		
Quadratic	0.9962	0.9312	0.9215	2.37	3.02	Suggested

Table 4. ANOVA of quadratic model for responses of developed BF-NLCs.

ANOVA Results	Particle Size (Y_1, nm)	Entrapment Efficiency (%) (Y_2)
	Regression	
Some of square	83149.71	494.61
Degree of freedom	9	9
Mean square	9238.86	54.96
F-value	363.48	9.75
P	<0.0001	<0.0033
Influence	Significant	Significant
	Lack of fit-test	
Some of square	143.43	28.23
Degree of freedom	3	3
Mean square	47.81	9.51
F-value	5.54	3.36
P	0.0658	0.1363
Influence	Non-significant	Non-significant
	Residual	
Some of square	177.93	39.44
Degree of freedom	7	7
Mean square	25.42	5.63

3.4. *Effect of Independent Variables on Particle Size (Y_1)*

The particle size of different batches was found to be in the range of 113.47 (BF11) to 311.75 nm (BF8). A significant difference in particle size was observed due to the variation in the composition of lipid (A), surfactant (B), and homogenization cycle (C). The polynomial Equation (1) and 3D response plot (Figure 2) showed the effect of independent variables on particle size:

$$Y_1 = +117.81 + 8.59A - 26.51B - 21.65C - 41.19AB + 0.48AC - 12.36BC + 19.04A^2 + 116.02B^2 + 25.14C^2 \quad (1)$$

Lipid showed a positive effect on particle size. So, the optimum concentration of lipid is necessary to obtain the minimum size. The particle size increases with an increase in the lipid concentration due to the availability of higher concentration lipid. This might be due to the lack of sufficient surfactant which solubilizes the drug and reduces the interfacial tension. At a low concentration of surfactant, the aggregation of particles takes place and the particle size increased [33]. The surfactant (B) showed the oppositive effect on the particle size. It influences particle size by giving stability [23]. As the concentration of surfactant increased, the particle size decreases. The surfactant reduces the available space to accommodate the drug [20]. Like surfactants, the homogenization cycle (C) also showed a negative effect on particle size. As the homogenization cycle increases, the particle size decreases due to the generation of high force which is responsible for the breaking and hence reducing the particle size.

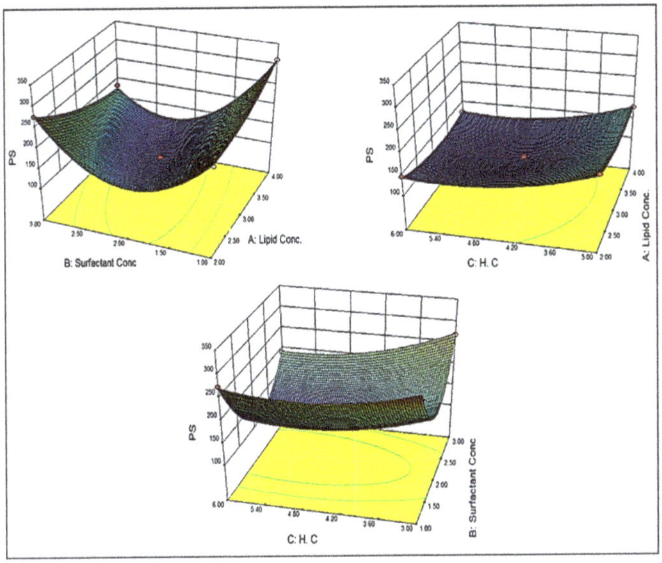

Figure 2. 3D response surface plot showing the effect of independent variables on the particle size.

Equation (1) also showed the combined effect of various variables on the particle size. The combination of lipid (A) and surfactant (B) showed a negative effect on particle size. So, from the equation, we can say that surfactant (B) has a more prominent effect on particle size than lipid (A). In the case of factors lipid (A) and homogenization cycle (C), it showed a positive effect. As the concentration of lipid and homogenization cycle increases the particle size also increases. At a high homogenization cycle, the lipid may form an aggregate due to the lack of sufficient surfactant (B) concentration. The combined effect of factors surfactant (B) and homogenization cycle (C) showed a negative effect on the size. The combined effect was also found to be similar to the individual effect of surfactant (B) and homogenization cycle (C).

3.5. *Effect of Independent Variables on Entrapment Efficiency (Y_2)*

The entrapment efficiency of different compositions was found to be in the range of 66.74 (BF1) to 87.42% (BF9). A significant difference in entrapment efficiency was observed due to the variation in the composition of lipid (A), surfactant (B), and homogenization cycle (C). The polynomial Equation (2) concern to entrapment efficiency is given below:

$$Y_2 = +84.85 + 3.03A - 2.069B - 1.01C + 0.61AB + 0.030AC - 1.93BC - 3.28A^2 - 8.69B^2 - 1.50\,C^2 \quad (2)$$

The polynomial equation and 3D response plot (Figure 3) showed a significant effect of lipid (A), surfactant (B), and homogenization cycle (C) on entrapment efficiency. The increase in the concentration of lipid (A) leads to an increase in the BF entrapment in the NLCs. The use of lipid blends (solid and liquid lipid) gives more space to accommodate more drugs. From the solubility study data, the BF showed good solubility in solid as well liquid lipid, so more amount of drug solubilize and gives greater entrapment. The second-factor surfactant (B) showed an overall negative effect on entrapment efficiency. An increase in surfactant concentration decreases the entrapment efficiency. At the initial stage, with the increase in surfactant concentration, the entrapment efficiency was increased; however, with further increase, there was a measurable reduction in entrapment efficiency, which might be due to drug leakage into the external environment (BF1 70.59% and BF3 66.74%) [20,34]. At a high concentration of surfactant, the drug may leach out from the lipid blend. The homogenization cycle (C) also depicted the negative effect on entrapment efficiency. At a higher homogenization cycle, the high shear force is generated which is responsible for the breakdown of particles and hence leaching of drug from lipid nanoparticles takes place and reduces the drug entrapment. The combination of lipid (A) and surfactant (B) showed a positive effect on drug entrapment. So, from the equation, we can say that surfactant (B) has a less prominent effect than the lipid (A). In the case of factor lipid (A) and homogenization cycle (C), it showed a positive effect. As the concentration of lipid and homogenization cycle increases the entrapment also increases. At a high lipid concentration and homogenization cycle, the amount of lipid breakdown is higher and a higher amount of BF is entrapped. The combined effect of factors surfactant (B) and homogenization cycle (C) showed a negative effect on the size. The combined effect was also found to be similar to the individual effect of surfactant (B) and homogenization cycle (C).

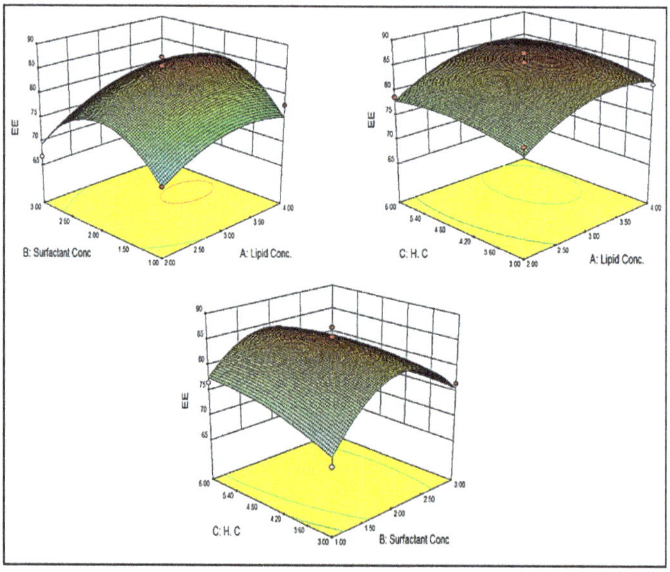

Figure 3. 3D response surface plot showing the effect of independent variables on entrapment efficiency.

3.6. Point Prediction

The formulation was further optimized by the software-generated point prediction method. The point prediction was made by making minute changes in the value of independent parameters to get the more close and accurate value of these parameters and their effect on the responses. The main objective of this step to get more accurate results and validate the resulted model. The optimized formulation (BF-NLCopt) shows the composition of lipid concentration 2.6%, surfactant concentration 2.4%, and the homogenization cycle of 4. This composition showed a particle size of 119.29 nm and an entrapment efficiency of 86.35%. The predicted value was found to be very closer to the practical value. The predicted particle size was found to be 112.14 nm with an entrapment efficiency of 88.56%. There was a non-significant difference in the actual and predicted particle size, and entrapment efficiency was observed (Figure 4). It indicates that the model was well fitted between them.

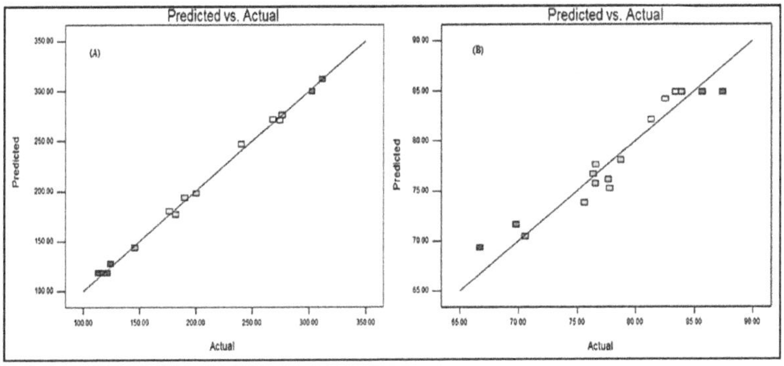

Figure 4. Actual and predicted value of particle size (**A**) and entrapment efficiency (**B**).

3.7. Particle Characterization

The particle size of the prepared BF-NLCs formulations was analyzed and is given in Table 2. The particle size of BF-NLCs was found in the range of 113.47 (BF11) to 311.75 nm (BF8). The selected formulation BF-NLCopt showed a particle size of 119.29 ± 3.23 nm (Figure 5), PDI of 0.29 with a ZP of −28 mV. The low PDI (<0.5) and zeta potential value (−28 mV near to optimum, ±30 mV) support the physical stability of the prepared formulation. The particle size plays an important role in the permeation of the drug through the skin and helps in the retention of the drug in the epidermis. Particle size more than 100 nm will not penetrate the lower skin layers. Nanoparticles (>100 nm) preferentially permeate into the follicles, but not the dermis, enabling high accumulation within the follicular reservoir. The epidermis releases the drug for a prolonged duration to treat the topical fungal infection [35,36]. As previously reported that lipid nanoparticles lower than 100 nm were capable to penetrate the deeper layer of skin and bigger size nanoparticles were incapable to do so [37]. Moreover, NLCs provide greater drug permeation and therapeutic efficacy by offering close contact with the stratum corneum. Due to the lipid, NLCs have bioadhesion properties and form an occlusive film which decreases the water loss from the transepidermal layer. NLCs also facilitate the penetration of the drug into deeper skin [19]. The morphology of BF-NLCs-opt was found to be spherical and is given in Figure 6.

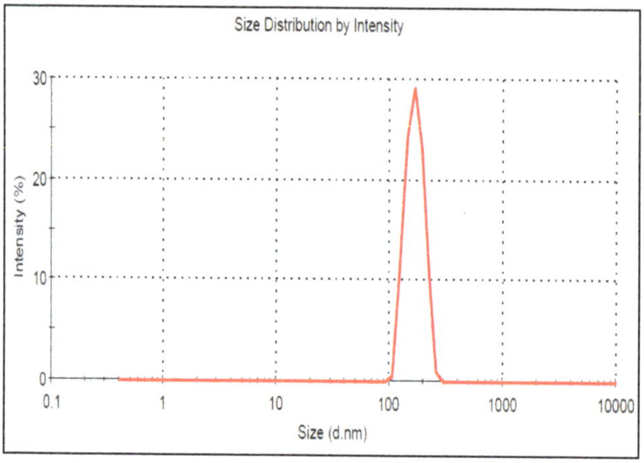

Figure 5. Particle size image of BF-NLCopt.

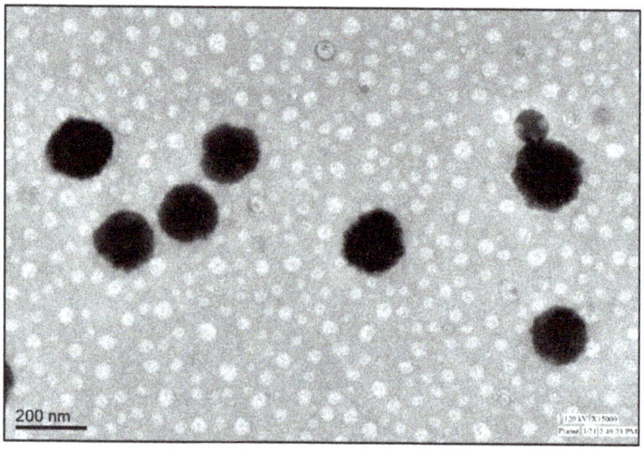

Figure 6. Surface morphology image of BF-NLCopt.

3.8. Entrapment Efficiency

The entrapment efficiency of the prepared BF-NLCs was determined by the indirect method and the data are given in Table 2. The maximum entrapment efficiency was found in the formulation BF1 (66.74%) to BF7 (87.42%). The selected optimized formulation BF-NLCopt showed an entrapment efficiency of 86.35 ± 4.11%. The selected formulation showed a drug load of 11.21 ± 0.67%. NLCs were prepared with a blend of solid lipid and liquid lipid. The solid lipid encloses the oil droplets of liquid lipid which solubilizes the greater amount of drug and helps to get high entrapment and loading. The presence of liquid lipid in the NLCs affects the loading and entrapment by creating imperfections in the order crystal structure and gives more space to accommodate a greater amount of drug [23].

3.9. Formulation of BF-NLCopt Gel

Due to the low viscosity of NLCs dispersion, it displays unsuitable rheological properties so it is difficult to apply to the skin layer. To overcome these issues, the low viscosity dispersion system was converted into a hydrogel system to provide precise spatial and temporal control of active ingredient release. Carbopol was selected as the gelling agent due to its compatibility with nanoparticulate delivery, easy formulation, thermal stability, optimum rheological properties. The use of triethanolamine (organic base) as neutralizing agent improves the stability of lipid nanoparticles [38]. BF-NLCs-opt was successfully transformed into gel using Carbopol in three different concentration (0.75, 1, 1.25% *w*/*v*) to select the optimum concentration. Among the three concentrations, BF-NLCopt gel prepared with Carbopol (1% *w*/*v*) showed good spreadability for the gel formulation. The other tested concentration (0.75 and 1.25%) results do not show a satisfactory result.

3.10. BF-NLCs Gel Characterization

The drug content indicates the % of BF existing in the gel formulation. It is the ratio (%) of the drug present in formulation to the actual amount of drug taken for gel formulation development. The high drug content is a desirable character of pharmaceutical formulation. Here, the drug content of the prepared BF-NLCopt gel was found to be 99.45 ± 2.11%. The high drug content approved the authenticity of the adopted method to develop the gel formulation. The pH of BF-NLCopt gel was determined by pH meter at room temperature and the value was found to be 6.5 ± 0.2. The pH was found to be within the limit of topical gel and does not produce any toxicity [7]. The viscosity of BF-NLCopt gel was evaluated by Brookfield viscometer and the result was found to be 519 ± 1.43 CPs. The value of viscosity is significantly affected by the particle size and PDI. The higher value gives a more viscous formulation [39]. In this formulation, the particle size and PDI value both are lower and support the findings. Generally, the viscosity of gel formulations reflects consistency. The viscosity decreases with increasing the rate of shear rate (shear thinning, Non-Newtonian flow) in the case of gel formulation. The viscosity must be optimum to show better adherence to the skin and spreadability [36,40]. Spreadability is a very important parameter for the determination of the spreading capacity of gel on the skin. The good spreadability of topical formulation gives better application to inflamed or diseased skin. It was determined by application of weight and expressed as the area of spread of the gel. The area of the spread gel was found to be 6.35 ± 0.56 cm in low spreading time (1.5 s, visually observed). The viscosity decreases when applied with shear (might be due to the characteristic pseudoplastic behavior of gel), which confirm the characteristic of high spreadability due to the decrease in viscosity when applying a certain force. Moreover, the gel retained the property of remaining at the application site without draining, indicating the adherence ability of the developed gel formulation.

3.11. In Vitro Release Study

The in vitro drug release study of BF-NLCopt, BF-NLCopt gel, and BF-gel was done by using a dialysis membrane. BF-NLCopt, BF-NLCopt gel, and BF-gel showed the cumulative BF release of 88.09 ± 3.01%, 65.09 ± 4.37%, and 34.54 ± 3.87% in 24 h, respectively (Figure 7). BF-NLCopt gel showed a significant ($p < 0.05$) low release of BF than BF-NLCopt dispersion due to the presence of Carbopol gel matrix. It hindered the release of BF from the gel matrix and slowly released due to the extra barrier. BF-NLCopt gel exhibited a significant ($p < 0.05$) higher release of BF than BF gel (conventional). The presence of lipid and surfactant in BF-NLCopt gel reduces the surface interfacial tension, which enhances the solubility and dissolution of BF. BF-NLCopt and BF-NLCopt gel exhibited initial burst release due to the diffusion of unentrapped BF quickly to release media. Later, sustained BF release was observed from the lipids (solid and liquid lipid) for BF-NLCs-opt and lipid as well as Carbopol matrix for BF-NLCopt gel. These release patterns agreed with the previous research study reported by [19,26]. The biphasic release behavior is ideal for the topical formulation because the burst release gives the amount of drug at the target site to achieve therapeutic concentration and the slow release helps to maintain the therapeutic concentration [37,41]. The release data fitted to the different kinetic models, i.e., zero-order, first-order, Higuchi, Korsmeyer-Peppas, Hixson-Crowell. The R^2 value for each model was found to be 0.7399 (zero-order), 0.834 (First order), 0.892 (Higuchi), 0.9506 (Korsmeyer-Peppas), and Hixson-Crowell (0.897). The maximum R^2 (0.9506) value was found to be for Korsmeyer-Peppas, so this model was selected as the best kinetic release model. The n value was found to be 0.48 (<0.5), indicating the release of the drug by diffusion (Fickain type) [42].

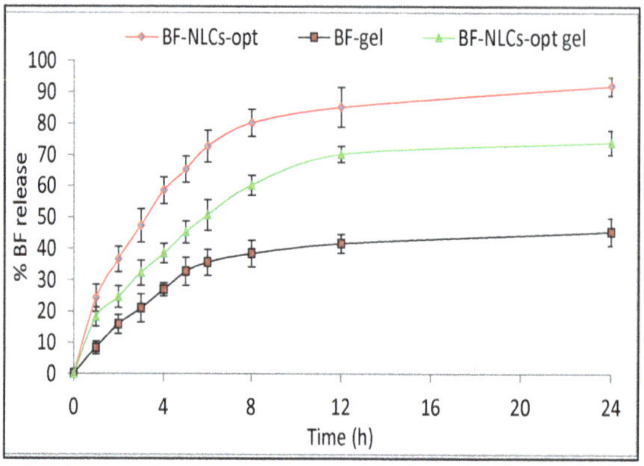

Figure 7. In vitro release study profile of BF-NLCopt, BF-NLCopt gel, and BF-gel. Each study was performed in triplicate and data are shown as mean ± SD. Comparison of permeation study of BF gel with BF-NLCsopt gel and BF-NLCs-opt. The significant differences observed at $p < 0.001$.

3.12. Permeation Study

The in vitro permeation study of BF-NLCopt, BF-NLCopt-gel, BF-gel was performed by diffusion shell using an eggshell membrane and data are shown in Figure 8. The amount of BF permeated from BF-NLCs-opt, BF-NLCs-opt-gel, and BF-gel through the membrane was found to be 845.52 ± 44.59 μg/cm^2 (42.28 ± 2.22%), 641.37 ± 46.59 μg/m^2 (32.07 ± 2.32%), and 285.52 ± 38.59 μg/cm^2 (14.28 ± 1.92%). There was a highly significant ($p < 0.001$) enhancement in the BF permeation from BF-NLCs-opt and BF-NLCsopt gel than BF gel. The greater permeation was found due to the nanometric size and presence of surfactant and lipid which act as permeation enhancers. The presence of surfactant helps solubilize the drug as well as helps open the pores of the membrane. In addition, There was a significant ($p < 0.001$) difference in the permeation between BF-NLCs-opt and BF-NLCopt gel. A lesser amount permeated from the gel formulation due to the slow diffusion of BF from the Carbopol gel matrix. The slow permeation from BF-NLCopt gel is ideal for topical preparation. The drug slowly permeates the skin and retains on the skin for a greater period. The presence of the drug at the target area gives a more prominent effect than the prepared NLCs formulation.

3.13. Anti-Fungal Study

The anti-fungal activity of BF-NLCopt gel, BF-gel, and BF dispersion was analyzed on *Candida albicans*, and *Aspergillus fumigatus* by agar diffusion method and data are shown in Figure 9. BF-NLCopt gel showed a ZOI of 14.1 ± 1.6 mm and 18.5 ± 2.1 mm against *Candida albicans*, whereas 15.6 ± 2.1 mm and 16.3 ± 1.6 mm against *Aspergillus fumigatus* in 12 h and 24 h, respectively. BF-gel showed ZOI of 10.3 ± 0.8 mm and 8.2 ± 0.9 mm against *Candida albicans* at 12 h and 24 h. Whereas, the observed ZOI was found to be 10.9 ± 1.1 mm and 7.1 ± 1.2 mm against *Aspergillus fumigatus* at the same time point, respectively. BF-dispersion showed good ZOI at 12 h (11.6 ± 1.3 mm) and very low ZOI at 24 h (6.9 ± 0.9 mm) against *Candida albicans*. In the case of *Aspergillus fumigatus*, BF dispersion showed a similar result as high ZOI at 12 h (14.1 ± 1.4 mm) and low ZOI at 24 h (6.6 ± 0.7 mm). There was a significant ($p < 0.05$) difference observed between the evaluated ZOI between BF-NLCopt gel and BF gel and BF dispersion. BF-NLCopt gel showed a marked enhancement in the ZOI at both time points. The higher activity is due to the higher solubility of BF in the used lipids and surfactant; and higher concentration of BF is available at the target area. The nanometric-sized BF-NLC particles slowly diffuse and produce antifungal activity.

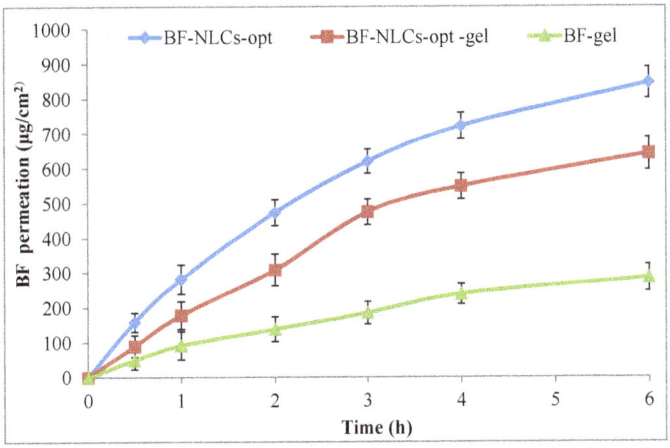

Figure 8. Permeation study profile of BF-NLCopt, BF-NLCopt gel, and BF-gel. Each study was performed in triplicate and data shown as mean ± SD. Comparison of permeation study of BF gel with BF-NLCsopt gel and BF-NLCs-opt. The significant differences observed at $p < 0.001$.

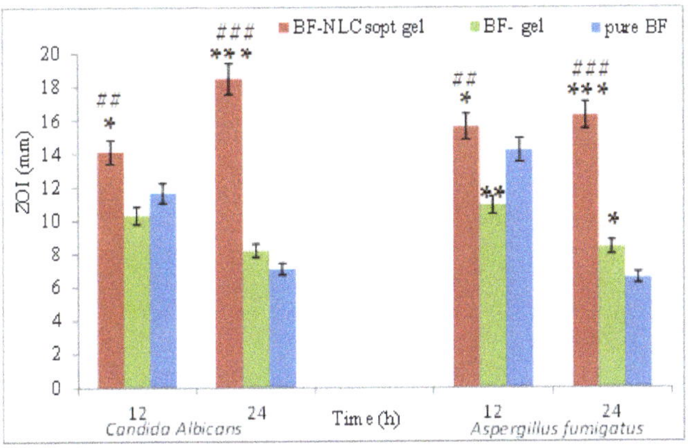

Figure 9. Antifungal study profile of Butenafine NLCopt gel, Butenafine gel, and Butenafine dispersion. Each study was performed in triplicate and data are shown as mean ± SD. * Comparison of antifungal activity of pure BF with BF-NLCsopt gel and BF-gel, Comparison of antifungal activity of BF-gel with BF-NLCsopt gel. *, **, ***, ##, ### show the significant differences at $p < 0.05$, $p < 0.01$, $p < 0.001$; respectively.

3.14. Irritation Study

The irritation study was performed by HET- CAM using the egg membrane and the irritation score was calculated. The experiments were observed for any sign of lysis, hemorrhage, coagulation of blood vessels for a maximum duration of 5 min. A test is considered acceptable if the negative and positive controls each induce a response that falls within the classification of nonirritating and severely irritating, respectively. The comparison of the study was performed between the positive control (1% *w*/*v*, SLS), negative control (0.9% NaCl), and BF-NLCopt-gel. The positive control-treated CAM showed a cumulative score of 17 (indicates severe irritation). The negative control-treated chorioallantoic membrane showed a cumulative score of 0, which is considered as non-irritant. BF-NLCopt-gel-treated CAM membrane exhibited a cumulative score of 0.17 with no signs of lysis, hemorrhage, coagulation of blood vessels after 5 min indicating that the optimized gel formulation was nonirritant and nontoxic. The overall score of the study of BF-NLCopt gel showed a score between 0 and 0.9 considered as non-irritant and safe. The finding of the study agreed with previously published research work [43].

4. Conclusions

The prepared BF-NLCs were optimized using three factors at three-level Box Behnken design. The optimized formulation was converted to carbopol gel to enhance skin contact. The prepared formulations showed nanometric size with high drug entrapment efficiency. The results showed an enhanced drug release pattern from BF-NLCopt than BF-NLCopt-gel. The permeation data showed higher permeation from BF-NLCopt and BF-NLCopt-gel across the tested membrane. Furthermore, the formulation showed optimum viscosity, pH, and spreadability. The irritation and antifungal result studies showed no irritation to the tested CAM as well as BF-NLCopt gel showed a marked enhancement in the ZOI at both time points. The overall study revealed that the butenafine nano lipid carrier-based gel system acts as a potential delivery system in the treatment of anti-fungal disease.

Author Contributions: Conceptualization, M.Y. and A.Z.; methodology and project administration M.Y.; visualization and software, A.Z.; validation, M.Y. and A.Z.; formal analysis, S.S.I.; investigation, M.Y.; resources, M.Y. and A.Z.; data curation, A.Z.; writing—S.S.I. and S.I.B.; writing—review and editing, S.A.; W.A.M.; S.I.B.; funding acquisition, W.A.M. All authors have read and agreed to the published version of the manuscript.

Funding: Deanship of Scientific Research at King Saud University for support of the present research work through research group project number (RG-1441-443).

Institutional Review Board Statement: Not applicable.

Informed Consent Statement: Not applicable.

Data Availability Statement: Not applicable.

Acknowledgments: The authors extend their appreciation to the Deanship of Scientific Research at King Saud University for funding this work through research group No (RG-1441-443).

Conflicts of Interest: The authors declare no conflict of interest.

References

1. Rao, S.; Barot, T.; Rajesh, K.S.; Jha, L.L. Formulation, optimization and evaluation of microemulsion based gel of Butenafine Hydrochloride for topical delivery by using simplex lattice mixture design. *J. Pharm. Investig.* **2016**, *46*, 1–12. [CrossRef]
2. Bezerra-Souza, A.; de Jesus, J.A.; Laurenti, M.D.; Lalatsa, A.; Serrano, D.R.; Passero, L.F.D. Nanoemulsified Butenafine for Enhanced Performance against Experimental Cutaneous Leishmaniasis. *J. Immunol. Res.* **2021**, *2021*, 1–13. [CrossRef] [PubMed]
3. Porras, A.M.G.; Terra, B.S.; Braga, T.C.; Magalhães, T.F.F.; Martins, C.V.B.; Da Silva, D.L.; Baltazar, L.M.; Gouveia, L.F.; De Freitas, G.J.C.; Santos, D.A.; et al. Butenafine and analogues: An expeditious synthesis and cytotoxicity and antifungal activities. *J. Adv. Res.* **2018**, *14*, 81–91. [CrossRef] [PubMed]
4. Syed, T.A.; Hadi, S.M.; Qureshi, Z.A.; Ali, S.M.; Ahmad, S.A. Butenafine 1% versus terbinafine 1% in cream for the treatment of Tinea Pedis. *Clin. Drug Investig.* **2000**, *196*, 393–397. [CrossRef]
5. Zhang, Z.; Tsai, P.-C.; Ramezanli, T.; Michniak-Kohn, B.B. Polymeric nanoparticles-based topical delivery systems for the treatment of dermatological diseases. *Wiley Interdiscip. Rev. Nanomed. Nanobiotechnol.* **2013**, *5*, 205–218. [CrossRef]
6. Dar, M.J.; Din, F.U.; Khan, G.M. Sodium stibogluconate loaded nano-deformable liposomes for topical treatment of leishmaniasis: Macrophage as a target cell. *Drug Deliv.* **2018**, *25*, 1595–1606. [CrossRef] [PubMed]
7. Riaz, A.; Hendrickx, S.; Elbrink, K.; Caljon, G.; Maes, L.; Ahmed, N.; Kiekens, F.; Khan, G.M. Preparation and Characterization of Nanostructured Lipid Carriers for Improved Topical Drug Delivery: Evaluation in Cutaneous Leishmaniasis and Vaginal Candidiasis Animal Models. *AAPS PharmSciTech* **2020**, *21*, 1–14. [CrossRef]
8. Souto, E.B.; Baldim, I.; Oliveira, W.P.; Rao, R.; Yadav, N.; Gama, F.M.; Mahant, S. SLN and NLC for topical, dermal, and transdermal drug delivery. *Expert Opin. Drug Deliv.* **2020**, *17*, 357–377. [CrossRef]
9. Pardeike, J.; Hommoss, A.; Müller, R.H. Lipid nanoparticles (SLN, NLC) in cosmetic and pharmaceutical dermal products. *Int. J. Pharm.* **2009**, *366*, 170–184. [CrossRef]

10. Wissing, S.A.; Müller, R.H. A novel sunscreen system based on tocopherol acetate incorporated into solid lipid nanoparticles. *Int. J. Cosmet. Sci.* **2001**, *23*, 233–243. [CrossRef]
11. Joshi, M.; Patravale, V. Formulation and Evaluation of Nanostructured Lipid Carrier (NLC)—Based Gel of valdecoxib. *Drug Dev. Ind. Pharm.* **2006**, *32*, 911–918. [CrossRef]
12. Patel, D.; Dasgupta, S.; Dey, S.; Ramani, Y.R.; Ray, S.; Mazumder, B. Nanostructured Lipid Carriers (NLC)—Based Gel for Topical Delivery of Aceclofenac: Preparation, Characterization and In Vivo Evaluation. *Sci. Pharm.* **2012**, *80*, 749–764. [CrossRef] [PubMed]
13. Madan, J.R.; Khobaragade, S.; Dua, K.; Awasthi, R. Formulation, optimization, and in vitro evaluation of nanostructured lipid carriers for topical delivery of Apremilast. *Dermatol. Ther.* **2020**, *33*, e13370. [CrossRef]
14. Bharadwaj, S.; Gupta, G.; Sharma, V. Topical gel: A novel approach for drug delivery. *J. Chem. Biol. Phys. Sci.* **2012**, *2*, 856–867.
15. Singla, V.; Saini, S.; Joshi, B.; Rana, A. Emulgel: A new platform for topical drug delivery. *Int. J. Pharma Biol. Sci.* **2012**, *3*, 485–498.
16. Mahmood, A.; Rapalli, V.K.; Gorantla, S.; Waghule, T.; Singhvi, G. Dermatokinetic assessment of luliconazole-loaded nanostructured lipid carriers (NLCs) for topical delivery: QbD-driven design, optimization, and in vitro and ex vivo evaluations. *Drug Deliv. Transl. Res.* **2021**, 1–18. [CrossRef]
17. Baghel, S.; Nair, V.S.; Pirani, A.; Sravani, A.B.; Bhemisetty, B.; Ananthamurthy, K.; Aranjani, J.M.; Lewis, S.A. Luliconazole-loaded nanostructured lipid carriers for topical treatment of superficial Tinea infections. *Dermatol. Ther.* **2020**, *33*, e13959. [CrossRef] [PubMed]
18. Ranpise, H.A.; Gujar, K.N.; Mathure, D.; Satpute, P.P.; Awasthi, R.; Dua, K.; Madan, J. Skin Targeting of Oxiconazole Nitrate Loaded Nanostructured Lipid- Carrier Gel for Fungal Infections. *Pharm. Nanotechnol.* **2018**, *6*, 192–200. [CrossRef]
19. Qumber, M.; Alruwaili, N.K.; Bukhari, S.N.A.; Alharbi, K.S.; Imam, S.S.; Afzal, M.; Alsuwayt, B.; Mujtaba, A.; Ali, A. BBD-Based Development of Itraconazole Loaded Nanostructured Lipid Carrier for Topical Delivery: In Vitro Evaluation and Antimicrobial Assessment. *J. Pharm. Innov.* **2021**, *16*, 85–98. [CrossRef]
20. Gaba, B.; Fazil, M.; Ali, A.; Baboota, S.; Sahni, J.K.; Ali, J. Nanostructured lipid (NLCs) carriers as a bioavailability enhancement tool for oral administration. *Drug Deliv.* **2014**, *22*, 1–10. [CrossRef]
21. Yasir, M.; Sara, U. Preparation and optimization of haloperidol loaded solid lipid nanoparticles by Box–Behnken design. *J. Pharm. Res.* **2013**, *7*, 551–558. [CrossRef]
22. Ali, J.; Khan, N.; Ali, A. Development and Optimization of Carteolol Loaded Carboxymethyl Tamarind Kernel Polysaccharide Nanoparticles for Ophthalmic Delivery: Box-Behnken Design, In Vitro, Ex Vivo Assessment. *Sci. Adv. Mater.* **2014**, *6*, 63–75. [CrossRef]
23. Muller, R.; Mader, K.; Gohla, S. Solid lipid nanoparticles (SLN) for controlled drug delivery—A review of the state of the art. *Eur. J. Pharm. Biopharm.* **2000**, *50*, 161–177. [CrossRef]
24. Barot, B.S.; Parejiya, P.B.; Patel, H.K.; Gohel, H.C.; Shelat, P.K. Microemulsion-based gel of terbinafine for the treatment of onychomycosis: Optimization of formulation using D-optimal design. *AAPS PharmSciTech* **2012**, *13*, 184–192. [CrossRef]
25. Nautiyal, U.; Mohammed, J.; Kazmi, I. Preparation and Evaluation of Antifungal Micro-Emulsion/Gel using reduce Dose of Silver, Supported by Ciprofloxacin. *Int. Pharm. Sci.* **2012**, *2*, 72–87.
26. Gaba, B.; Fazil, M.; Khan, S.; Ali, A.; Baboota, S.; Ali, J. Nanostructured lipid carrier system for topical delivery of terbinafine hydrochloride. *Bull. Fac. Pharm. Cairo Univ.* **2015**, *53*, 147–159. [CrossRef]
27. Saleem, M.N.; Idris, M. Formulation Design and Development of a Unani Transdermal Patch for Antiemetic Therapy and Its Pharmaceutical Evaluation. *Scientifica* **2016**, *2016*, 7602347. [CrossRef] [PubMed]
28. Shah, V.; Raval, S.; Peer, S.; Upadhyay, U.M. A comparative evaluation of different membranes for their diffusion efficiency: An in-vitro study. *Pharma Sci. Monit.* **2010**, *1*, 41–49.
29. Ansari, M.J.; Ahmed, M.M.; Anwer, K.; Aldawsari, M.F.; Al Shahrani, S.M.; Ahmad, N. Development and Validation of Simple, Rapid and Sensitive High- Performance Liquid Chromatographic Method for the Determination of Butenafine Hydrochloride. *J. Pharm. Res. Int.* **2020**, *32*, 116–125. [CrossRef]
30. Asmerom, D.; Kalay, T.H.; Tafere, G.G. Antibacterial and Antifungal Activities of the Leaf Exudate of Aloe megalacantha Baker. *Int. J. Microbiol.* **2020**, *2020*, 1–6. [CrossRef]
31. Mehling, A.; Kleber, M.; Hensen, H. Comparative studies on the ocular and dermal irritation potential of surfactants. *Food Chem. Toxicol.* **2007**, *45*, 747–758. [CrossRef]
32. Irimia, T.; Dinu-Pîrvu, C.-E.; Ghica, M.V.; Lupuleasa, D.; Muntean, D.-L.; Udeanu, D.I.; Popa, L. Chitosan-Based In Situ Gels for Ocular Delivery of Therapeutics: A State-of-the-Art Review. *Mar. Drugs* **2018**, *16*, 373. [CrossRef]
33. Yasir, M.; Sara, U.V.S.; Som, I.; Gaur, P.; Singh, M. Bentham Science Publisher Ameeduzzafar Nose to Brain Drug Delivery: A Novel Approach Through Solid Lipid Nanoparticles. *Curr. Nanomed.* **2016**, *6*, 105–132. [CrossRef]
34. Gupta, S.; Kesarla, R.; Omri, A. Formulation Strategies to Improve the Bioavailability of Poorly Absorbed Drugs with Special Emphasis on Self-Emulsifying Systems. *ISRN Pharm.* **2013**, *2013*, 1–16. [CrossRef] [PubMed]
35. Jenning, V.; Schafer-Korting, M.; Gohla, S. Vitamin aloaded solid lipid nanoparticles for topical use: Drug release properties. *J. Control. Release* **2000**, *66*, 115–126. [CrossRef]
36. Abdel-Mottaleb, M.M.A.; Mortada, N.D.; El-Shamy, A.A.; A Awad, G. Physically Cross-Linked Polyvinyl Alcohol for the Topical Delivery of Fluconazole. *Drug Dev. Ind. Pharm.* **2009**, *35*, 311–320. [CrossRef] [PubMed]

37. Adib, Z.M.; Ghanbarzadeh, S.; Kouhsoltani, M.; Khosroshahi, A.Y.; Hamishehkar, H. The Effect of Particle Size on the Deposition of Solid Lipid Nanoparticles in Different Skin Layers: A Histological Study. *Adv. Pharm. Bull.* **2016**, *6*, 31–36. [CrossRef]
38. Patel, D.K.; Kesharwani, R.; Kumar, V. Etodolac loaded solid lipid nanoparticle based topical gel for enhanced skin delivery. *Biocatal. Agric. Biotechnol.* **2020**, *29*, 101810. [CrossRef]
39. Koca, H.D.; Doganay, S.; Turgut, A.; Tavman, I.H.; Saidur, R.; Mahbubul, I.M. Effect of particle size on the viscosity of nanofluids: A review. *Renew. Sustain. Energy Rev.* **2018**, *82*, 1664–1674. [CrossRef]
40. Motawea, T.B.; El-Gawad, A. Topical phenytoin nanostructured lipid carriers: Design and development. *Drug Dev. Ind. Pharm.* **2018**, *44*, 144–151. [CrossRef]
41. Uprit, S.; Sahu, R.K.; Roy, A.; Pare, A. Preparation and characterization of minoxidil loaded nanostructured lipid carrier gel for effective treatment of alopecia. *Saudi Pharm. J.* **2013**, *21*, 379–385. [CrossRef]
42. Permanadewi, I.; Kumoro, A.C.; Wardhani, D.H.; Aryanti, N. Modelling of controlled drug release in gastrointestinal tract simulation. *IOP Conf. Ser. J. Phys. Conf. Ser.* **2019**, *1295*, 012063. [CrossRef]
43. Sandeep, D.S.; Mahitha, M.; Meghna, S. Development, Characterization, and In vitro Evaluation of Aceclofenac Emulgel. *Asian J. Pharm.* **2020**, *14*, 330.

pharmaceutics

MDPI

Review

Can Plant Materials Be Valuable in the Treatment of Periodontal Diseases? Practical Review

Anna Gościniak [1], Magdalena Paczkowska-Walendowska [1], Agnieszka Skotnicka [2], Marek A. Ruchała [3] and Judyta Cielecka-Piontek [1,*]

1 Department of Pharmacognosy, Faculty of Pharmacy, Poznań University of Medical Sciences, Rokietnicka 3, 60-806 Poznań, Poland; anna.gosciniak@student.ump.edu.pl (A.G.); mpaczkowska@ump.edu.pl (M.P.-W.)
2 Department of Pharmaceutical Technology, Faculty of Pharmacy, Poznan University of Medical Sciences, Grunwaldzka 6, 60-780 Poznań, Poland; askotnic@ump.edu.pl
3 Department of Conservative Dentistry and Endodontics, Poznan University of Medical Sciences, Bukowska 70, 60-812 Poznań, Poland; maruchala@ump.edu.pl
* Correspondence: jpiontek@ump.edu.pl

Abstract: Periodontal diseases are one of the most significant challenges in dental health. It is estimated that only a few percent of the worldwide population have entirely healthy teeth, and according to WHO, oral diseases may affect up to 3.5 billion people worldwide. One of the most serious oral diseases is periodontitis, an inflammatory disease affecting periodontal tissues, caused by pathogenic bacteria and environmental factors such as the ageing population, abuse of tobacco products, and lack of adequate oral hygiene due low public awareness. Plant materials are widely and successfully used in the management of many conditions, including periodontitis. Plant materials for periodontitis exhibit antibacterial, anti-inflammatory, antioxidant activities and affect the periodontium structure. Numerous studies demonstrate the advantages of phytotherapy for periodontitis relief and indicate the usefulness of Baikal skullcap root, Pomegranate fruit peel and root cortex, Tea leaves, Chamomile flowers, Magnolia bark, Blackberry leaves and fruits, Cranberry fruits and Lippia sidoides essential oil. This review aims to analyze the use and applicability of selected plant materials in periodontitis management since it is of paramount importance to evaluate the evidence of the traditionally used plant materials in light of continuously growing interest in phytotherapy and its adjuvant role in the treatment of periodontitis.

Keywords: periodontitis; herbal drugs; plant extracts; oromucosal route

1. Introduction

Poor oral health and its harmful consequences confer a substantial burden worldwide on people, healthcare systems and societies. The demographic shift and population ageing have significant implications for public health, mainly by increasing the prevalence of chronic diseases and disabilities [1,2]. As the burden of chronic diseases increases, healthcare systems' priorities focus on the effective management of chronic conditions and the continuous optimization of the elderly population's functioning [1]. Despite the global focus on chronic health problems of an ageing population, still, there is an inadequate emphasis on oral health [3]. According to The Global Burden of Disease 2010 Study, oral health problems accounted for 15 million disability-adjusted life years, implying an average health loss of 224 years per 100,000 population [4]. According to World Health Organization (WHO), oral diseases constitute a significant problem for numerous countries and their healthcare systems. Oral health conditions cause pain, discomfort, lead to a substantial decrease in the quality of life, and also determine growing care needs in later life [5]. It is estimated that oral diseases may affect even up to 3.5 billion people worldwide [2]. The most common oral conditions are tooth decay, periodontitis, loss of dentition, cancers of the lips and mouth [6], and dry mouth. Untreated infections can lead to oral cancers, one of the three most commonly diagnosed malignancies in the Asia Pacific, and in Europe, with the incidence ranging from 5 to 10 cases per 100,000 people [7]. WHO statistics on oral diseases present the continuous increase in the prevalence of oral health deterioration in both children and adults. Over 530 million children globally suffer from decay in primary teeth, approximately 60–90% of schoolchildren have experienced caries, with the disease being most prevalent in Asian and Latin American countries [8]. Despite the natural ageing process, common factors leading to tooth loss in adults are caries and severe periodontal diseases [5]. About 50% of Europeans suffer from at least one type of periodontitis, and in the 60–65 age group, the prevalence increases up to 70–85% [9]. About 30% of Europeans aged between 65 and 74 years no longer have permanent teeth, what significantly affects their functioning at physiological and neuromuscular levels as well as in psychological and social aspects of wellbeing [10,11] and consequently, their quality of life [12].

2. Periodontitis

Periodontitis is an infectious tissue disease with an inflammatory response. Periodontitis is characterized by microbially-associated, host-mediated inflammation resulting in periodontal attachment loss [13]. Periodontal tissues include the structures surrounding the tooth such as the gum, the mucosa of the alveolar process, alveolar bone, periodontium and root cementum. The periodontium supports the tooth, protects against oral microflora, and makes the attachment to the bone possible. The inflammatory process leads to the loss of connective tissue and alveolar bone, resulting in bone support loss and pathological tooth mobility [14].

There are three types of periodontitis [13]:

1. Necrotizing periodontitis
    a. Necrotizing gingivitis
    b. Necrotizing periodontitis
    c. Necrotizing stomatitis
2. Periodontitis as a manifestation of Systemic Disease
3. Periodontitis.

The more recent classification differentiates periodontitis also into aggressive and chronic forms [13].

Initially the disease is asymptomatic. The early symptoms may exhibit as gingivitis, later progressing to periodontitis. Patients may notice gum bleeding, soreness, redness and swelling of the gums, during tooth brushing and when chewing the food [15]. The disease may become chronic with the increase in the amount of plaque. Initially, the cavities are mild [14]. The symptoms presenting later are abscess in the gingival tissue, unpleasant taste and smell in the mouth. Along with the disease progression, the alveolar and periodontium bones are resorbed and, as a consequence, gingival recession leads to loosening and ultimately the loss of the tooth [16,17].

Etiopathogenesis of Periodontitis

The exact mechanism of periodontal pathogenesis is not fully understood. Still, the onset and progression of the disease are invariably associated with the imbalance in the periodontal and microbial homeostasis [18]. In periodontitis, the microorganisms colonizing the oral cavity dominate bacteria called a red-complex. Red-complex is formed by gram-negative pathological anaerobic bacteria such as *Porphyromonas gingivalis, Tannerella forsythia, Treponema denticola* [19,20]. It is believed that the presence of the red-complex bacteria is responsible for weakening the innate periodontal defence functions, stimulating untoward interaction between the host bacteria and dental plaque leading to tissue damage [21,22].

The immune inflammation of tissues, gingivitis, and periodontitis, is a response to the chronic presence of plaque bacteria producing damaging factors such as chemotoxins, endotoxins and enzymes. The immune reaction aims to eliminate the damaging factors following the release of pro-inflammatory cytokines and other mediators of inflammation. However, prolonged exposure to pathogens suppresses this compensative process converting it to harmful, leading to a significant tissue damage. Therefore, the progression of periodontitis is the amalgamation of external factors such as pathogens and the host immune response, a direct cause of periodontal tissue damage. The long-term exposure to the factor causing the humoral response leads to the activation of osteoclasts, which begin the bone damage and activate the matrix metalloproteinase damaging the ligaments [17]. A graphical representation of the mechanism is shown in Figure 1.

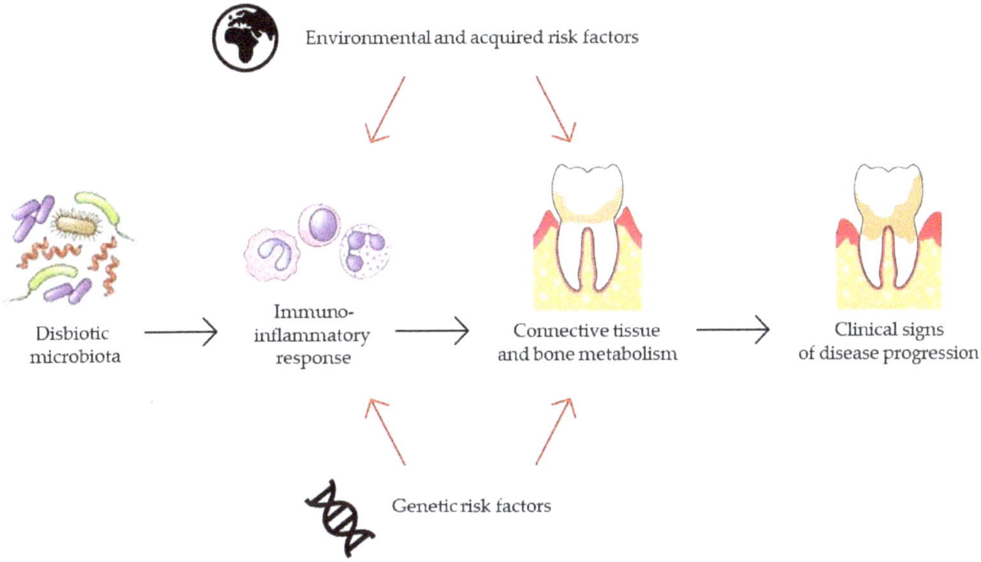

Figure 1. Model of pathogenesis of periodontitis.

Additionally, it has been noted that the oxidative stress caused by the antimicrobial reaction during periodontitis may be an additional cause of tissue damage. When inflammation occurs, the production of reactive oxygen species by the immune system (i.e., neutrophils and macrophages) increases dramatically [23]. They can directly cause tissue damage by lipid peroxidation, DNA damage, protein damage, and oxidation of essential enzymes, and in the meantime, act as signalling molecules or mediators of the inflammation process. In addition, reactive oxygen species can directly attack fibroblast and osteoblast cells and reduce collagen production in cells leading to periodontitis [24].

Understanding the pathogenesis of periodontitis is fundamental element for discovering effective targeted treatment. Currently, the preferred management strategies include using antioxidant agents, modifying the process of elimination of reactive oxygen species and supporting the bioprocesses protecting cells [25].

3. The Role of Herbs in the Treatment of Periodontitis

Plant materials are commonly and widely used to treat several medical conditions. Herbal products are often preferred over conventional drugs due to their complex biological activity, often favourable safety profile, lower therapy costs, biocompatibility, and low impact on the environment. Moreover, conventional, synthetic compounds usually cause more side effects, intolerance, and irresponsible use, for example, not evidence-based antibiotic therapy, leads to long-term global threat such as antibiotic resistance [26]. For example, chlorhexidine-based mouthwashes for periodontitis are associated with an unpleasant aftertaste, allergic reactions, tongue numbness discouraging use [27]. Concurrently, the interest in using herbal medicines is growing continuously to combat or prevent diseases, including periodontal diseases [28,29].

Due to the multifactorial aetiology and complex disease pathogenesis, the management of periodontitis is multidirectional [30]. The primary therapy goal is the reduction in bleeding and inhibition of the disease progression [31]. The ideal formulation should exhibit antibacterial, antioxidant, and anti-inflammatory properties to reduce dental plaque and limit the periodontal tissue damage caused by the progressive inflammatory process. The therapeutic outlook is the activity against the bacteria mentioned above, i.e., red-complex and the inhibition of plaque formation through the anti-adhesive effect. Reducing inflammation can be manifested by inhibiting both pro-inflammatory cytokines and proteinases such as collagenase [32]. The inhibition of the inflammatory processes minimizes the risk of periodontal tissues damage. The concept of advantageous effects in managing periodontitis includes the antioxidant effect of plant materials protecting tissues against the damaging effects of free radicals [33]. Bone resorption in the course of periodontitis is associated with the hyperactivity of RANKL (Receptor Activator for Nuclear Factor κB Ligand), a protein responsible for the maturation and activation of osteoclast cells [34]. The inhibition of bone resorption by the plant materials mentioned in the literature illustrates the purposefulness of their use.

Selected plant materials evaluated with proven effectiveness in the treatment and prevention of periodontitis are presented below.

3.1. *Sage Leaves (Salvia officinalis)*

Salvia officinalis is a shrub belonging to the *Lamiaceae* family. Active compounds of sage leaves are flavonoids and their glycosides, especially rosemary acid, ellagic acid, and luteolin-7-glucoside. The plant material contains the terpenes and terpenoids bornyl acetate, camphene, camphor, humulene limonene and ursolic acid [35]. Sage leaf is widely used in oral disorders for rinses or as one of the ingredients in complex preparations. One of the traditional uses stated in a monograph of the European Medicines Agency includes relief of mouth and throat inflammations (Table 1) [36]. Sage leaf has an anti-inflammatory effect helpful for the topical treatment of inflammatory diseases. Many studies have been carried out on compounds obtained from sage. Mansourabadi et al. [37] evidenced activity of flavanoid–salvigenin, Baricevic et al. [38] of ursolic acid—triterpenoid and Rodrigues et al. [39] of hydroalcoholic extract. Essential oils isolated from sage are a promising source of antibacterial agents. They may prevent oral infectious diseases, as confirmed in in vitro study by Tambur et al. [40]. However, the antibacterial activity of methanol extract against *Enterococcus faecalis* infection in root canals is lower than chlorhexidine activity. A randomized clinical trial aiming to evaluate the clinical effects of a mouthwash containing *Salvia officinalis* extracts on *S. mutans* (bacteria causing dental plaque) demonstrated the effectivness of sage mouthwash in a reduction in the count of *S. mutans* in dental plaque [41]. In contrast, the study by Mendes et al. [42] concluded that the extract has a moderate effect on periopathogenic bacteria and proved the activity against *S. mutans* in the combination of dichloromethane-soluble fraction with chlorhexidine only.

Extensive research on the progression of periodontitis has also been conducted. Anti-inflammatory effect of ethanolic extract of *Salvia sclarea* L., a plant belonging to the same genus as *Salvia officinalis* was investigated. Use of *S. sclarea* extracts significantly inhibits the inflammatory process by lowering IL-1β, IL-6 and TNF-α levels, reducing gingival tissue changes, and preserving alveolar bone resorption in lipopolysaccharide-induced periodontitis in rats [43]. A randomized, placebo-controlled, double-blind did not confirm the superior beneficial effect of the mouthwash with *S. officinalis* on inflammatory parameters and plaque indices compared with the placebo after 6 weeks [44]. On the other hand, *S. officinalis* gel has a potential anti-inflammatory role in the management of chronic periodontitis. The study by Aljuboori et al. [45] investigated both clinical and immunological parameters depending on the application of a gel containing sage extract preceded by scaling and root planning procedure and concluded a significant difference between the control and test group.

Combining sage with other raw materials with anti-periodontitis activity is also clinically relevant. Toothpaste with Swiss Medicinal Herbs (*Chamomilla recutita*, *Arnica montana*, *Echinacea purpurea*, and *Salvia officinalis*) diminished bacterial load specific for gingivitis/periodontitis in comparison to a control toothpaste [46]. preparations shown In Table 2 presented a variety of preparations containing sage leaves with proven efficacy in the management of periodontitis.

3.2. *Oak Bark (Quercus spp.)*

Oak bark is a dried bark obtained from several oak species found in Europe—*Quercus robur*, *Quercus petraea*, *Quercus pubescens* belonging to a family of *Fagaceae*. Plant materials are rich in tannins and effectively treat periodontitis. The main active compounds found in oak bark are hydrolyzable and condensed (8–20%) tannins [47]. The European Pharmacopoeia provides the minimal tannins level reference value as 3% in the raw plant material of *Qercus* spp. expressed as pyrogallol and calculated per dried herbal material [48]. The phenolic acids—ellagic acid, gallic acid, protocatechuic acid, and vanillic acid are also present. The predominant compound is ellagic acid [47]. The bark of *Q. robur*, *Q. petraea*, *Q. pubescens* is on the official database the European Medicine Agency of pharmaco-therapeutic plants are presented in Table 1 [49] The traditional use in periodontitis described in the monograph targets symptomatic treatment of a low-level inflammation of the oral mucosa or skin [49]. Oak bark extract as an ingredient in dermatological medicines can be used to treat skin conditions [50]. Oak bark tannins have strong astringent, bactericidal, antioxidant, and anti-inflammatory properties [51,52]. Antibacterial properties of tannins from oak bark result from inhibition and deactivation of proteins through forming stable, insoluble complexes. The astringent effect prevents minor bleeding in the mouth associated with periodontitis progression. Studies conducted by Tsubanova et al. confirm the effectiveness of gel containing *Aloe vera* and oak bark extracts in periodontitis treatment [53–55]. The proven properties of oak bark extract permits the presumption that it could be an excellent adjunctive agent in periodontitis treatment (Table 2).

3.3. *Peppermint Leaves (Mentha piperita)*

Peppermint leaf is a plant raw material valued for its proven activity. *Mentha piperita* (*Lamiaceae*) is a cultivated natural hybrid of *Mentha aquatica* and *Mentha spicata*. In addition to the essential oil peppermint leaves typically contain 1.2–3.9% (*v/w*) of essential oil, various flavonoids such as luteolin and its 7-glycoside, rutin, hesperidin, and eriocitrin (eriodictyol 7-*O*-rutinoside). The main constituents of the monoterpenes fraction are menthol and menthone [56]. Other components include phenolic acids, e.g., rosemary acid [57]. The European Medicines Agency has published monographs on peppermint leaves and the essential oil extracted from peppermint (Table 1) [58]. It is well documented that the essential oil and extracts of *Mentha* species have antimicrobial, anti-inflammatory, and antioxidant properties [59–62] with antimicrobial activity against pathogens involved in the development of periodontal disease [63]. Peppermint oil or peppermint leaf extract are often present in oral preparations for their antiseptic properties and flavour correction value. Essential oils of *Mentha piperita* showed good antibacterial and antibiofilm activities in the study conducted by Karicheri et al. [64] concerning *Aggregatibacter actinomycetemcomitans*, a strain identified in periodontopathogenesis. Peppermint oil is also used in novel formulations, for example as an anti-ulcer and anti-inflammatory agent in self-emulsifying drug delivery system for the improvement of meloxicam's solubility used in periodontal disease [65]. Furthermore, the study by Phaechamud et al. [66] successfully used peppermint oil to modify the release of doxycycline from the in situ formed gel systems with Eudragit RS. Above studies demonstrate a modern trend on the development of innovative drug forms by using plant-based raw materials.

3.4. Calamus Rhizome (*Acorus calamus*)

Acorus calamus, also known as a sweet flag, cradle root or muskrat, is a plant belonging to *Acoraceae*, common in central Asia and eastern Europe [67]. A total of fifty-three volatile organic compounds were isolated and identified. The main bioactive compound is acorenone followed by isocalamendiol whose content varies according to variety and region [68]. Monoterpene hydrocarbons, sequestrine ketones, α- and β-asarone, and eugenol were also identified [69]. Due to the toxicity of α- and β-asarone, the European Medicines Agency recommends that their concentration in herbal medicinal products should be kept to a minimum and that diploid variety lacking these compounds should always be preferred [70]. The traditional use by many Native American tribes utilized the calamus products as an anaesthetic for toothaches and to alleviate headaches [71]. Both calamus rhizome extract and essential oil can be used to manage periodontitis for their antibacterial, antioxidant, and anti-inflammatory properties. Antibacterial activity against peripathogenic bacteria such as *Actinomyces odontolyticus*, *Eikenella corrodens*, *Fusobacterium nucleatum*, although present, is weaker than with ethanolic extracts of sage or chamomile. The essential oil exhibits anti-inflammatory and antioxidant effects [72,73]. The methanol extract also shows antioxidant and antifungal activity, as demonstrated by Dinev et al. [74]. The active compound presented in calami rhizoma is asaroaldehyde [75], which according to Hwang et al. promotes osteogenic activity via the p38/extracellular-signal-regulated kinase signalling pathway. Thus, it may be used in the future as a compound of natural origin adjuvant in periodontal regeneration [76]. Additionally, anti-inflammatory properties of ursolic acid present in the calamus were confirmed by Baricevic et al. [38].

3.5. Baikal Skullcap Root (*Scutellaria baicalensis*)

Scutellaria baicalensis, also known as a Baikal skullcap or Chinese skullcap, and is a perennial herbaceous plant from the *Lamiaceae* (*Labiatae*) family. *Scutellaria baicalensis* is traditionally used in Chinese medicine as an anti-inflammatory agent [77]. Baicalin, a flavonoid with valued properties, is obtained from the root and is used to treat periodontitis. It can be used for treating periodontitis in oral gels and rinses for its antibacterial and antifungal properties. *S. baicalensis* has a strong antibacterial activity against oral pathogens, including *Streptococcus salivarius* or *Bacteroides gingivalis* [78]. A synergistic antibacterial effect has also been reported when used along with chlorhexidine, an agent known as the current gold standard for periodontitis [79]. Baicalin exhibits anti-inflammatory activity by inhibiting inflammatory process mediators such as arachidonic acid metabolites of nitric oxide [80,81]. In the animal periodontitis model, baicalin in 200 mg/kg dose regulates COX-2 expression in macrophages, plasma cells, fibroblasts, and gingival tissues [82].

Baicalin inhibits toll-like receptors expression and downstream signalling and mitigates inflammatory responses and the alveolar bone loss in experimental rat periodontitis [83]. *Scutellaria baicalensis* extract significantly inhibits alveolar bone resorption, reduces the production of pro-inflammatory cytokines expression in gingival tissues, and also improves the recovery of periodontal structures [84]. Baicalein attenuates the inflammatory response and restores osteogenesis in vitro in lipopolysaccharide-treated periodontal ligament cells, suggesting its potential use as a modulator of the host response in the treatment of periodontitis. [85]. Baicalin may inhibit metalloproteinase, e.g., collagenase activity, by more than 30%, which significantly moderates the destruction process of collagen in the gum tissue [86]. Baicalin has proven to protect the periodontal tissue in induced inflammation in the animal model [82,87]. Baicalin can significantly increase bone mineral density and height of the alveolar bone, and histological observations have shown that it may promote bone repair and regeneration [88]. Herbal toothpaste containing 0.5% Baikal skullcap extract and fluoride reduces plaque and biofilm viability more effectively than a conventional toothpaste containing fluoride only [89].

3.6. Pomegranate (Punica granatum)

Punica granatum L. known as pomegranate belongs to the family *Punicaceae*. As raw materials, fruit peel and root cortex are used. The main active compounds are flavonoids, ellagitannin, punicalagin, ellagic acid, vitamins and minerals, and alkaloids such as pelletierine [90,91]. Pomegranate extract exhibited anti-inflammatory activity through inhibition of NF-κB, decreasing NO level and PGE2 synthesis [92]. Pomegranate can potentially be used in the treatment of periodontitis due to its antioxidant activity. Pomegranate extract scavenges free radicals and reduces macrophage oxidative stress and lipid peroxidation [93]. Active compounds found in the raw material exhibit antibacterial action against gingivitis-causing microbes. Therefore, pomegranate mouthwash has an antiplaque effect and is effective against *Aggregatibacter actinomycetemcomitans*, *Porphyromonas gingivalis*, and *Prevotella intermedia* [94]. In addition, pomegranate glycolic extract has antimicrobial activity against *P. gingivalis* in vivo using *Galleria mellonella* [95]. Pomegranate juice is effective against dental plaque forming microorganisms decreasing the Colony Forming Units by 32% [96]. Pomegranate extract in gel form used in addition to mechanical cleaning exhibits the antigingivitis activity measured by Papillary Bleeding Index [97]. *Punica granatum* mouthwash showed similar improvement in bleeding and gingivitis score compared to chlorhexidine, with no impact on the reduction of plaque scores [98]. 5% *Punica granatum* peel extract decreases chronic gingivitis' clinical signs and IL-1β levels without any adverse effect. *Punica granatum* chip may increase the effectiveness of conventional therapies and used in adjunction to scaling and root planning proves to be more effective than traditional management alone [99].

3.7. Tea Leaves (Camellia sinensis)

Camelia sinensis, belonging to the *Theaceae* family, is a plant with a proven healing potential mainly due to strong antioxidant activity. The main components of the leaves extract are catechins, including epigallocatechin gallate, epicatechin, epigallocatechin, and epicatechin gallate. Tea polyphenols inhibit the proliferation of *P. gingivalis*, and the minimum inhibitory concentration (MIC) for epigallocatechin gallate ranges from 250 μm/mL to 500 μm/mL pending on the strain tested. In addition, tea polyphenols at a concentration of 100 μg/mL reduced *P. gingivalis* adhesion on epithelial cells by 70% [100]. The catechins in green tea reduce the expression of peroxidation indicators and the pro-inflammatory cytokine index in rats after a topical application [101]. Green tea extract delays the disease progression of induced periodontitis in the animal model. The green tea application lowered as well the RANKL expression [102]. *Camelia sinensis* extract can reduce alveolar bone resorption by reducing cytokines and TRAP-positive multinucleated osteoclasts inhibiting periodontal inflammation [103]. Adjunctive topical extract therapy with a green tea gel reduces inflammation after 4 weeks of daily use [104]. In addition, a reduction of gingival inflammation and improvement in periodontal parameters was observed [105]. The clinical effect is also observed following oral supplementation of green tea. A significant improvement in clinical parameters and higher antioxidant capacity was observed [106]. Daily consumption of green tea has positively affected the probing depth and bleeding index [107]. The effectiveness against the plaque, gingivitis, oral hygiene, and salivary pH was higher following use of 0.5% tea mouthwashes compared to 0,2% chlorhexidine [108].

Traditionally used herbal medicinal products relieve fatigue and weakness (Table 1) [109].

3.8. Aloe vera Gel

Aloe is a plant belonging to the *Asphodelaceae* family. The following potentially active compounds are present in the leaves: vitamins soluble in water and lipids, complex or simple polysaccharides, especially acemannan—acetylated glucomannan, a mixture of polymers of various lengths, minerals, organic acids, and phenolic compounds [110,111]. Clinical study results proved antimicrobial activity against various periodontal pathogens—*Actinobacillus actinomycetemcomitans*, *Clostridium bacilli*, *Streptococcus mutans* and *Staphylo-*

coccus aureus. The efficacy was similar to ofloxacin (5 mcg) and ciprofloxacin (30 mcg) [112]. The effectiveness of aloe vera extract was analysed in various formulations: toothpaste, mouthwash, and gel, including methods of obtaining the extracts. Bhat et al. [113] compared scaling and root planning with and without following of intra-pocket application of *Aloe vera* gel. The results confirmed that the oral application of gel improved patient's oral condition. The gel seems to be more effective than scaling and root planning alone. Moghaddam et al. [114] run single-blind clinical trial confirmed that the adjunction to scaling improved clinical parameters such as plaque index, gingival index, and probing depth. The effectiveness of subgingivally applied *Aloe vera* gel was comparable to 1% ornidazole, 0.25% chlorhexidine, gluconate (Ornigreat™ gel) [115]. Likewise, the effect of aloe vera toothpaste on plaque index, the gingival index was similar to fluoride toothpaste alone [116], and its effectiveness was comparable to chlorhexidine [117]. Similarly, Penmetsa et al. has shown that *Ocimum sanctum* mouthwash, *Aloe vera* mouthwash, and chlorhexidine mouthwash are equally effective in reducing plaque, gingival, and bleeding 30 days apart. A study conducted by Karim et al. on 345 patients also showed that *Aloe vera* mouth rinse is equally effective in reducing periodontal indices as chlorhexidine. *Aloe vera* mouthwash was proven to be as effective on the 4-day de novo plaque formation as 0.2% chlorhexidine (the current gold standard) [118]. It improved bleeding index, plaque index, probing depth and clinical attachment level in patients with comorbidities such as type 2 diabetes mellitus and chronic periodontitis [119].

3.9. Chamomile Flowers (Matricaria chamomilla)

Chamomile flower, a representative of the *Asteraceae* family, is a plant material rich in essential oils, sesquiterpenes such as α-bisabolol and chamazulene, and flavonoid compounds. Traditionally, a herbal medicinal product is used to treat minor mouth and throat ulcerations and inflammations (Table 1) [120]. Chamomile flower extract is a complex of blends with topical anti-inflammatory and antioxidant properties [121]. The scope of common chamomile activity formed the basis for the use in periodontitis. Hans et al. [122] and Saderi et al. [123] confirmed the activity of essential oil against *P. gingivalis*. The essential oil inhibits the formation of a biofilm formed by *A. actinomycetemcomitans* and *T. denticola* bacteria. In Guimarães et al. study [124] dry extract of chamomile prevented inflammation and alveolar bone resorption in rats with ligature-induced periodontitis. The effectiveness of chamomile mouth rinses in reducing gingival bleeding was similar to 0.12% chlorhexidine solution [125]. Moreover, Agarwal et al. [126] showed that 1% chamomile extract improves periodontal patients' clinical and microbiological presentation.

3.10. Magnolia Bark (Magnolia officinalis)

Magnolia officinalis, a plant from the *Magnoliaceae* family is a source of magnolol and honokiol—polyphenolic, compounds with anti-neurodegenerative, antioxidant, and anti-inflammatory activity [127]. Magnolia bark extract also has an advantageous impact on periodontitis management. Honokiol and magnolol have antimicrobial activity against *A. actinomycetemcomitans*, *P. gingivalis*, *P. intermedia*, *M. luteus*, and *B. subtilis* [128]. Magnolol may also be successfully used in the treatment of caries. The minimal inhibitory concentrations of magnolol, honokiol and chlorhexidine for *S. mutans*, a primary pathogen in caries, were 10, 10, and 0.25 μg/mL, respectively [129]. Walker et al. [130] demonstrated that magnolia bark extract inhibited the release of pro-inflammatory cytokines and the activity of metalloproteinases. Kim et al.'s in vitro study [131] proved that 75% ethanol magnolia cortex extract in the concentration of 60 μg/mL in 1% DMSO exhibited an anti-inflammatory effect on *P. gingivalis*-stimulated cells. Magnolol and honokiol exhibit anti-inflammatory and antioxidant activity [132,133]. In the animal model study, magnolol moderates and alleviates bone resorption and improves the structure of the periodontium [134]. In addition, clinical studies by Hellström et al. [135] confirmed the effectiveness of magnolia extract in reducing gingivitis when used for six-month in a dentifrice, in the concentration of 0.3%.

3.11. *Blackberry Leaves and Fruits (Rubus fruticosus)*

Blackberries are one of the many representatives of the *Rosaceae* family. Due to the content of astringent tannins, blackberry leaves are used to manage oral cavity disorders. Blackberry leaf rinses have been used to treat pharyngitis, relieve pain and for mouth ulcers [136]. The raw material is effective in treating oral thrush. Chewing blackberry leaves strengthens the gums and may be used as a dressing to heal abscesses [137]. Leonti et al. [138] reported the use of blackberry leaves, imported from Italy for gingivitis and abscess treatment. The active compounds present in the raw material have anti-inflammatory properties. Epicatechin, ellagic acid and quercetin derivatives inhibit the secretion of pro-inflammatory cytokines [139–141]. In addition, hyperoside, high in content in the raw plant material, inhibits the activation of the nuclear factor-kappa B (NF-κB) signalling pathway, suggesting the potential application of blackberry leaves in the treatment of inflammatory diseases [142]. In addition to the leaves, blackberry fruits are also effective in the treatment of stomatitis. Active compounds present in blackberry fruit exhibit anti-inflammatory, antioxidant and antiviral activity [143–145]. Antibacterial activity against pathogenic periodontal pathogens and *S. mutans* has been noted by González et al. [146].

3.12. *Cranberry Fruit (Vaccinum macrocarpon)*

Vaccinum macrocarpon is a plant that belongs to the *Ericaceae* family. The main active compounds present in the cranberry fruit are proanthocyanidins—compounds with advantageous properties in periodontitis treatment. Proanthocyanidins found in cranberries mainly consist of epicatechin subunits with at least one A-type bond. The main properties are antiadhesive, antibacterial activities and inhibition of collagenase and proteinase activity. One of the common use of cranberry fruits is in the treatment of urinary tract infections (Table 1) [147] The likely mechanism of its advantageous effect is the inhibition of bacterial adhesion to the urinary tract wall. The anti-adhesive impact is also being investigated in dental diseases. The study of cranberry juice effect on *Streptococcus* bacteria biofilm formation confirmed significant inhibition of bacteria adhesion and, consequently, prevention of dental plaque formation [131]. The results of Weiss et al. study proved a reduction in *Streptococcus* count in saliva after 6 week-use of mouthwash containing extracts from cranberry fruit [132]. Cranberries have proven antibacterial activity against red-complex bacteria. The active components from cranberries inhibit biofilm formation [148]. In addition, products obtained from cranberry juice inhibit the proteinases of *P. gingivalis*, *T. forsythia* and *T. denticola* and type I collagen and transferrin degradation by *P. gingivalis* [149]. Moreover, Tipton et al. [150] suggested the role in the inhibition of NF-κB and MMP-3, through the regulation of fibroblast inflammatory response. A-type cranberry proanthocyanidins can interfere with osteoclastic cell maturation and physiology and thus prevent bone resorption [151]. Furthermore, the consumption of cranberry functional beverages improves gingival and plaque indexes with no risk of developing caries [152].

3.13. *Lippia Sidoides*

Lippia sidoides, also known as *Alecrim pimento* and Pepper-Rosmarin, is an aromatic and medicinal plant species of the family *Verbenaceae,* originally from South America. The aromatic species of Lippia are commonly used as brews and inhalations for allergic rhinitis and the treatment of vaginal, oral, and throat infections. Additionally, the plant has potential use in periodontitis treatment. The main constituents of the essential oil obtained from *Lippia* are thymol, carvacrol, and eugenol [153]. These compounds exhibit potent antimicrobial activity against the typical oral pathogens, i.e., *Streptococcus* and *Candida albicans* [154]. In a study conducted by Botelho et al. [155], a standard 0.12% chlorhexidine mouth rinse reduced plaque index, gingival bleeding, and the number of *S. mutans* colonies more efficiently than *Lippia sidoides* essential oil 1% mouth rinse but did not reach a statistical significance. Both types of mouthwashes were equally effective in reducing microbial plaque and gingival inflammation. A gel containing 10% *Lippia sidoides* essential oil significantly reduced plaque and gingivitis after three months of treatment, similarily to chlorhexidine mouthwash [156]. There was a significant difference in the gingival index after 21 days of treatment with 10% *Lippia sidoides* gel [157]. In an animal model study with stimulated periodontal disease, the effect of topical herbal gel from 0.5% *Lippia sidoides* and 5% *Myracrodruon urundeuva* was investigated. The results showed that topical herbal gel decreased myeloperoxidase activity and significantly inhibited tumour necrosis factor-alpha and interleukin-1beta concentration in gingival tissue. Alveolar bone loss and tissue lesion with histopathological changes were considerably reduced [158]. Botelho et al. investigated thymol nanogel (1.2 mg/g) containing *Lippia sidoides* in acute periodontitis in rats. The treatment with nanogel reduced both lesions and myeloperoxidase activity in gingival tissue [159].

Table 1. Traditionally-used pharmaceutical dosage forms and therapeutic indications of selected plant materials according to European Union herbal monographs.

Plant Material	Pharmaceutical Dosage Form	Therapeutic Indications	References
Salvia officinalis L., *folium*	- Comminuted herbal substance as herbal tea for oral use. - Comminuted herbal substance for infusion preparation for oromucosal or cutaneous use. - Herbal preparations in liquid and solid dosage forms for oral use. - Herbal preparations in liquid or semi-solid dosage forms for cutaneous use or oromucosal use.	- Traditional herbal medicinal product for relief of inflammations in the mouth or the throat. - Traditional herbal medicinal product for relief of minor skin inflammations.	[36]
Quercus robur L., *Quercus petraea* (Matt.) Liebl., *Quercus pubescens* Willd., *cortex*	- Comminuted herbal substance as herbal tea for oral use or as a decoction preparation for oromucosal or cutaneous use. - Herbal preparation in solid or liquid dosage forms for oral use.	- Traditional herbal medicinal product for symptomatic treatment of minor inflammation of the oral mucosa or skin.	[49]
Mentha x piperita L., *folium*	- Herbal substance and comminuted herbal substance a herbal tea for oral use. - Herbal preparations in solid or liquid dosage forms for oral use.	No traditional use in the oral cavity.	[58]
Camellia sinensis (L.) Kuntze, non fermentatum *folium*	- Herbal substance or comminuted herbal substance as a herbal tea for oral use. - Herbal preparations in solid dosage forms for oral use.	No traditional use in the oral cavity	[109]
Matricaria recutita L., *flos*	- Herbal substance or comminuted herbal substance as an herbal tea for oral use and inhalation. - Herbal preparations in liquid dosage forms for oral use. - Herbal substance or comminuted herbal substance for infusion preparation for oromucosal or cutaneous use. - Herbal preparations in liquid dosage forms for preparation of dilutions for oromucosal or cutaneous use.	- Traditional herbal medicinal product for the treatment of minor ulcers and inflammations of the mouth and throat.	[120]
Vaccinium macrocarpon Aiton, *fructus*	- Herbal preparations in liquid dosage forms for oral use.	No traditional use in the oral cavity.	[147]

4. Oromucosal Formulations in Periodontal Diseases

4.1. Oromucosal Route

Oral mucosa and mouth have been positioned as one of the most common drug delivery routes, designed for both systemic and local drug action targets. The oromucosal route may grant effective absorption at the site of administration (local administration with systemic drug action), or the action may be limited to the administration site. Within the oral mucosal cavity, delivery of drugs is classified into three categories: (1) sublingual delivery (systemic delivery of drugs through the mucosal membranes lining the floor of the mouth), (2) buccal delivery (drug administration through the mucosal membranes lining the cheeks—buccal mucosa) and (3) local delivery (drug delivery into the oral cavity). Oral mucosa is highly vascularised, permeable, so active substances are well and fast absorbed to directly enter the systemic circulation, bypassing the gastrointestinal tract and first-pass metabolism in the liver, which can be manifested by fast onset of action [160]. The systemic effect following administration depends on the physicochemical properties of active substances. The hydrophilic and lipophilic properties of a drug mainly influence the success of the transport and absorption [161]. The absorption of hydrophilic substances depends on the size of the molecule, and with increasing molecular size, its oromucosal absorption decreases. Furthermore, pH value plays an important role; depending on the acid dissociation constant (pKa) and the saliva's pH value, altering the drug ionization may occur.

The oromucosal route is convenient and non-invasive. This route encompasses both alternative to oral or parenteral and patient-oriented approaches. Currently, European Pharmacopoeia (Ph. Eur.) lists two main categories of preparations: oromucosal preparations and preparations for dental use. Drugs administered via the sublingual and buccal routes are traditionally formulated as solid dosage forms (e.g., tablets, wafers, films, inserts, and patches, mucoadhesive tablet), liquid dosage forms (e.g., sprays and drops), and semi-solid dosage forms (e.g., gels). Specifically for dental use, preparations are formulated as gargle (concentrate, powder, or tablet for gargle), gingival solution, gel or a paste, lozenge or pastille, mouth wash (or tablet or powder for mouth wash), dental paste, stick or cream [162].

4.2. Oromucosal Preparations and Formulations in Periodontitis

Systemic administration of drugs for the treatment and prevention of oral diseases leads to therapeutic concentrations at the site of infection, but for short periods, forcing repeated dosing and, most importantly, exposing the system to medicines. Local delivery of medicines has been investigated for the possibility of overcoming the limitations of conventional therapy. Oral local drug delivery systems (OLDDSs) have been used both as immediate and sustained release forms. Drug dosage systems (DDS) are considered important adjuvant therapy for treating and preventing oral diseases [163]. Both already marketed and novel OLDDSs are formulated on a biopolymers matrix (biodegradable and biocompatible), which does not disturb the periodontal tissue regeneration process. Amongst various natural polymers, chitosan, a deacetylated product of chitin is widely used in drug delivery devices. Drug carriers investigated in OLDDSs are liposomes, micelles, and other copolymer nanoparticles and dendrimers [163]. In terms of polymers investigated for biodegradable matrixes the potential agents are the natural polymers such as chitosan, cellulose, alginate, and synthetic polymers such as poly(ε-caprolactone) (PCL), poly(d,l-lactide) (PLA), poly-(d,l-lactide-co-glycolide) (PLGA), poly- (vinylpyrrolidone) (PVP) and poly(vinyl alcohol) (PVAL) [164]. Currently, various novel approaches have been investigated. A wide array of OLDDSs is being investigated as treatment options such as fibres, stripes, films, microparticulate systems.

Despite the achievement of effective concentration of the drug at the sites of microbial infection and periodontal pockets, the above systems have specific disadvantages, e.g., the application of fibres causes patient's discomfort, and the removal of fibre various degrees of gingival redness. Microparticulate systems have poor retention of the system into the periodontal pocket [165]. A very promising OLDDSs are in situ forming gels containing alginates, hyaluronic acid, and gellan gum or gelrite, which by partial adhesion to the surrounding tissue resulting in a significant residence time of the system at the site of action, prolonged residence times in periodontal pockets and increased exposure to the drug [166].

One of the novel OLDDSs type is mucoadhesive tablets, which, due to mucoadhesive polymers, deposit at the site of application, consequently, extend the duration of action in the affected area [167]. This form can be used for accurate and local dosing of the drug, with a flexible and controlled dosing schedule, and what's more, this formulation is easy to use, improving patient's compliance [168]. Numerous literature data on delivering polyphenols, especially resveratrol by this route is available. It has been proven that by the use of mucoadhesive polymers, a controlled and sustained release of resveratrol at the affected site can be achieved [169,170], which in combination with antioxidant and antibacterial properties can be a potential alternative in the treatment of periodontal diseases.

Table 2 presents herbal preparations available in the European Union and registered as non-prescription medicines for gingivitis and periodontitis. The availability of these preparations and the fact that they are commonly used confirm the interest in using herbal medicines in the treatment of periodontitis. The use of novel preparations, such as the aforementioned polymers or new plant materials, is a drive for the development of innovative dosage forms, required for the ageing population affected by periodontitis.

Table 2. Selected preparations marketed as non-prescription medicines in the European Union [171].

Marketed Product	Active Ingredient(s)	Manufacturer	Pharmaceutical Form	Registered Indications
Aperisan® 20%	Liquid extract of sage leaves (*Salvia officinalis* L.),	Dentinox® Gesellschaft für pharmazeutische Präparate Lenk and Schuppan KG, Berlin, Germany	oral gel (topical)	The symptomatic treatment of oral inflammation
Argol Essenza Balsamica®	Menthol, melissa oil, cinnamon oil, clove oil (*Syzygium aromaticum* L.), lemon oil (*Citrus limon* L.), nutmeg oil (*Myristica fragrans*), thyme oil (*Thymus* spp.), coriander oil (*Coriandrum sativum* L.), peppermint oil (*Mentha piperita* L.),	Alba Thyment Sp. z o.o., Suchy Las, Poland	oral solution (topical)	For aphthous and inflamed gums
Baikadent®	A complex of flavones isolated from the root of Baikal Scullcap (*Scutellaria baicalensis*)	Herbapol Wrocław S.A., Wrocław, Poland	oral gel (topical)	In the complementary treatment of superficial and deep periodontopathies; in the prophylaxis of periodontal diseases; in chronic inflammatory conditions of the oral cavity mucosa (also in case of injuries caused by dentures)

Table 2. *Cont.*

Marketed Product	Active Ingredient(s)	Manufacturer	Pharmaceutical Form	Registered Indications
Dentosept®	A complex liquid extract of camomile basket (*Matricaria recutita* L.), oak bark (*Quercus* spp.), sage leaf (*Salvia officinalis* L.), arnica herb (*Arnica* spp.), calamus rhizome (*Acorus calamus* L.), peppermint herb (*Mentha piperita* L.), thyme (*Thymus* spp.)	Phytopharm Klęka S.A., Klęka, Poland	oral solution (topical)	Anti-inflammatory, anti-bacterial, disinfectant and astringent, in inflammation of the oral and pharyngeal mucosa, gingivitis and stomatitis (including inflammation of the tongue); superficial periodontitis; bleeding gums; adjunctive in periodontosis
Dentosept A®	A complex liquid extract of Chamomile basket (*Matricaria recutita* L.), oak bark (*Quercus* spp.), sage leaf (*Salvia officinalis* L.), arnica herb (*Arnica* spp.), calamus rhizome (*Acorus calamus* L.), peppermint herb (*Mentha piperita* L.), thyme herb (*Thymus* spp.)	Phytopharm Klęka S.A., Klęka, Poland	oral solution (topical)	In the inflammations of oral cavity and gums, aphthae, mouth sores (after dentures), as an aid in periodontosis
Herbadent®	Benzocaine, Salicylic Acid, Herbal Liquid Extract for Herbadent	Herbai a.s., Prague, Czech Republic	oral topical solution	To massage the gums, especially in periodontitis, gingivitis and periodontitis
Kamistad® Gel	Matricaria liquid extract (*Matricaria recutita* L.), lidocaine hydrochloridum monohydricum	STADA Arzneimittel AG, Berlin, Germany	oral topical gel	Traditionally used as a remedy for mild inflammation of the gums and oral mucosa
Kamistad Senzitiv®	Matricaria Liquid Extract (*Matricaria recutita* L.) lidocaine hydrochloridum monohydricum	STADA Arzneimittel AG, Berlin, Germany	oral topical gel	Kamistad Sensitive is indicated for the treatment of minor infections of the gums and oral mucosa in adults and adolescents above 12 years old
Kamillosan Konzentrat®	Chamomile flower extract (*Matricaria recutita* L.)	MEDA Pharma GmbH and Co. KG, Bad Homburg, Germany	oral topical solution	As an adjunctive treatment for moist compresses, rinses or washes for inflammatory skin and mucous membrane disorders, including the oral cavity and gums
Kamillosan® Mund- und Rachenspray®	Chamomile flower extract (*Matricaria recutita* L.)	MEDA Pharma GmbH and Co. KG, Bad Homburg, Germany	oral spray	Kamillosan mouth and throat spray is used in inflammation of the throat (tonsillitis) in colds without fever, inflammation of the mucous membranes of the mouth and gums

Table 2. *Cont.*

Marketed Product	Active Ingredient(s)	Manufacturer	Pharmaceutical Form	Registered Indications
Mucosit®	Extract from chamomile (*Matricaria recutita* L.), calendula (*Calendula officinalis* L.) coltsfoot leaf (*Tussilago farfara*), oak bark (*Quercus* spp.), sage leaf (*Salvia officinalis* L.), thyme herb (*Thymus* spp.); essential oils (mint and chamomile), allantoin	Herbapol Poznań S.A., Poznań, Poland	oral topical gel	Topically as an astringent, anti-inflammatory and antimicrobial agent and as a local anaesthetic, accelerating granulation and wound healing. The drug for topical application on oral mucosa as a traditionally used supportive agent in the treatment of periodontal diseases and inflammatory conditions of the oral cavity
Septosan fix®	Peppermint herb (*Mentha piperita* L.), thyme (*Thymus* spp.), sage leaf (*Salvia officinalis* L.),	Pharmaceutical Works POLPHARMA S.A., Starogard Gdański, Poland	herbs for infusion in sachets	Traditionally used as a disinfectant in acute and chronic inflammations of the oral and pharyngeal mucosa, and mouth, throat and gum inflammation

5. Conclusions and Future Perspectives

The presented summary has confirmed that plant materials could be successfully used in the management of periodontitis. Generally, plant materials therapy has a favourable safety profile with fewer side effects compared to conventional agents such as chlorhexidine. It is crucial, especially in polypharmacy and chronic illnesses, when side effects may become a reason for therapy termination. The consequences of periodontitis involve tooth loss and related ailments as well as other conditions predisposing to their occurrence with a more severe course or a need for aggressive treatment. It is confirmed that the relationship between periodontitis and diabetes manifests in a more severe course of the disease and the more extensive destruction of periodontal tissues in diabetic patients and, at the same time, poorer glycemic control in diabetic patients with periodontal diseases [172]. Periodontitis also predisposes to the occurrence of cardiovascular diseases. A greater arterial stiffness, a biomarker of cardiovascular disease risk, is observed [173]. Moreover, the risk of cardiovascular and thrombotic stroke doubles in patients with periodontitis [174]. Therefore, it is crucial to introduce the prevention and effective and safe management of periodontitis from the initial developmental stages.

So far, herbal ingredients are promising, treatment option adjuvant to conventional management such as scaling, leading to improved results and reduction of the aggressive therapies burden. Plant raw materials show promising activity in preventing periodontitis [175]; however, further research is essential to evaluate their effectiveness and long-term safety. The medicinal potential of plant products can be exploited by increasing their bioavailability, for example, through nanoencapsulation, which has been applied successfully for baicalin [176]. The use of nanotechnology is a promising current strategy for periodontitis treatment. Encapsulation of active compounds in nanoparticles enables sustained release at specific sites at the specified designed profile and improves absorption [177]. Due to the high potential of novel OLDDSs as effective active pharmaceutical ingredient delivery systems both in treatment and prevention of development of localized periodontitis or in areas that do not respond to the standard mechanical therapy, further research is required to investigate the effectiveness and applicability of formulations containing plant materials as well as to compare various novel systems. The potential use of polymers such as chitosan is worth noting due to their biocompatibility, biodegradability, and adhesion ability. Option of great interest and potential is a local delivery system obtained from biomaterials of plants origin. The current data confirm that using hydrogels as adjunctive therapy to improve periodontal healing shows promising results [178–180]. In addition to biocompatibility and bio-friendliness plant, biomaterials are usually associated with a lower cost, better safety profile, and solid evidence may be used to substitute conventional therapies or adjuvant, increasing their effectiveness.

Author Contributions: Conceptualization, M.P.-W. and J.C.-P.; methodology, A.G. and M.P.-W.; formal analysis, J.C.-P.; investigation, A.G.; writing—original draft preparation, A.G.; writing—review and editing, M.P.-W., A.G., A.S., M.A.R. and J.C.-P.; supervision, J.C.-P. All authors have read and agreed to the published version of the manuscript.

Funding: This research was funded by National Science Center, Poland, under the Preludium BIS grant (number 2020/39/O/NZ7/03441).

Institutional Review Board Statement: Not applicable.

Informed Consent Statement: Not applicable.

Conflicts of Interest: The authors declare no conflict of interest.

References

1. Beard, J.R.; Bloom, D.E. Towards a comprehensive public health response to population ageing. *Lancet* **2015**, *385*, 658–661. [CrossRef]
2. GBD 2017 Disease and Injury Incidence and Prevalence Collaborators. Global, regional, and national incidence, prevalence, and years lived with disability for 354 Diseases and Injuries for 195 countries and territories, 1990–2017: A systematic analysis for the Global Burden of Disease Study 2017. *Lancet* **2018**, *392*, 1789–1858. [CrossRef]
3. Petersen, P.E.; Yamamoto, T. Improving the oral health of older people: The approach of the WHO Global Oral Health Programme. *Community Dent. Oral Epidemiol.* **2005**, *33*, 81–92. [CrossRef] [PubMed]
4. Marcenes, W.; Kassebaum, N.J.; Bernabé, E.; Flaxman, A.; Naghavi, M.; Lopez, A.; Murray, C.J.L. Global burden of oral conditions in 1990–2010: A systematic analysis. *J. Dent. Res.* **2013**, *92*, 592–597. [CrossRef] [PubMed]
5. WHO. Europe Disease Prevention—Data and Statistics. 2018. Available online: https://www.euro.who.int/en/health-topics/disease-prevention/oral-health/data-and-statistics (accessed on 9 December 2020).
6. Peres, M.A.; Macpherson, L.M.D.; Weyant, R.J.; Daly, B.; Venturelli, R.; Mathur, M.R.; Listl, S.; Celeste, R.K.; Guarnizo-Herreño, C.C.; Kearns, C.; et al. Oral diseases: A global public health challenge. *Lancet* **2019**, *394*, 249–260. [CrossRef]
7. Ferlay, J.; Colombet, M.; Soerjomataram, I.; Mathers, C.; Parkin, D.; Piñeros, M.; Znaor, A.; Bray, F. Estimating the global cancer incidence and mortality in 2018: GLOBOCAN sources and methods. *Int. J. Cancer* **2019**, *144*, 1941–1953. [CrossRef] [PubMed]
8. Prasai Dixit, L.; Shakya, A.; Shrestha, M.; Shrestha, A. Dental caries prevalence, oral health knowledge and practice among indigenous Chepang school children of Nepal. *BMC Oral Health* **2013**, *13*, 1–5. [CrossRef] [PubMed]
9. Patel, R. The State of Oral Health in Europe-Report Commissioned by the Platform for Better Oral Health in Europe. 2012. Available online: http://www.oralhealthplatform.eu/our-work/the-state-of-oral-health-in-europe (accessed on 15 December 2020).
10. Alves, A.C.; Cavalcanti, R.V.; Calderon, P.S.; Pernambuco, L.; Alchieri, J.C. Quality of life related to complete denture. *Acta Odontol. Latinoam.* **2018**, *31*, 91–96.
11. Murthy, V.; Sethuraman, K.R.; Rajaram, S.; Choudhury, S. Predicting denture satisfaction and quality of life in completely edentulous: A mixed-mode study. *J. Indian Prosthodont. Soc.* **2021**, *21*, 88–98. [CrossRef]
12. Gerritsen, A.E.; Allen, P.F.; Witter, D.J.; Bronkhorst, E.M.; Creugers, N.H.J. Tooth loss and oral health-related quality of life: A systematic review and meta-analysis. *Health Qual. Life Outcomes* **2010**, *8*, 126. [CrossRef]
13. Caton, J.G.; Armitage, G.; Berglundh, T.; Chapple, I.L.C.; Jepsen, S.; Kornman, K.S.; Mealey, B.L.; Papapanou, P.N.; Sanz, M.; Tonetti, M.S. A new classification scheme for periodontal and peri-implant diseases and conditions—Introduction and key changes from the 1999 classification. *J. Clin. Periodontol.* **2018**, *45*, S1–S8. [CrossRef] [PubMed]
14. Stawińska, N.; Ziętek, M.; Kochanowska, I. Molekularne procesy resorpcji kości i ich potencjał terapeutyczny w leczeniu chorób przyzębia i osteoporozy. *Dent. Med. Probl.* **2005**, *42*, 627–635.
15. Lange, L.; Thiele, G.M.; McCracken, C.; Wang, G.; Ponder, L.A.; Angeles-Han, S.T.; Rouster-Stevens, K.A.; Hersh, A.O.; Vogler, L.B.; Bohnsack, J.F.; et al. Symptoms of periodontitis and antibody responses to Porphyromonas gingivalis in juvenile idiopathic arthritis. *Pediatr. Rheumatol.* **2016**, *14*, 8. [CrossRef]
16. Page, R.C.; Engel, L.D.; Narayanan, A.S.; Clagett, J.A. Chronic Inflammatory Gingival and Periodontal Disease. *J. Am. Med. Assoc.* **1978**, *240*, 545–550. [CrossRef]

17. Pihlstrom, B.L.; Michalowicz, B.S.; Johnson, N.W. Periodontal diseases. *Lancet* **2005**, *366*, 1809–1820. [CrossRef]
18. Hajishengallis, G. The inflammophilic character of the periodontitis-associated microbiota. *Mol. Oral Microbiol.* **2014**, *29*, 248–257. [CrossRef]
19. Suzuki, N.; Yoneda, M.; Hirofuji, T. Mixed Red-Complex Bacterial Infection in Periodontitis. *Int. J. Dent.* **2013**, *2013*, 587279. [CrossRef] [PubMed]
20. Socransky, S.S.; Haffajee, A.D.; Smith, C.; Martin, L.; Haffajee, J.A.; Uzel, N.G.; Goodson, J.M. Use of checkerboard DNA-DNA hybridization to study complex microbial ecosystems. *Oral Microbiol. Immunol.* **2004**, *19*, 352–362. [CrossRef]
21. Curtis, M.A.; Zenobia, C.; Darveau, R.P. The Relationship of the Oral Microbiotia to Periodontal Health and Disease. *Cell Host Microbe* **2011**, *10*, 302–306. [CrossRef]
22. Holt, S.C.; Ebersole, J.L. Porphyromonas gingivalis, Treponema denticola, and Tannerella forsythia: The "red complex", a prototype polybacterial pathogenic consortium in periodontitis. *Periodontol. 2000* **2005**, *38*, 72–122. [CrossRef] [PubMed]
23. Miricescu, D.; Totan, A.; Calenic, B.; Mocanu, B.; Didilescu, A.; Mohora, M.; Spinu, T.; Greabu, M. Salivary biomarkers: Relationship between oxidative stress and alveolar bone loss in chronic periodontitis. *Acta Odontol. Scand.* **2014**, *72*, 42–47. [CrossRef] [PubMed]
24. Çanakçi, C.F.; Çiçek, Y.; Çanakçi, V. Reactive oxygen species and human inflammatory periodontal diseases. *Biochemistry* **2005**, *70*, 619–628. [CrossRef]
25. Scannapieco, F.A.; Gershovich, E. The prevention of periodontal disease—An overview. *Periodontol. 2000* **2020**, *84*, 9–13. [CrossRef] [PubMed]
26. Sabtu, N.; Enoch, D.A.; Brown, N.M. Antibiotic resistance: What, why, where, when and how? *Br. Med. Bull.* **2015**, *116*, 105–113. [CrossRef] [PubMed]
27. Pemberton, M.N.; Gibson, J. Chlorhexidine and hypersensitivity reactions in dentistry. *Br. Dent. J.* **2012**, *213*, 547–550. [CrossRef]
28. Li, F.S.; Weng, J.K. Demystifying traditional herbal medicine with modern approach. *Nat. Plants* **2017**, *3*, 17109. [CrossRef]
29. Cruz Martínez, C.; Diaz Gómez, M.; Oh, M.S. Use of traditional herbal medicine as an alternative in dental treatment in Mexican dentistry: A review. *Pharm. Biol.* **2017**, *55*, 1992–1998. [CrossRef] [PubMed]
30. Graziani, F.; Karapetsa, D.; Alonso, B.; Herrera, D. Nonsurgical and surgical treatment of periodontitis: How many options for one disease? *Periodontol. 2000* **2017**, *75*, 152–188. [CrossRef]
31. Teughels, W.; Dhondt, R.; Dekeyser, C.; Quirynen, M. Treatment of aggressive periodontitis. *Periodontol. 2000* **2014**, *65*, 107–133. [CrossRef]
32. Całkosiński, I.; Dobrzyński, M.; Całkosińska, M.; Seweryn, E.; Bronowicka-Szydełko, A.; Dzierzba, K.; Ceremuga, I.; Gamian, A. Characterization of an inflammatory response. *Postepy Hig. Med. Dosw.* **2009**, *63*, 395–408.
33. Machlin, L.J.; Bendich, A. Free radical tissue damage: Protective role of antioxidant nutrients 1. *FASEB J.* **1987**, *1*, 441–445. [CrossRef]
34. Chen, B.; Wu, W.; Sun, W.; Zhang, Q.; Yan, F.; Xiao, Y. RANKL Expression in Periodontal Disease: Where Does RANKL Come from? *BioMed Res. Int.* **2014**, *2014*, 731039. [CrossRef]
35. Ghorbani, A.; Esmaeilizadeh, M. Pharmacological properties of Salvia officinalis and its components. *J. Tradit. Complement. Med.* **2017**, *7*, 433–440. [CrossRef]
36. Committee on Herbal Medicinal Products (HMPC). European Union Herbal Monograph on Salvia Officinalis L., Folium. 2016. Available online: https://www.ema.europa.eu/en/medicines/herbal/salviae-officinalis-folium (accessed on 2 December 2021).
37. Mansourabadi, A.H.; Sadeghi, H.M.; Razavi, N.; Rezvani, E. Anti-inflammatory and analgesic properties of salvigenin, Salvia officinalis flavonoid extracted. *Adv. Herb. Med.* **2016**, *2*, 31–41.
38. Baricevic, D.; Sosa, S.; Della Loggia, R.; Tubaro, A.; Simonovska, B.; Krasna, A.; Zupancic, A. Topical anti-inflammatory activity of Salvia officinalis L. leaves: The relevance of ursolic acid. *J. Ethnopharmacol.* **2001**, *75*, 125–132. [CrossRef]
39. Rodrigues, M.R.A.; Kanazawa, L.K.S.; das Neves, T.L.M.; da Silva, C.F.; Horst, H.; Pizzolatti, M.G.; Santos, A.R.S.; Baggio, C.H.; de Paula Werner, M.F. Antinociceptive and anti-inflammatory potential of extract and isolated compounds from the leaves of Salvia officinalis in mice. *J. Ethnopharmacol.* **2012**, *139*, 519–526. [CrossRef] [PubMed]
40. Tambur, Z.; Miljković-Selimović, B.; Opačić, D.; Vuković, B.; Malešević, A.; Ivančajić, L.; Aleksić, E. Inhibitory effects of propolis and essential oils on oral bacteria. *J. Infect. Dev. Ctries.* **2021**, *15*, 1027–1031. [CrossRef] [PubMed]
41. Beheshti-Rouy, M.; Azarsina, M.; Rezaie-Soufi, L.; Alikhani, M.Y.; Roshanaie, G.; Komaki, S. The antibacterial effect of sage extract (Salvia officinalis) mouthwash against Streptococcus mutans in dental plaque: A randomized clinical trial. *Iran. J. Microbiol.* **2015**, *7*, 173–177. [PubMed]
42. Mendes, F.S.F.; Garcia, L.M.; da Silva Moraes, T.; Casemiro, L.A.; de Alcantara, C.B.; Ambrósio, S.R.; Veneziani, R.C.S.; Miranda, M.L.D.; Martins , C.H.G. Antibacterial activity of salvia officinalis L. against periodontopathogens: An in vitro study. *Anaerobe* **2020**, *63*, 102194. [CrossRef]
43. Kostić, M.; Kitić, D.; Petrović, M.B.; Jevtović-Stoimenov, T.; Jović, M.; Petrović, A.; Živanović, S. Anti-inflammatory effect of the Salvia sclarea L. ethanolic extract on lipopolysaccharide-induced periodontitis in rats. *J. Ethnopharmacol.* **2017**, *199*, 52–59. [CrossRef]
44. Jünger, H.; Jaun-Ventrice, A.; Guldener, K.; Ramseier, C.A.; Reissmann, D.R.; Schimmel, M. Anti-inflammatory potential of an essential oil-containing mouthwash in elderly subjects enrolled in supportive periodontal therapy: A 6-week randomised controlled clinical trial. *Clin. Oral Investig.* **2020**, *24*, 3203–3211. [CrossRef] [PubMed]

45. Aljuboori, I.W.; Mahmood, M.S. The Effects of Salvia officinalis Gel as an Adjunct to Scaling and Root Planning in Patients with Periodontitis (Clinical and Immunological Study). *Int. J. Drug Deliv. Technol.* **2020**, *10*, 232–237. [CrossRef]
46. Kharaeva, Z.F.; Mustafaev, M.S.; Khazhmetov, A.V.; Gazaev, I.H.; Blieva, L.Z.; Steiner, L.; Mayer, W.; De Luca, C.; Korkina, L.G. Anti-Bacterial and Anti-Inflammatory Effects of Toothpaste with Swiss Medicinal Herbs towards Patients Suffering from Gingivitis and Initial Stage of Periodontitis: From Clinical Efficacy to Mechanisms. *Dent. J.* **2020**, *8*, 10. [CrossRef] [PubMed]
47. Elansary, H.O.; Szopa, A.; Kubica, P.; Ekiert, H.; Mattar, M.A.; Al-Yafrasi, M.A.; El-Ansary, D.O.; Zin El-Abedin, T.K.; Yessoufou, K. Polyphenol Profile and Pharmaceutical Potential of Quercus spp. Bark Extracts. *Plants* **2019**, *8*, 486. [CrossRef]
48. Council of Europe. *European Pharmacopoeia*, 6th ed.; Council of Europe: Strasbourg, France, 2008.
49. Committee on Herbal Medicinal Products (HMPC). Community Herbal Monograph on Quercus Robur L., Quercus Petraea (Matt.) Liebl., Quercus Pubescens Willd., Cortex. Available online: https://www.ema.europa.eu/en/medicines/herbal/quercus-cortex (accessed on 2 December 2021).
50. Bhatia, N.; Friedman, A.; Del Rosso, J. Applications of Topical Oak Bark Extract in Dermatology: Clinical Examples and Discussion. *J. Drugs Dermatol.* **2019**, *18*, 203–206.
51. Deryabin, D.G.; Tolmacheva, A.A. Antibacterial and Anti-Quorum Sensing Molecular Composition Derived from Quercus cortex (Oak bark) Extract. *Molecules* **2015**, *20*, 17093. [CrossRef]
52. Dróżdż, P.; Pyrzynska, K. Assessment of polyphenol content and antioxidant activity of oak bark extracts. *Eur. J. Wood Wood Prod.* **2018**, *76*, 793–795. [CrossRef]
53. Tsubanova, N. Screening study for finding the optimal combination gel composition for the treatment of periodontal disease, which contains extracts of Aloe vera and oak bark. *Asian J. Pharm.* **2017**, *11*, S353–S357.
54. Tsubanova, N.A.; Zhurenko, D.S. A pharmacodynamic study of a new gel containing an extract of Aloe vera and an extract of oak bark for potential treatment of periodontal diseases. *Ceska Slov. Farm.* **2020**, *69*, 143–148.
55. Tsubanova, N.A.; Zhurenko, D.S.; Sakharova, T.S. The study of the effect of a new gel containing the extract of oak bark and aloe extract on the mucous membrane of the oral cavity under the conditions of the experimental stomatitis. *Clin. Pharm.* **2018**, *3*, 4–10. [CrossRef]
56. Riachi, L.G.; De Maria, C.A.B. Peppermint antioxidants revisited. *Food Chem.* **2015**, *176*, 72–81. [CrossRef] [PubMed]
57. Brahmi, F.; Khodir, M.; Mohamed, C.; Pierre, D.; Brahmi, F. Chemical Composition and Biological Activities of Mentha Species. *Aromat. Med. Plants Back Nat.* **2017**, *10*, 47–79. [CrossRef]
58. Committee on Herbal Medicinal Products. European Union Herbal Monograph on Mentha x Piperita L., Folium. 2017. Available online: https://www.ema.europa.eu/en/medicines/herbal/menthae-piperitae-aetheroleum (accessed on 3 December 2021).
59. Snoussi, M.; Noumi, E.; Trabelsi, N.; Flamini, G.; Papetti, A.; De Feo, V. Mentha spicata essential oil: Chemical composition, antioxidant and antibacterial activities against planktonic and biofilm cultures of Vibrio spp. strains. *Molecules* **2015**, *20*, 14402–14424. [CrossRef]
60. Camele, I.; Grul'ová, D.; Elshafie, H.S. Chemical Composition and Antimicrobial Properties of Mentha× piperita cv.'Kristinka'Essential Oil. *Plants* **2021**, *10*, 1567. [CrossRef]
61. Kizil, S.; Hasimi, N.; Tolan, V.; Kilinc, E.; Yuksel, U. Mineral content, essential oil components and biological activity of two mentha species (M. piperita L., M. spicata L.). *Turkish J. Field Crop.* **2010**, *15*, 148–153.
62. İşcan, G.; Kirimer, N.; Kürkcüoğlu, M.; Başer, H.C.; Demirci, F. Antimicrobial screening of Mentha piperita essential oils. *J. Agric. Food Chem.* **2002**, *50*, 3943–3946. [CrossRef] [PubMed]
63. Peixoto, I.T.A.; Furlanetti, V.F.; Anibal, P.C.; Duarte, M.C.T.; Höfling, J.F. Potential pharmacological and toxicological basis of the essential oil from *Mentha* spp. *Rev. Ciências Farm. Básica e Apl.* **2009**, *30*, 235–239.
64. Karicheri, R.; Antony, B. Antibacterial and antibiofilm activities of peppermint (Mentha piperita Linn) and menthol mint (Mentha arvensis Linn) essential oils on Aggregatibacter Actinomycetemcomitans isolated from orodental infections. *Eur. J. Pharm. Med. Res.* **2016**, *3*, 577–581.
65. Sindi, A.M.; Hosny, K.M.; Alharbi, W.S. Lyophilized Composite Loaded with Meloxicam-Peppermint oil Nanoemulsion for Periodontal Pain. *Polymers* **2021**, *13*, 2317. [CrossRef]
66. Phaechamud, T.; Mahadlek, J.; Tuntarawongsa, S. Peppermint oil/doxycycline hyclate-loaded Eudragit RS in situ forming gel for periodontitis treatment. *J. Pharm. Investig.* **2018**, *48*, 451–464. [CrossRef]
67. Gupta, R. Survey record of medicinal and aromatic plants of Chamba forest division, Himachal Pradesh. *Indian For.* **1964**, *90*, 454–468.
68. Marongiu, B.; Piras, A.; Porcedda, S.; Scorciapino, A. Chemical Composition of the Essential Oil and Supercritical CO 2 Extract of Commiphora myrrha (Nees) Engl. And of Acorus calamus L. *J. Agric. Food Chem.* **2005**, *53*, 7939–7943. [CrossRef] [PubMed]
69. Raja, A.E.; Vijayalakshmi, M.; Devalarao, G. Acorus calamus Linn.: Chemistry and biology. *Res. J. Pharm. Technol.* **2009**, *2*, 256–261.
70. Committee on Herbal Medicinal Products (HMPC). Public Statement on the Use of Herbal Medicinal Products Containing Asarone. 2005. Available online: https://www.ema.europa.eu/en/use-herbal-medicinal-products-containing-asarone (accessed on 3 December 2021).
71. Rajput, S.B.; Tonge, M.B.; Karuppayil, S.M. An overview on traditional uses and pharmacological profile of Acorus calamus Linn. (Sweet flag) and other Acorus species. *Phytomedicine* **2014**, *21*, 268–276. [CrossRef]

72. Loying, R.; Gogoi, R.; Sarma, N.; Borah, A.; Munda, S.; Pandey, S.K.; Lal, M. Chemical Compositions, In-vitro Antioxidant, Anti-microbial, Anti-inflammatory and Cytotoxic Activities of Essential Oil of Acorus calamus L. Rhizome from North-East India. *J. Essent. Oil-Bear. Plants* **2019**, *22*, 1299–1312. [CrossRef]
73. Nanda, B.L. Determination of phytochemicals and antioxidant activity of Acorus calamus rhizome. *J. Drug Deliv. Ther.* **2014**, *4*, 117–121. [CrossRef]
74. Dinev, T.; Tzanova, M.; Velichkova, K.; Dermendzhieva, D.; Beev, G. Antifungal and Antioxidant Potential of Methanolic Extracts from Acorus calamus L., Chlorella vulgaris Beijerinck, Lemna minuta Kunth and Scenedesmus dimorphus (Turpin) Kützing. *Appl. Sci.* **2021**, *11*. [CrossRef]
75. Khwairakpam, A.D.; Damayenti, Y.D.; Deka, A.; Monisha, J.; Roy, N.K.; Padmavathi, G.; Kunnumakkara, A.B. Acorus calamus: A bio-reserve of medicinal values. *J. Basic Clin. Physiol. Pharmacol.* **2018**, *29*, 107–122. [CrossRef] [PubMed]
76. Hwang, J.W.; Park, W.J.; Han, Y. Asarylaldehyde enhances osteogenic differentiation of human periodontal ligament stem cells through the ERK/p38 MAPK signaling pathway. *Biochem. Biophys. Res. Commun.* **2021**, *545*, 27–32. [CrossRef]
77. Zhang, X.-W.; Li, W.-F.; Li, W.-W.; Ren, K.-H.; Fan, C.-M.; Chen, Y.-Y.; Shen, Y.-L. Protective effects of the aqueous extract of *Scutellaria baicalensis* against acrolein-induced oxidative stress in cultured human umbilical vein endothelial cells. *Pharm. Biol.* **2011**, *49*, 256–261. [CrossRef]
78. Tsao, T.F.; Newman, M.G.; Kwok, Y.Y.; Horikoshi, A.K. Effect of Chinese and Western Antimicrobial Agents on Selected Oral Bacteria. *J. Dent. Res.* **1982**, *61*, 1103–1106. [CrossRef]
79. Leung, K.; Seneviratne, C.; Li, X.; Leung, P.; Lau, C.; Wong, C.-H.; Pang, K.; Wong, C.; Wat, E.; Jin, L. Synergistic Antibacterial Effects of Nanoparticles Encapsulated with Scutellaria baicalensis and Pure Chlorhexidine on Oral Bacterial Biofilms. *Nanomaterials* **2016**, *6*, 61. [CrossRef]
80. Cui, L.; Feng, L.; Zhang, Z.H.; Jia, X. Bin The anti-inflammation effect of baicalin on experimental colitis through inhibiting TLR4/NF-κB pathway activation. *Int. Immunopharmacol.* **2014**, *23*, 294–303. [CrossRef]
81. De Oliveira, J.S.; Pinto, M.e.S.C.; Santana, L.d.A.d.B.; Pinto, A.S.B.; di Lenardo, D.; Vasconcelos, D.F.P. Biological effects of medicinal plants on induced periodontitis: A systematic review. *Int. J. Dent.* **2016**, *2016*, 3719879. [CrossRef] [PubMed]
82. Cai, X.; Li, C.; Du, G.; Cao, Z. Protective effects of baicalin on ligature-induced periodontitis in rats. *J. Periodontal Res.* **2008**, *43*, 14–21. [CrossRef]
83. Sun, J.Y.; Li, D.L.; Dong, Y.; Zhu, C.H.; Liu, J.; Zhou, T.; Gou, J.Z.; Li, A.; Zang, W.J. Baicalin inhibits toll-like receptor 2/4 expression and downstream signaling in rat experimental periodontitis. *Int. Immunopharmacol.* **2016**, *36*, 86–93. [CrossRef] [PubMed]
84. Kim, M.H.; Lee, H.; Choi, Y.Y.; Lee, D.H.; Yang, W.M. Scutellaria baicalensis ameliorates the destruction of periodontal ligament via inhibition of inflammatory cytokine expression. *J. Chin. Med. Assoc.* **2018**, *81*, 141–146. [CrossRef]
85. Ren, M.; Zhao, Y.; He, Z.; Lin, J.; Xu, C.; Liu, F.; Hu, R.; Deng, H.; Wang, Y. Baicalein inhibits inflammatory response and promotes osteogenic activity in periodontal ligament cells challenged with lipopolysaccharides. *BMC Complement. Med. Ther.* **2021**, *21*, 1–11. [CrossRef] [PubMed]
86. Chung, C.P.; Park, J.B.; Bae, K.H. Pharmacological effects of methanolic extract from the root of Scutellaria baicalensis and its flavonoids on human gingival fibroblast. *Planta Med.* **1995**, *61*, 150–153. [CrossRef]
87. Cai, X.; Li, C.; Cao, Z.; Du, G.; Liu, L. Protective effect of baicalin on experimental periodontitis in rats and its possible mechanisms. *Zhonghua Kou Qiang Yi Xue Za Zhi* **2008**, *43*, 281–285.
88. Zeng, H.; Li, F.; Wei, H.; Shi, J.; Rao, G.; Li, A.; Gou, J. Preliminary study of the dual release baicalin and rhBMP-2 system to improve periodontal tissue regeneration in minipigs. *Shanghai Kou Qiang Yi Xue* **2013**, *22*, 126–131. [PubMed]
89. Arweiler, N.B.; Pergola, G.; Kuenz, J.; Hellwig, E.; Sculean, A.; Auschill, T.M. Clinical and antibacterial effect of an anti-inflammatory toothpaste formulation with Scutellaria baicalensis extract on experimental gingivitis. *Clin. Oral Investig.* **2011**, *15*, 909–913. [CrossRef] [PubMed]
90. Farhan, M.; Ariffin, M.; Fauzi, N.; Venkata, C.; Prakash, S.; Prakash, I. Bioactive Chemical Constituents from Pomegranate (Punica granatum) Juice, Seed and Peel-A Review. *Int. J. Res. Chem. Environ.* **2011**, *1*, 1–18.
91. Rahmani, A.H.; Alsahli, M.A.; Almatroodi, S.A. Active constituents of pomegranates (Punica granatum) as potential candidates in the management of health through modulation of biological activities. *Pharmacogn. J.* **2017**, *9*, 689–695. [CrossRef]
92. Arun, N.; Singh, D.P. Punica granatum: A review on pharmogological and therapeutic properties. *J. Pharm. Sci. Res.* **2012**, *3*, 5.
93. Jurenka, J. Therapeutic Applications of Pomegranate (Punica granatum L.): A Review. *Altern. Med. Rev.* **2008**, *13*, 128–144.
94. Bhadbhade, S.J.; Acharya, A.B.; Rodrigues, S.V.; Thakur, S.L. The antiplaque efficacy of pomegranate mouthrinse. *Quintessence Int.* **2011**, *42*, 29–36.
95. Aparecida Procópio Gomes, L.; Alves Figueiredo, L.M.; Luiza do Rosário Palma, A.; Corrêa Geraldo, B.M.; Isler Castro, K.C.; Ruano de Oliveira Fugisaki, L.; Cardoso Jorge, A.O.; Dias de Oliveira, L.; Junqueira, J.C. Punica granatum L. (Pomegranate) Extract: In Vivo Study of Antimicrobial Activity against Porphyromonas gingivalis in Galleria mellonella Model. *Sci. World J.* **2016**, *2016*, 8626987. [CrossRef]
96. Kote, D.S.; Kote, D.S.; Nagesh, D.L. Effect of Pomegranate Juice on Dental Plaque Microorganisms (Streptococci and Lactobacilli). *Anc. Sci. Life* **2011**, *31*, 49–51.
97. Somu, C.A.; Ravindra, S.; Ajith, S.; Ahamed, M.G. Efficacy of a herbal extract gel in the treatment of gingivitis: A clinical study. *J. Ayurveda Integr. Med.* **2012**, *3*, 85–90. [CrossRef]

98. Ahuja, S.; Dodwad, V.; Kukreja, B.J.; Mehra, P.; Kukreltayeltaeja, P. A comparative evaluation of efficacy of Punica granatum and chlorhexidine on plaque and gingivitis. *J. Int. Clin. Dent. Res. Organ.* **2011**, *3*, 29–32. [CrossRef]
99. Eltay, E.G.; Gismalla, B.G.; Mukhtar, M.M.; Awadelkarim, M. Punica granatum peel extract as adjunct irrigation to nonsurgical treatment of chronic gingivitis. *Ther. Clin. Pract.* **2021**, *43*, 101383. [CrossRef]
100. Sakanaka, S.; Aizawa, M.; Kim, M.; Yamamoto, T. Inhibitory Effects of Green Tea Polyphenols on Growth and Cellular Adherence of an Oral Bacterium, Porphyromonas gingivalis. *Biosci. Biotechnol. Biochem.* **1996**, *60*, 745–749. [CrossRef]
101. Maruyama, T.; Tomofuji, T.; Endo, Y.; Irie, K.; Azuma, T.; Ekuni, D.; Tamaki, N.; Yamamoto, T.; Morita, M. Supplementation of green tea catechins in dentifrices suppresses gingival oxidative stress and periodontal inflammation. *Arch. Oral Biol.* **2011**, *56*, 48–53. [CrossRef] [PubMed]
102. Yoshinaga, Y.; Ukai, T.; Nakatsu, S.; Kuramoto, A.; Nagano, F.; Yoshinaga, M.; Montenegro, J.L.; Shiraishi, C.; Hara, Y. Green tea extract inhibits the onset of periodontal destruction in rat experimental periodontitis. *J. Periodontal Res.* **2014**, *49*, 652–659. [CrossRef] [PubMed]
103. De Almeida, J.M.; Marques, B.M.; Novaes, V.C.N.; de Oliveira, F.L.P.; Matheus, H.R.; Fiorin, L.G.; Ervolino, E. Influence of adjuvant therapy with green tea extract in the treatment of experimental periodontitis. *Arch. Oral Biol.* **2019**, *102*, 65–73. [CrossRef] [PubMed]
104. Chava, V.K.; Vedula, B.D. Thermo-Reversible Green Tea Catechin Gel for Local Application in Chronic Periodontitis: A 4-Week Clinical Trial. *J. Periodontol.* **2013**, *84*, 1290–1296. [CrossRef]
105. Hrishi, T.S.; Kundapur, P.P.; Naha, A.; Thomas, B.S.; Kamath, S.; Bhat, G.S. Effect of adjunctive use of green tea dentifrice in periodontitis patients—A Randomized Controlled Pilot Study. *Int. J. Dent. Hyg.* **2016**, *14*, 178–183. [CrossRef]
106. Chopra, A.; Thomas, B.S.; Sivaraman, K.; Prasad, H.K.; Kamath, S.U. Green tea intake as an adjunct to mechanical periodontal therapy for the management of mild to moderate chronic periodontitis: A randomised controlled clinical trial. *Oral Health Prev. Dent.* **2016**, *14*, 293–303. [CrossRef] [PubMed]
107. Taleghani, F.; Rezvani, G.; Birjandi, M.; Valizadeh, M. Impact of green tea intake on clinical improvement in chronic periodontitis: A randomized clinical trial. *J. Stomatol. Oral Maxillofac. Surg.* **2018**, *119*, 365–368. [CrossRef] [PubMed]
108. Balappanavar, A.Y.; Sardana, V.; Singh, M. Comparison of the effectiveness of 0.5% tea, 2% neem and 0.2% chlorhexidine mouthwashes on oral health: A randomized control trial. *Indian J. Dent. Res.* **2013**, *24*, 26. [CrossRef] [PubMed]
109. Committee on Herbal Medicinal Products (HMPC). Community Herbal Monograph on Camellia sinensis (L.) Kuntze, Non Fermentatum Folium. 2014. Available online: https://www.ema.europa.eu/en/medicines/herbal/camelliae-sinensis-non-fermentatum-folium (accessed on 5 June 2021).
110. Radha, M.H.; Laxmipriya, N.P. Evaluation of biological properties and clinical effectiveness of Aloe vera: A systematic review. *J. Tradit. Complement. Med.* **2015**, *5*, 21–26. [CrossRef] [PubMed]
111. Krzysztof Gołąb, J.G.M.W. Acemannan—Properties and medical utility. *Postępy Fitoter.* **2016**, *1*, 268–273.
112. Jain, S.; Rathod, N.; Nagi, R.; Sur, J.; Laheji, A.; Gupta, N.; Agrawal, P.; Prasad, S. Antibacterial Effect of Aloe Vera Gel against Oral Pathogens: An In-vitro Study. *J. Clin. Diagn. Res.* **2016**, *10*, ZC41. [CrossRef] [PubMed]
113. Bhat, G.; Kudva, P.; Dodwad, V. Aloe vera: Nature's soothing healer to periodontal disease. *J. Indian Soc. Periodontol.* **2011**, *15*, 205. [CrossRef]
114. Moghaddam, A.A.; Radafshar, G.; Jahandideh, Y.; Kakaei, N. Clinical Evaluation of Effects of Local Application of Aloe vera Gel as an Adjunct to Scaling and Root Planning in Patients with Chronic Periodontitis. *J. Dent.* **2017**, *18*, 165–172.
115. Penmetsa, G.S.; Subbareddy, B.; Mopidevi, A.; Arunbhupathi, P.; Baipalli, V.; Pitta, S. Comparing the Effect of Combination of 1% Ornidazole and 0.25% Chlorhexidine Gluconate (Ornigreat™) Gel and Aloe vera Gel in the Treatment of Chronic Periodontitis: A Randomized, Single-Blind, Split-Mouth Study. *Contemp. Clin. Dent.* **2019**, *10*, 226. [CrossRef]
116. Sayar, F.; Farahmand, A.H.; Rezazadeh, M. Clinical Efficacy of Aloe Vera Toothpaste on Periodontal Parameters of Patients with Gingivitis-A Randomized, Controlled, Single-masked Clinical Trial. *J. Contemp. Dent. Pract.* **2021**, *22*, 242–247.
117. Vangipuram, S.; Jha, A.; Bhashyam, M. Comparative efficacy of aloe vera mouthwash and chlorhexidine on periodontal health: A randomized controlled trial. *J. Clin. Exp. Dent.* **2016**, *8*, e442–e447. [CrossRef]
118. Chhina, S.; Singh, A.; Menon, I.; Singh, R.; Sharma, A.; Aggarwal, V. A randomized clinical study for comparative evaluation of Aloe Vera and 0.2% chlorhexidine gluconate mouthwash efficacy on de-novo plaque formation. *J. Int. Soc. Prev. Community Dent.* **2016**, *6*, 251–255. [CrossRef]
119. Pradeep, A.R.; Garg, V.; Raju, A.; Singh, P. Adjunctive Local Delivery of Aloe Vera Gel in Patients With Type 2 Diabetes and Chronic Periodontitis: A Randomized, Controlled Clinical Trial. *J. Periodontol.* **2016**, *87*, 268–274. [CrossRef] [PubMed]
120. Committee on Herbal Medicinal Products (HPMC). European Union Herbal Monograph on Matricaria Recutita L., Flos. 2015. Available online: https://www.ema.europa.eu/en/committees/committee-herbal-medicinal-products-hmpc (accessed on 2 December 2021).
121. Murti, K.; Panchal, M.A.; Gajera, V.; Solanki, J. Pharmacological properties of Matricaria recutita: A review. *Pharmacologia* **2012**, *3*, 348–351. [CrossRef]
122. Hans, V.M.; Grover, H.S.; Deswal, H.; Agarwal, P. Antimicrobial efficacy of various essential oils at varying concentrations against periopathogen Porphyromonas gingivalis. *J. Clin. Diagnostic Res.* **2016**, *10*, ZC16. [CrossRef] [PubMed]
123. Saderi, H.; Owlia, P.; Hosseini, A.; Semiyari, H. Antimicrobial effects of chamomile extract and essential oil on clinically isolated Porphyromonas gingivalis from periodontitis. *Acta Hortic.* **2003**, *680*, 145–146. [CrossRef]

124. Guimarães, M.V.; Melo, I.M.; Adriano Araújo, V.M.; Tenazoa Wong, D.V.; Roriz Fonteles, C.S.; Moreira Leal, L.K.; Ribeiro, R.A.; Lima, V. Dry Extract of Matricaria recutita L. (Chamomile) Prevents Ligature-Induced Alveolar Bone Resorption in Rats via Inhibition of Tumor Necrosis Factor-α and Interleukin-1β. *J. Periodontol.* **2016**, *87*, 706–715. [CrossRef]
125. Batista, A.L.A.; Diógenes Alves UchÔa Lins, R.; de Souza Coelho, R.; do Nascimento Barbosa, D.; Moura Belém, N.; Alves Celestino, F.J. Clinical efficacy analysis of the mouth rinsing with pomegranate and chamomile plant extracts in the gingival bleeding reduction. *Complement. Ther. Clin. Pract.* **2014**, *20*, 93–98. [CrossRef]
126. Agarwal, A.; Chaudhary, B. Clinical and microbiological effects of 1% Matricaria chamomilla mouth rinse on chronic periodontitis: A double-blind randomized placebo controlled trial. *J. Indian Soc. Periodontol.* **2020**, *24*, 354–361. [CrossRef]
127. Shen, J.L.; Man, K.M.; Huang, P.H.; Chen, W.C.; Chen, D.C.; Cheng, Y.W.; Liu, P.L.; Chou, M.C.; Chen, Y.H. Honokiol and magnolol as multifunctional antioxidative molecules for dermatologic disorders. *Molecules* **2010**, *15*, 6452–6465. [CrossRef] [PubMed]
128. Ho, K.Y.; Tsai, C.C.; Chen, C.P.; Huang, J.S.; Lin, C.C. Antimicrobial activity of honokiol and magnolol isolated from Magnolia officinalis. *Phyther. Res.* **2001**, *15*, 139–141. [CrossRef]
129. Sakaue, Y.; Domon, H.; Oda, M.; Takenaka, S.; Kubo, M.; Fukuyama, Y.; Okiji, T.; Terao, Y. Anti-biofilm and bactericidal effects of magnolia bark-derived magnolol and honokiol on Streptococcus mutans. *Microbiol. Immunol.* **2016**, *60*, 10–16. [CrossRef]
130. Walker, J.M.; Maitra, A.; Walker, J.; Ehrnhoefer-Ressler, M.M.; Inui, T.; Somoza, V. Identification of magnolia officinalis L. Bark extract as the most potent anti-inflammatory of four plant extracts. *Am. J. Chin. Med.* **2013**, *41*, 531–544. [CrossRef] [PubMed]
131. Kim, J.Y.; Kim, K.H.; Kwag, E.H.; Seol, Y.J.; Lee, Y.M.; Ku, Y.; Rhyu, I.C. Magnoliae Cortex and maize modulate Porphyromonas gingivalis-induced inflammatory reactions. *J. Periodontal Implant Sci.* **2018**, *48*, 70–83. [CrossRef] [PubMed]
132. Park, J.; Lee, J.; Jung, E.; Park, Y.; Kim, K.; Park, B.; Jung, K.; Park, E.; Kim, J.; Park, D. In vitro antibacterial and anti-inflammatory effects of honokiol and magnolol against Propionibacterium sp. *Eur. J. Pharmacol.* **2004**, *496*, 189–195. [CrossRef]
133. Lee, J.; Jung, E.; Park, J.; Jung, K.; Lee, S.; Hong, S.; Park, J.; Park, E.; Kim, J.; Park, S.; et al. Anti-inflammatory effects of magnolol and honokiol are mediated through inhibition of the downstream pathway of MEKK-1 in NF-κB activation signaling. *Planta Med.* **2005**, *71*, 338–343. [CrossRef] [PubMed]
134. Lu, S.-H.; Huang, R.-Y.; Chou, T.-C. Magnolol Ameliorates Ligature-Induced Periodontitis in Rats and Osteoclastogenesis: In Vivo and In Vitro Study. *Evidence-Based Complement. Altern. Med.* **2013**, *2013*, 634095. [CrossRef] [PubMed]
135. Hellström, M.K.; Ramberg, P. The effect of a dentifrice containing Magnolia extract on established plaque and gingivitis in man: A six-month clinical study. *Int. J. Dent. Hyg.* **2014**, *12*, 96–102. [CrossRef]
136. Hamedi, S.; Sadeghpour, O.; Shamsardekani, M.R.; Amin, G.; Hajighasemali, D.; Feyzabadi, Z. The Most Common Herbs to Cure the Most Common Oral Disease: Stomatitis Recurrent Aphthous Ulcer (RAU). *Iran. Red Crescent Med. J.* **2016**, *18*, 21694. [CrossRef]
137. Verma, R.; Gangrade, T.; Punasiya, R.; Ghulaxe, C. Rubus fruticosus (blackberry) use as an herbal medicine. *Pharmacogn. Rev.* **2014**, *8*, 101–104. [CrossRef]
138. Leonti, M.; Casu, L.; Sanna, F.; Bonsignore, L. A comparison of medicinal plant use in Sardinia and Sicily-De Materia Medica revisited? *J. Ethnopharmacol.* **2009**, *121*, 255–267. [CrossRef]
139. Wang, H.; Cao, Z.-R. Anti-inflammatory Effects of (-)-Epicatechin in Lipopolysaccharide-Stimulated Raw 264.7 Macrophages. *Trop. J. Pharm. Res.* **2014**, *13*, 1415–1419. [CrossRef]
140. El-Shitany, N.A.; El-Bastawissy, E.A.; El-Desoky, K. Ellagic acid protects against carrageenan-induced acute inflammation through inhibition of nuclear factor kappa B, inducible cyclooxygenase and proinflammatory cytokines and enhancement of interleukin-10 via an antioxidant mechanism. *Int. Immunopharmacol.* **2014**, *19*, 290–299. [CrossRef] [PubMed]
141. Lesjak, M.; Beara, I.; Simin, N.; Pintać, D.; Majkić, T.; Bekvalac, K.; Orčić, D.; Mimica-Dukić, N. Antioxidant and anti-inflammatory activities of quercetin and its derivatives. *J. Funct. Foods* **2018**, *40*, 68–75. [CrossRef]
142. Ku, S.K.; Zhou, W.; Lee, W.; Han, M.S.; Na, M.K.; Bae, J.S. Anti-Inflammatory Effects of Hyperoside in Human Endothelial Cells and in Mice. *Inflammation* **2015**, *38*, 784–799. [CrossRef] [PubMed]
143. Bowen-Forbes, C.S.; Zhang, Y.; Nair, M.G. Anthocyanin content, antioxidant, anti-inflammatory and anticancer properties of blackberry and raspberry fruits. *J. Food Compos. Anal.* **2010**, *23*, 554–560. [CrossRef]
144. Dai, J.; Gupte, A.; Gates, L.; Mumper, R.J. A comprehensive study of anthocyanin-containing extracts from selected blackberry cultivars: Extraction methods, stability, anticancer properties and mechanisms. *Food Chem. Toxicol.* **2009**, *47*, 837–847. [CrossRef]
145. Danaher, R.J.; Wang, C.; Dai, J.; Mumper, R.J.; Miller, C.S. Antiviral effects of blackberry extract against herpes simplex virus type 1. *Oral Surg. Oral Med. Oral Pathol. Oral Radiol. Endod.* **2011**, *112*, e31–e35. [CrossRef] [PubMed]
146. González, O.A.; Escamilla, C.; Danaher, R.J.; Dai, J.; Ebersole, J.L.; Mumper, R.J.; Miller, C.S. Antibacterial effects of blackberry extract target periodontopathogens. *J. Periodontal Res.* **2013**, *48*, 80–86. [CrossRef]
147. Committee on Herbal Medicinal Products (HMPC). European Union Herbal Monograph on Vaccinium Macrocarpon Aiton, Fructus. 2021. Available online: https://www.ema.europa.eu/en/medicines/herbal/vaccinii-macrocarpi-fructus (accessed on 5 June 2021).
148. Labrecque, J.; Bodet, C.; Chandad, F.; Grenier, D. Effects of a high-molecular-weight cranberry fraction on growth, biofilm formation and adherence of Porphyromonas gingivalis. *J. Antimicrob. Chemother.* **2006**, *58*, 439–443. [CrossRef]
149. Bodet, C.C.; Piché, M.; Chandad, F.; Grenier, D. Inhibition of periodontopathogen-derived proteolytic enzymes by a high-molecular-weight fraction isolated from cranberry. *J. Antimicrob. Chemother.* **2006**, *57*, 685–690. [CrossRef]

150. Tipton, D.A.; Babu, J.P.; Dabbous, M. Effects of cranberry components on human aggressive periodontitis gingival fibroblasts. *J. Periodontal Res.* **2013**, *48*, 433–442. [CrossRef]
151. Tanabe, S.; Santos, J.; La, V.D.; Howell, A.B.; Grenier, D. A-type cranberry proanthocyanidins inhibit the RANKL-dependent differentiation and function of human osteoclasts. *Molecules* **2011**, *16*, 2365–2374. [CrossRef]
152. Woźniewicz, M.; Nowaczyk, P.M.; Kurhańska-Flisykowska, A.; Wyganowska-Świątkowska, M.; Lasik-Kurdyś, M.; Walkowiak, J.; Bajerska, J. Consumption of cranberry functional beverage reduces gingival index and plaque index in patients with gingivitis. *Nutr. Res.* **2018**, *58*, 36–45. [CrossRef] [PubMed]
153. Fontenelle, R.O.S.; Morais, S.M.; Brito, E.H.S.; Kerntopf, M.R.; Brilhante, R.S.N.; Cordeiro, R.A.; Tomé, A.R.; Queiroz, M.G.R.; Nascimento, N.R.F.; Sidrim, J.J.C.; et al. Chemical composition, toxicological aspects and antifungal activity of essential oil from Lippia sidoides Cham. *J. Antimicrob. Chemother.* **2007**, *59*, 934–940. [CrossRef] [PubMed]
154. Botelho, M.A.; Nogueira, N.A.; Bastos, G.M.; Fonseca, S.G.; Lemos, T.L.; Matos, F.J.; Montenegro, D.; Heukelbach, J.; Rao, V.S.; Brito, G.A. Antimicrobial activity of the essential oil from Lippia sidoides, carvacrol and thymol against oral pathogens. *Braz. J. Med. Biol. Res. Rev. Bras. Pesqui. Medicas Biol.* **2007**, *40*, 349–356. [CrossRef]
155. Botelho, M.A.; dos Santos, R.A.; Martins, J.G.; Carvalho, C.O.; Paz, M.C.; Azenha, C.; Ruela, R.S.; Queiroz, D.B.; Ruela, W.S.; Marinho, G.; et al. Comparative effect of an essential oil mouthrinse on plaque, gingivitis and salivary Streptococcus mutans levels: A double blind randomized study. *Phytother. Res.* **2009**, *23*, 1214–1219. [CrossRef]
156. da Silva, P.S.L.; Machado, P.Y.C.; Catunda, B.T.; Barbosa, A.P.N. Clinical effect of a gel containing Lippia sidoides on plaque and gingivitis control. *Eur. J. Dent.* **2013**, *7*, 28–34.
157. Rodrigues, I.S.; Tavares, V.N.; Pereira, S.L.; Costa, F.N. Antiplaque and antigingivitis effect of Lippia Sidoides: A double-blind clinical study in humans. *J. Appl. Oral Sci.* **2009**, *17*, 404–407. [CrossRef]
158. Botelho, M.A.; Rao, V.S.; Carvalho, C.B.; Bezerra-Filho, J.G.; Fonseca, S.G.; Vale, M.L.; Montenegro, D.; Cunha, F.; Ribeiro, R.A.; Brito, G.A. Lippia sidoides and Myracrodruon urundeuva gel prevents alveolar bone resorption in experimental periodontitis in rats. *J. Ethnopharmacol.* **2007**, *113*, 471–478. [CrossRef]
159. Botelho, M.A.; Barros, G.; Queiroz, D.B.; Carvalho, C.F.; Gouvea, J.; Patrus, L.; Bannet, M.; Patrus, D.; Rego, A.; Silva, I.; et al. Nanotechnology in Phytotherapy: Antiinflammatory Effect of a Nanostructured Thymol Gel from Lippia sidoides in Acute Periodontitis in Rats. *Phyther. Res.* **2016**, *30*, 152–159. [CrossRef] [PubMed]
160. Rathbone, M.; Senel, S.; Pather, I. *Oral Mucosal Drug Delivery and Therapy*; Springer: Boston, MA, USA, 2015; pp. 1–3, ISBN 978-1-4899-7557-7.
161. Harris, D.; Robinson, J.R. Drug delivery via the mucous membranes of the oral cavity. *J. Pharm. Sci.* **1992**, *81*, 1–10. [CrossRef] [PubMed]
162. Council of Europe. *European Pharmacopoeia*, 10th ed.; Council of Europe: Strasbourg, France, 2019.
163. Liang, J.; Peng, X.; Zhou, X.; Zou, J.; Cheng, L. Emerging Applications of Drug Delivery Systems in Oral Infectious Diseases Prevention and Treatment. *Molecules* **2020**, *25*, 516. [CrossRef]
164. Zięba, M.; Chaber, P.; Duale, K.; Martinka Maksymiak, M.; Basczok, M.; Kowalczuk, M.; Adamus, G. Polymeric Carriers for Delivery Systems in the Treatment of Chronic Periodontal Disease. *Polymers* **2020**, *12*, 1574. [CrossRef] [PubMed]
165. Sholapurkar, A.; Sharma, D.; Glass, B.; Miller, C.; Nimmo, A.; Jennings, E. Professionally delivered local antimicrobials in the treatment of patients with periodontitis-a narrative review. *Dent. J.* **2021**, *9*, 2. [CrossRef]
166. Yadav, R.; Kanwar, I.L.; Haider, T.; Pandey, V.; Gour, V.; Soni, V. In situ gel drug delivery system for periodontitis: An insight review. *Futur. J. Pharm. Sci.* **2020**, *6*, 33. [CrossRef]
167. Abruzzo, A.; Cerchiara, T.; Bigucci, F.; Gallucci, M.C.; Luppi, B. Mucoadhesive buccal tablets based on chitosan/gelatin microparticles for delivery of propranolol hydrochloride. *J. Pharm. Sci.* **2015**, *104*, 4365–4372. [CrossRef]
168. Shaikh, R.; Singh, T.R.R.; Garland, M.J.; Woolfson, A.D.; Donnelly, R.F. Mucoadhesive drug delivery systems. *J. Pharm. Bioallied Sci.* **2011**, *3*, 89–100. [CrossRef]
169. Paczkowska-Walendowska, M.; Dvořák, J.; Rosiak, N.; Tykarska, E.; Szymańska, E.; Winnicka, K.; Ruchała, M.A.; Cielecka-Piontek, J. Buccal Resveratrol Delivery System as a Potential New Concept for the Periodontitis Treatment. *Pharmaceutics* **2021**, *13*, 417. [CrossRef]
170. Paczkowska-Walendowska, M.; Szymańska, E.; Winnicka, K.; Szwajgier, D.; Baranowska-Wójcik, E.; Ruchała, M.A.; Simon, M.; Cielecka-Piontek, J. Cyclodextrin as functional carrier in development of mucoadhesive tablets containing Polygoni cuspidati extract with potential for dental applications. *Pharmaceutics* **2021**, *13*, 1916. [CrossRef] [PubMed]
171. European Medicines Agency (EMA). 2021 National Registers of Authorised Medicines. 2021. Available online: https://www.ema.europa.eu/en/medicines/national-registers-authorised-medicines (accessed on 23 October 2021).
172. Llambés, F. Relationship between diabetes and periodontal infection. *World J. Diabetes* **2015**, *6*, 927. [CrossRef] [PubMed]
173. Schmitt, A.; Carra, M.C.; Boutouyrie, P.; Bouchard, P. Periodontitis and arterial stiffness: A systematic review and meta-analysis. *J. Clin. Periodontol.* **2015**, *42*, 977–987. [CrossRef]
174. Sen, S.; Giamberardino, L.D.; Moss, K.; Morelli, T.; Rosamond, W.D.; Gottesman, R.F.; Beck, J.; Offenbacher, S. Periodontal Disease, Regular Dental Care Use, and Incident Ischemic Stroke. *Stroke* **2018**, *49*, 355–362. [CrossRef] [PubMed]
175. Salinas-Peña, E.A.; Mendoza-Rodríguez, M.; Velázquez-González, C.; Medina-Solis, C.E.; Pontigo-Loyola, A.P.; Márquez-Corona, M.; Rodríguez-Hernández, A.P.; Ximénez-Fyvie, L.A. Antibacterial properties in-vitro of Mexican serviceberry extracts against dental biofilm species. *J. Berry Res.* **2021**, *11*, 431–446. [CrossRef]

176. Liu, X.; Chen, Y.; Chen, X.; Su, J.; Huang, C. Enhanced efficacy of baicalin-loaded TPGS polymeric micelles against periodontitis. *Mater. Sci. Eng. C. Mater. Biol. Appl.* **2019**, *101*, 387–395. [CrossRef] [PubMed]
177. Mercado, N.; Bhatt, P.; Sutariya, V.; Florez, F.L.E.; Pathak, Y.V. Application of Nanoparticles in Treating Periodontitis: Preclinical and Clinical Overview. In *Surface Modification of Nanoparticles for Targeted Drug Delivery*; Springer: Berlin/Heidelberg, Germany, 2019; pp. 467–480. [CrossRef]
178. Wang, B.; Booij-Vrieling, H.E.; Bronkhorst, E.M.; Shao, J.; Kouwer, P.H.J.; Jansen, J.A.; Walboomers, X.F.; Yang, F. Antimicrobial and anti-inflammatory thermo-reversible hydrogel for periodontal delivery. *Acta Biomater.* **2020**, *116*, 259–267. [CrossRef]
179. Ji, Q.X.; Deng, J.; Xing, X.M.; Yuan, C.Q.; Yu, X.B.; Xu, Q.C.; Yue, J. Biocompatibility of a chitosan-based injectable thermosensitive hydrogel and its effects on dog periodontal tissue regeneration. *Carbohydr. Polym.* **2010**, *82*, 1153–1160. [CrossRef]
180. Ji, Q.X.; Zhao, Q.S.; Deng, J.; Lü, R. A novel injectable chlorhexidine thermosensitive hydrogel for periodontal application: Preparation, antibacterial activity and toxicity evaluation. *J. Mater. Sci. Mater. Med.* **2010**, *21*, 2435–2442. [CrossRef]

pharmaceutics

MDPI

Article

Biocomposite Hydrogels for the Treatment of Bacterial Infections: Physicochemical Characterization and In Vitro Assessment

Delia Mihaela Rata [1], Anca Niculina Cadinoiu [1,*], Marcel Popa [1], Leonard Ionut Atanase [1], Oana Maria Daraba [1], Irina Popescu [2], Laura Ecaterina Romila [1] and Daniela Luminita Ichim [1]

[1] Faculty of Medical Dentistry, Apollonia University of Iasi, 700511 Iasi, Romania; delia.rata@univapollonia.ro (D.M.R.); marpopa@tuiasi.ro (M.P.); leonard.atanase@univapollonia.ro (L.I.A.); oana.daraba@univapollonia.ro (O.M.D.); laura.romila@univapollonia.ro (L.E.R.); daniela.ichim@univapollonia.ro (D.L.I.)

[2] Petru Poni Institute of Macromolecular Chemistry, Aleea Grigore Ghica Voda 41A, 700487 Iasi, Romania; ipopescu@icmpp.ro

* Correspondence: jancaniculina@yahoo.com or anca.n.cadinoiu@univapollonia.ro

Abstract: Hydrogels based on natural and synthetic polymers and inorganic nanoparticles proved to be a viable strategy in the fight against some Gram-positive and Gram-negative bacteria. Additionally, numerous studies have demonstrated the advantages of using ZnO nanoparticles in medicine due to their high antibacterial efficacy and relatively low cost. Consequently, the purpose of our study was to incorporate ZnO nanoparticles into chitosan/poly (vinyl alcohol)-based hydrogels in order to obtain a biocomposite with antimicrobial properties. These biocomposite hydrogels, prepared by a double crosslinking (covalent and ionic) were characterized from a structural, morphological, swelling degree, and mechanical point of view. FTIR spectroscopy demonstrated both the apparition of new imine and acetal bonds due to covalent crosslinking and the presence of the sulfate group following ionic crosslinking. The morphology, swelling degree, and mechanical properties of the obtained hydrogels were influenced by both the degree of covalent crosslinking and the amount of ZnO nanoparticles incorporated. In vitro cytotoxicity assessment showed that hydrogels without ZnONPs are non-cytotoxic while the biocomposite hydrogels are weak (with 3% ZnONPs) or moderately (with 4 and 5% ZnONPs) cytotoxic. Compared to nanoparticle-free hydrogels, the biocomposite hydrogels show significant antimicrobial activity against *S. aureus*, *E. coli*, and *K. pneumonia*.

Keywords: chitosan; poly (vinyl alcohol); ZnO nanoparticles; hydrogels; antimicrobial activities

1. Introduction

Bacterial infectious diseases represent a severe threat to human health and according to data from the World Health Organization (WHO), millions of patients around the world suffer from these bacterial infections each year [1]. The low effectiveness of antimicrobial agents on antibiotic-resistant bacterial infections can lead to increased mortality rates and prolonged hospitalization, but also to economic losses. Antibiotic abuse or incorrect use can lead to changes in the ability of microorganisms to resist antibacterial agents, leading to a decrease in their therapeutic efficacy [2,3]. Therefore, an increased number of researchers are involved in finding a viable strategy for the treatment of antibiotic-resistant bacterial infections by trying to explore new strategies without antibiotics or based on the administration of lower doses of drugs [1,4]. The development of new drug delivery systems that can extend the half-life, improve the bioavailability, optimize the pharmacokinetics, and decrease the dosing frequency of drugs may be a great solution to these issues [5]. Among various drug delivery systems, hydrogels can be a promising alternative due to their multiple advantages: nontoxicity, biocompatibility, biodegradability, mechanical properties, bioresorption ability, increased degree of flexibility, good permeability, easy loading, and controlled drug release but also due to their relatively low cost [6]. Hydrogels are three-dimensional hydrophilic polymer networks with a high capacity to absorb aqueous solutions or biological fluids, which have a degree of flexibility similar to natural tissue and that can generally be used for a wide range of bio-applications, such as drug delivery, artificial muscle manufacturing, enzyme and cell immobilization, contact lenses, biosensor membranes, artificial skin materials, and artificial heart liners [7,8]. Different types of hydrogels with potential antibacterial properties in the treatment of bacterial infectious (osteomyelitis [9], implant infections [10], skin wounds infectious [11], dental infectious, ocular infections [12]) were already reported in the literature.

Among all types of hydrogels, hydrogels from natural polymers (chitosan, dextran, gelatin, and hyaluronic acid) have great potential in medicine because of their biocompatibility, biodegradability, mechanical properties, bioresorption ability, and relatively low cost [13]. On the other hand, the synthetic polymers present a high chemical purity, reproducible synthesis process, controllable degradation rate, very good mechanical properties, and an important advantage to achieve a sustained release of the therapeutic agents over a longer period of time in comparison with the natural ones [14,15]. Hydrogels based on chitosan (CS) and poly (vinyl alcohol) (PVA) containing zinc oxide nanoparticles (ZnONPs) have already been studied in the literature by the freezing-thawing cycle [16,17] or solution casting method [18], but to the best of our knowledge, there are no studies on the obtaining of this type of biocomposite hydrogel using the double crosslinking method. In this context, the novelty of our study consists in the development of a new antibacterial biocomposite hydrogel based on CS and PVA containing ZnONPs using the double crosslinking method in order to reduce the quantity of the potential toxic covalent crosslinking agents.

Chitosan is a biocompatible, biodegradable, and non-toxic polysaccharide known to have antimicrobial properties and very commonly used to obtain various controlled release systems (hydrogels, nanoparticles, films, gels) with applications in medicine [19]. The main disadvantage of CS-based hydrogels is that they have low mechanical properties that restrict their practical application. However, in order to overcome this drawback, it is recommended to mix CS with synthetic polymers, such as PVA, which has good mechanical strength and is generally considered as a safe and biocompatible, non-toxic, and non-carcinogenic material that can be crosslinked by various physical and chemical methods to obtain hydrogels [20,21]. In our study, we used two crosslinkers for the preparation of hydrogels, the covalent one, glutaraldehyde (GA), and the ionic one, magnesium sulfate ($MgSO_4$). In our previous studies, it was demonstrated that this double crosslinking is a viable strategy for the preparation of hydrogels in the form of films or micro/nanoparticles [22–26]. An important advantage of this strategy is that the amount of covalent crosslinker, which can be potentially toxic for biomedical applications, is reduced but it still allows hydrogels with good mechanical properties to be obtained. Moreover, it is well known that the incorporation of inorganic antibacterial materials into hydrogels leads to the development of complex systems with superior mechanical strength, increases adhesion to surfaces (skin, soft tissues), and enhances antibacterial properties [27,28]. ZnONPs have been shown to possess great antibacterial activities on broad-spectrum pathogenic bacteria (*Staphylococcus aureus, Bacillus subtilis, Escherichia coli, Salmonella enteritidis, Salmonella typhimurium, Pseudomonas fluorescens, Pseudomonas aeruginosa*) [29–31]. The production of reactive oxygen species, loss of cellular integrity as a result of contact between ZnONPs and the cell wall, internalization of ZnONPs, release of Zn^{2+} ions, and photoconductivity are the most important antimicrobial mechanisms responsible for the destruction of microorganisms [32].

The urgent need to combat bacterial resistance to antibiotics has led to an increase in the research of alternative ways to control pathogenic attacks. Therefore, biocomposite hydrogels were obtained in this study and their physicochemical, morphological, and mechanical characterization and the in vitro cytotoxicity and antimicrobial activity against a variety of microorganisms that included Gram-positive bacteria (*Staphylococcus aureus*) and Gram-negative bacteria (*Escherichia coli, Klebsiella pneumoniae,* and *Pseudomonas aeruginosa*) were evaluated.

2. Materials and Methods

2.1. Materials

Medium molecular weight CS, degree of deacetylation~75% (molecular weight: 100,000–300,000 g/mol), and PVA, 13,000–23,000 g/mol, 87–89% degree of hydrolysis were purchased from Acros Organics BVBA (Geel, Belgium). Glutaraldehyde (GA) 50% in aqueous solution, anhydrous ≥98.0% magnesium sulfate ($MgSO_4$), and glycerine ≥99.7% were purchased from VWR International. Zinc oxide nanopowder (ZnO, 99+%, 10–30 nm) was obtained from US Research Nanomaterials, Inc. (Houston, TX, USA). Human dermal fibroblasts cell line (HDFa), Dulbeco's modified Eagle's Medium (DMEM), 10% fetal bovine serum (FBS), antibiotics (streptomycin/penicillin), non-essential amino acids, phosphate-buffered saline (PBS), and trypsin-EDTA, necessary for in vitro cytotoxicity assay, were purchased from Thermo Fisher Scientific (Waltham, MA, USA). 3-(4,5-Dimethyl-2-thia zolyl)-2,5-diphenyl-2H-tetrazolium bromide was obtained from Merck Millipore (Darmstadt, Germany). Freeze-dried stains (*Staphylococcus aureus*-ATCC 25923, *Escherichia coli*-ATCC 11775, *Klebsiella pneumonia*-ATCC BAA–1705, and *Pseudomonas aeruginosa*-ATCC 10145) were purchased from ATCC (Manassas, VA, USA). Chapman agar (mannitol salt agar) was purchased from Oxoid (Hampshire, United Kingdom) and MacConkey agar from G & M Procter Ltd. (Perth, UK). All other chemicals used were of analytical grade.

2.2. The Preparation of Biocomposite Hydrogels

Double crosslinked hydrogels were obtained according to the experimental program provided in Table 1.

Table 1. The experimental design for double crosslinked hydrogels preparation.

Sample	CS/PVA Ratio (mg/mg)	GA/Free NH_2 and OH Ratio (mol/mol)	$MgSO_4$ / Free NH_2 (mol/mol)	ZnO Nanoparticles in Relation to the Total Amount of Polymers (%)
C2P2.5	50/50	1/20	1/20	0
C2P2.10		1/10		
C2P2.15		3/20		
C2P2.20		1/5		
C2P4.5	50/100	1/20		
C2P4.10		1/10		0
C2P4.10.Z3				3
C2P4.10.Z4				4
C2P4.10.Z5				5
C2P4.15		3/20		0
C2P4.15.Z3				3
C2P4.15.Z4				4
C2P4.15.Z5				5
C2P4.20		1/5		0

Appropriate amounts of CS and PVA were dissolved separately, CS in 25 mL of lactic acid solution 1% (*w/v*) and PVA in 25 mL of ultrapure water. After their complete dissolution, the two solutions were mixed. The specific amount of GA required to crosslink the free amine of CS and hydroxyl groups of PVA was added dropwise under vigorous stirring over the mixed polymer solution. After 30 min, the $MgSO_4$ solution was added, stirring for a further 15 min. In order to increase the elasticity of hydrogels, 500 mg of glycerine were added to the polymer solution and the mixture was to stirred for another 15 min. After that, the obtained mixture was carefully transferred into a square silicone mold with a side of 6.5 cm. The samples were dried in the oven at 37 °C. After purification (washing with ultrapure water), the hydrogels were stored in the desiccator, at a constant humidity of 65%.

To impart antimicrobial properties to hydrogels, ZnO nanoparticles were added to their composition. Specific amounts of inorganic nanoparticles (Table 1) were first added to the PVA solution and allowed to stir overnight, at room temperature and in the dark. The next day, CS was added to this mixture and the same steps, as described above, were followed in order to prepare biocomposite hydrogels.

2.3. Characterization of Biocomposite Hydrogels

2.3.1. The Structural Characteristics

The structural characteristics of the obtained hydrogels, the ionic and covalent new bonds, and also the presence of ZnONPs were analyzed using a Bruker FT-IR Spectrometer (VERTEX 70) equipped with a DLaTGS detector in the ATR operating mode (ZnSe crystal: 600–4000 cm^{-1}).

2.3.2. The Morphological Characteristics

The morphological characteristics of the hydrogels were determined by scanning electron microscopy (SEM) using a Quanta 200 Scanning Electron Microscope (FEI Company, Bruno, Czech Republic). The hydrogels were analyzed in the dry state. This analysis gave us information on the porosity and homogeneity of hydrogels.

The shape and size of ZnONPs were assessed using a HITACHI-HT7700 Transmission Electron Microscope (Hitachi High-Technologies Corporation, Tokyo, Japan).

The elemental composition of the hydrogels and the presence of ZnONPs were demonstrated using a Verios G4 UC Scanning Electron Microscope from Thermo Scientific (Bruno, Czech Republic) equipped with an Octane Elect Super SDD detector (an energy-dispersive X-ray spectrometer from EDAX (Ametek, Mahwah, NJ, USA).

To provide electrical conductivity and to prevent charge buildup during exposure to the electron beam, the samples were coated with 6 nm platinum using a Leica EM ACE200 Sputter coater.

2.3.3. The Swelling Behavior

The swelling behavior of the obtained hydrogels was evaluated, using the gravimetric method, at 37 °C, in phosphate buffer solution (PBS) at pH 7.4, which mimics biological fluids [22]. The dried square-shaped hydrogel samples (1 cm side) were first weighed (W_d, mg) and then immersed in 20 mL of PBS. At predetermined times, the samples were easily removed from PBS and placed on a filter paper. The excess PBS on the surface of the hydrogels was easily absorbed with another filter paper and the swollen samples were weighed (W_{st}). This procedure was repeated until equilibrium was reached. The swelling degree at time t ($Q_t(\%)$) was calculated using the following equation:

$$Q_t(\%) = \frac{W_{st} - W_d}{W_d} \times 100 \tag{1}$$

2.3.4. Tensile Mechanical Measurements

Tensile mechanical measurements were performed using a Brookfield CT3 Texture Analyzer (Texture Brookfield Engineering Laboratories Inc., Middleboro, MA, USA) with a 50N load cell, at room temperature. The Roller Cam Accessory (TA-RCA) was used for this test. The films (dry hydrogels kept in an atmosphere with 65% humidity) were cut into rectangles (length 55 mm, width 10 mm) and were caught between two clamps positioned at a distance of 30 mm. The thickness of the films was determined by taking the average of the thickness, measured at five different places with a digital Caliper and found to be around 0.5 mm in all the samples. The speed of the tensile measurements was fixed at 0.5 mm/s. The stress (σ, N/m^2) was calculated from the load divided by the cross-sectional area of the undeformed sample, and the strain (ε) was determined as the clamp displacement divided by the initial distance between the two clamps. The ultimate strain and stress (from the point of failure) were determined, and the elastic modulus values were calculated from the slope of the linear climbing tract of the stress–strain plot within the fixed strain region 0.1–2%.

2.3.5. The In Vitro Cytotoxic Effects

The in vitro cytotoxic effects of the hydrogels were assessed using human dermal fibroblasts, adult, and cell line (HDFa). HDFa cells were cultured and prepared for testing as described in our previous paper [33]. The sterilized hydrogel samples were cut with a biopsy punch to obtain discs 4 mm in diameter and incubated with the HDFa cells. After the incubation time (24 and 48 h), the cell viability was determined using the MTT assay. All procedures were performed in a laminar flow hood (Lamil Plus 13, Kartusalan Metally Oy). Each hydrogel sample was tested in triplicate in order to get the average percentage and the standard error.

2.3.6. The Antimicrobial Activities

The antimicrobial activities of the obtained hydrogels without and with ZnONPs were evaluated using the disk diffusion assay against 4 reference strains: *S. aureus* (Gram-positive bacteria), *E. coli*, *K. pneumoniae*, and *P. aeruginosa* (Gram-negative bacteria). Commercially available antimicrobial disks with Chapman agar for Gram-positive bacteria, and MacConkey agar for Gram-negative bacteria were used for these tests. A suspension of microorganisms (0.5 McFarland density) was inoculated on a Petri dish with the substrate. The sterilized hydrogel discs (4 mm diameter) were firstly hydrated in sterile 0.9% saline solution and then placed on the agar plate and incubated for 24 h at 37 °C. The antimicrobial activities were evaluated by measuring the diameters of the inhibition zones (din, mm) according to the Kirby–Bauer method [34]. The experiments were repeated three times. In order to evaluate the significance of their antibacterial activity against *S. aureus*, *E. coli*, and *K. pneumonia* strains, as compared to samples without ZnONPs, the one-way ANOVA statistical test was processed.

3. Results and Discussions

3.1. Preparation of Biocomposite Hydrogels

Double crosslinked biocomposite hydrogels, based on CS and PVA, with antimicrobial properties were obtained by incorporating of ZnONPs into the polymer solution before crosslinking. The preparation method and the structure of the obtained biocomposite hydrogels is represented schematically in Figure 1. The reaction with GA takes place both at the amino groups from the CS chain, with the formation of imine bonds, and at the hydroxyl groups from PVA with the formation of new acetal-type bonds. Ionic crosslinking was performed using $MgSO_4$. Amino-type functional groups in the form of the ammonium ion (belonging to CS) with the sulfate groups of the ionic crosslinking agent participate in this reaction.

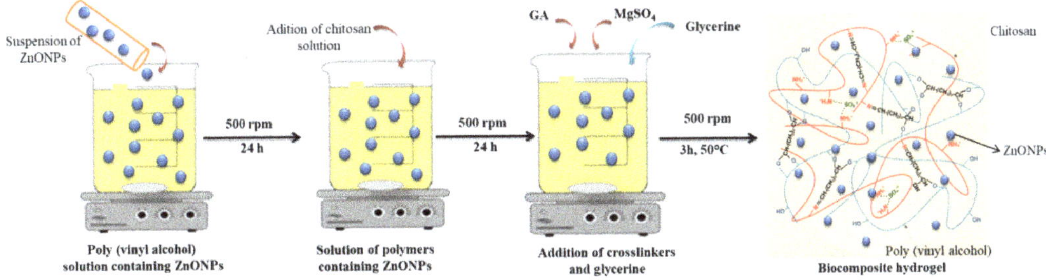

Figure 1. The schematic representation of the preparation method and the structure of the obtained biocomposite hydrogel.

The optimization of the obtaining conditions and the addition of the plasticizer (glycerine) led to the obtaining of hydrogels with improved mechanical properties. Additionally, the addition of ZnO nanoparticles to the polymer mixture yielded new biocomposite hydrogels with antimicrobial properties.

3.2. *Characterization of Biocomposite Hydrogels*

3.2.1. The Structural Characteristics

The structural characteristics of the obtained hydrogels were determined using FT-IR spectroscopy. The FT-IR spectra of hydrogel samples with different degrees of covalent crosslinking in the absence and in the presence of ZnONPs are shown in Figures 2 and 3, respectively.

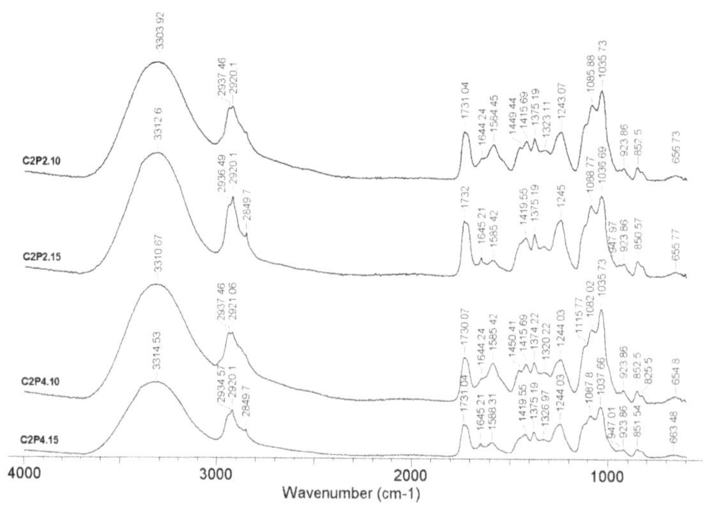

Figure 2. FT-IR spectra for hydrogel samples with different amounts of GA.

The obtained results confirmed, on the one hand, the reaction between the amino functional groups from CS and the hydroxyl functional groups of PVA with the covalent crosslinker (GA), and on the other hand, the reaction between the amino functional groups from CS with the $SO4^{2-}$ groups of the ionic crosslinker ($MgSO_4$). The characteristic peaks located around 3314, 3310, 3312, and 3303 cm^{-1}, respectively, that were found in the spectrum from Figure 2, and peaks located between 3305 and 3323 cm^{-1} in the spectrum from Figure 3 can be attributed to vibration stretching of secondary amine groups (NH) from CS or hydroxyl groups (OH) from PVA that are involved in the intra- and intermolecular bonds of hydrogen [35,36]. The absorption bands at about 2849 cm^{-1} and those between 2920 and 2940 cm^{-1} (Figures 2 and 3) correspond to the C-H bonds in the alkyl groups. In the case of samples with a higher amount of GA (C2P2.15, C2P4.15, C2P4.15.Z3, and C2P4.15.Z5), there is an intensification of the peak observed at 2849 cm^{-1}. Absorption bands around 1732 cm^{-1}, which are found in the spectra of all analyzed samples, can be associated with the stretching vibration of the —C=O group in PVA [23]. The peaks noticed for all the analyzed samples at 1629–1646 cm^{-1} can be attributed to the imine group (C=N), which is proof of the covalent crosslinking between GA and CS. The intense peak located at approximately 1590 cm^{-1} indicates the presence of amine groups (NH_2) specific for a polysaccharide structure [24]. The absorption bands at 1035–1040 and 1082–1088 cm^{-1} can be attributed to the acetal groups (C-O-C) formed by crosslinking of PVA with GA. The presence of ZnONPs into hydrogels was confirmed by the appearance of the peak at approximately 774 cm^{-1}, which corresponds to the ZnO stretching vibration [37]. The absorption peaks ranging from approximately 654 to 667 cm^{-1} that are found in the spectra of all analyzed samples can be attributed to the $SO4^{2-}$ group, which evidenced the ionic crosslinking between the functional groups of CS with $MgSO_4$ [25].

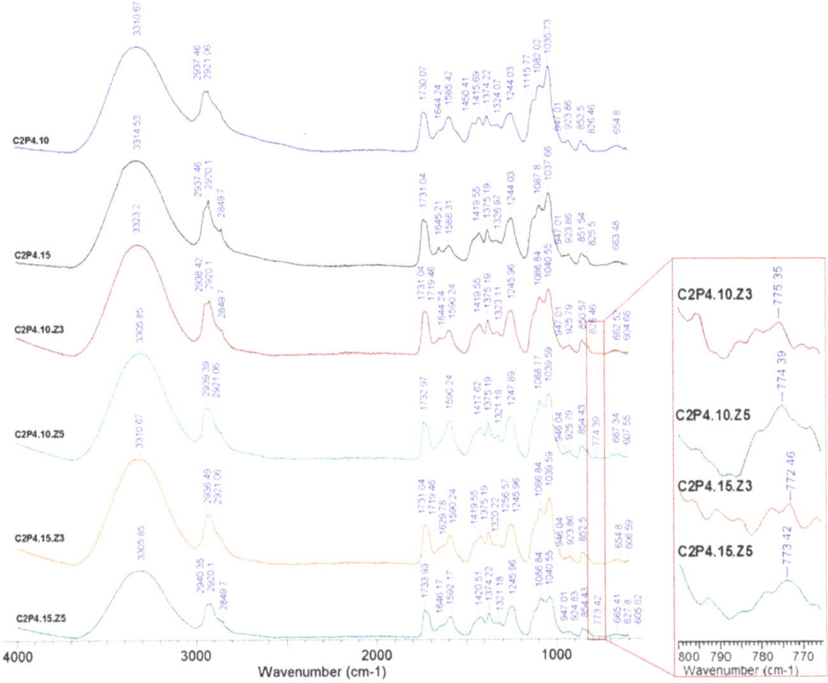

Figure 3. FT-IR spectra for hydrogel samples in the absence and in the presence of ZnONPs and different degrees of crosslinking.

3.2.2. Morphological Characteristics

ZnO Nanoparticles

Figure 4A shows the transmission electron microscopy (TEM) image of the ZnONPs. The TEM micrographs revealed that these nanoparticles, with a hexagonal shape, have dimensions of approximately 30 nm in a dry state.

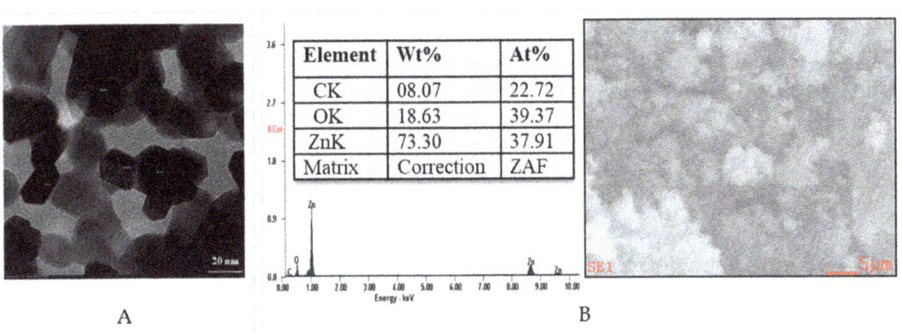

Figure 4. Transmission electron microscopy (TEM) (**A**) and EDAX spectrum (**B**) for ZnONPs.

EDAX elemental analysis of ZnONPs (Figure 4B) showed a large amount of Zn of approximately 73%. The amount of O present in the EDAX spectrum was approximately of 18%.

Biocomposite Hydrogels

The cross-section morphology of the hydrogel samples, with and without ZnONPs, was studied by scanning electron microscopy (SEM). The micrographs presented in Figure 5 reveal a fibrillar structure probably due to the presence of polysaccharide (CS) in the sample composition. It was also observed that the hydrogels become smoother when the amount of covalent crosslinker increases. Increasing the amount of covalent crosslinker increases the crosslinking density because an increased number of imine and acetal groups are formed, and this makes the structure of hydrogels smoother. This is in accordance with the swelling degree results, which reveal that as the amount of covalent crosslinker increases, the degree of swelling decreases. The hydrogels have a homogeneous structure, due to the strong covalent and ionic crosslinking, and the samples without ZnONPs did not show porosity. On the contrary, it has been observed that the hydrogels obtained in the presence of ZnONPs shows porosity, with the pores being small and irregular in size. At this point, it is very important to mention that with the increase of the amount of ZnONPs, a higher porosity was observed. Following the purifications, a part of the ZnONPs was removed and the place occupied by these nanoparticles inside the hydrogel matrix led to the appearance of pores and implicitly to the increase of the porosity. This effect is also illustrated by the results of the EDAX tests.

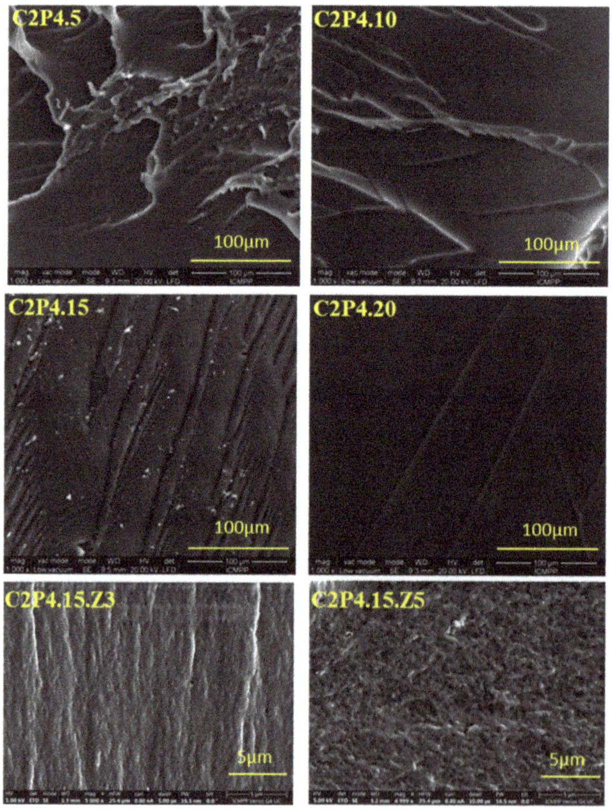

Figure 5. SEM micrographs for hydrogel samples in the cross-section with and without ZnONPs.

EDAX Analysis for Biocomposite Hydrogels

EDAX elemental analysis revealed the presence of ZnONPs in the analyzed biocomposite hydrogels (Figure 6).

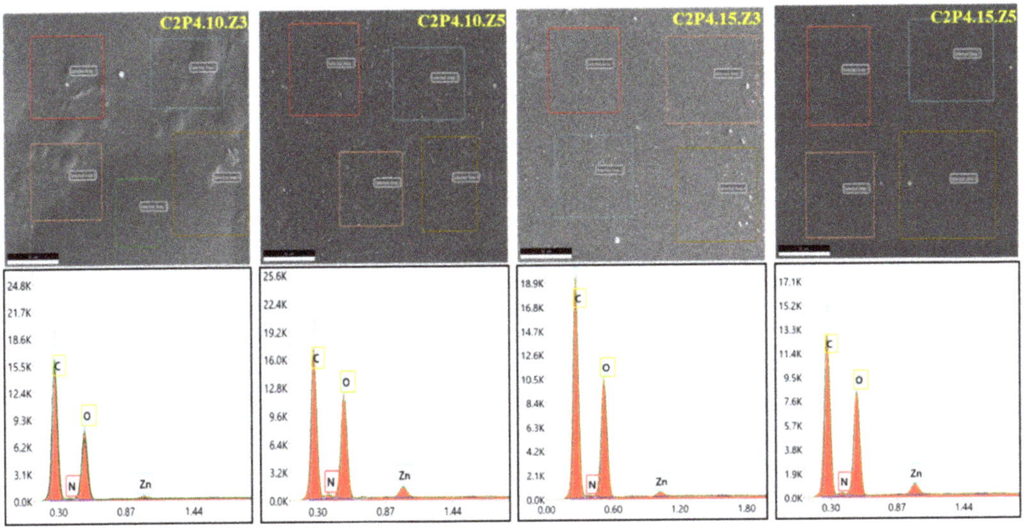

Figure 6. EDAX spectrum for samples C2P4.10.Z3, C2P4.10.Z5, C2P4.15.Z3, and C2P4.15.Z5.

The EDAX images also illustrated the presence of large amounts of C and O. The final content of ZnONPs in the tested hydrogels is presented in Table 2.

Following the calculations performed based on the EDAX analysis (Figure 4B, Table 2 and Table S1), it was found that the remaining percentages of ZnNPs in biocomposite hydrogels were approximately 70% in C2P4.10.Z3 hydrogel, approximately 93% in C2P4.10.Z5 hydrogel, approximately 71% in C2P4.15.Z3 hydrogel, and approximately 95% in C2P4.15.Z5 hydrogel. The obtained results demonstrate that the percentage of ZnONPs in the system is higher when the covalent crosslinking density increases (the amount of GA increases). After the purification step, a small amount of ZnONPs was removed from the biocomposite hydrogel network, thus explaining the decrease in the percentage of ZnONPs in the hydrogel compared to the initial one. The elemental composition of all tested samples is given in the Supplementary Materials (Table S1).

Table 2. Content of ZnONPs in tested hydrogels.

Sample	C2P4.10	C2P4.10.Z3	C2P4.10.Z5	C2P4.15	C2P4.15.Z3	C2P4.15.Z5
Initial content of ZnONPs (*w*/*w* %)	0	3	5	0	3	5
Final content of ZnONPs (*w*/*w* %)	0	2.09	4.63	0	2.14	4.74

3.2.3. Swelling Behavior

The swelling behavior of the obtained hydrogels was investigated in slightly alkaline medium (PBS at pH 7.4) in order to evaluate their potential to be used locally in biomedical applications for the treatment of bacterial infections. The swelling of the hydrogels was caused by the penetration of the aqueous solution, by diffusion, into the meshes of the polymer network. The amount of aqueous solution penetrating inside the hydrogel depends on the membrane elasticity, which can influence the diffusion rate. The swelling behavior of hydrogels was influenced by both the amount of covalent crosslinking density (Figure 7A) and the amount of ZnONPs (Figure 7B). As expected, it was found that the increase of the amount of covalent crosslinker (increase of the ratios between GA and the functional groups of the two polymers) led to a decrease of the swelling degree

(Figure 7A). This behavior can be attributed to the increase of the crosslinking density. Additionally, increasing the amount of ZnONPs in hydrogels also leads to a decrease of the swelling degree (Figure 7B). This behavior can be attributed to the decrease of the space inside the hydrogel network due to the increase of the amount of ZnONPs present in the system. Therefore, the amount of water that penetrated the hydrogel network was reduced.

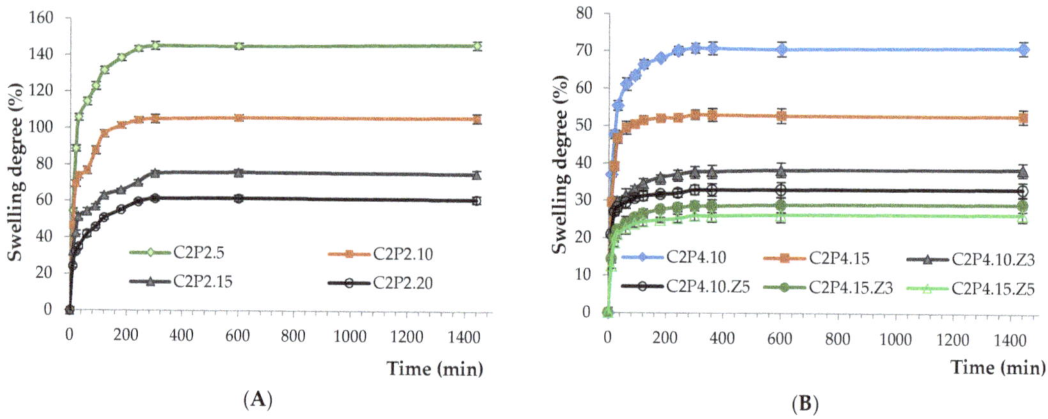

Figure 7. Swelling degree kinetics for 24 h in PBS (pH = 7.4) of (**A**) C2P2 hydrogels with different ratios between GA and the functional groups of the two polymers; (**B**) C2P4 hydrogels obtained in the absence and in the presence of ZnONPs.

It should also be mentioned that after 6 h, all the samples reached the maximum swelling degree. In fact, between 6 and 24 h, there were no changes in the samples' swelling degree as can be seen in Figure 7.

3.2.4. Tensile Mechanical Measurements

The mechanical properties of hydrogels are one of most important factors for establishing their future biomedical applications. Tensile properties of the dry hydrogels/films were investigated, and the obtained stress–strain curves are presented in Figure 8A,B. The tensile strength and modulus of elasticity are plotted in Figure 8C,D. Glycerine, a typical plasticizer for hydrophilic films, was used to obtain films that are not brittle, with an elongation at break between 20 and 50%. The tensile strength ranged from 2.3 to 5.6 MPa and the elastic modulus from 6 to 30 MPa, but the composition and the crosslinking density influenced the mechanical properties of the hydrogel network.

The increase of the crosslinking leads to a decrease in the free volume between the chains, with a consequent stiffening of the network. Therefore, the crosslinking determines an enhancement in the tensile strength and Young's modulus and a decrease of the elongation at break [38–41]. This behavior was also observed for the double crosslinked PVA/CS hydrogels: the modulus of elasticity and tensile strength increased in the C2P2 series (Figure 8C) and C2P4 series (Figure 8D) with the increase of the GA content. The increase of the PVA content in the C2P2 to C2P4 samples determined an increase of the tensile strength and Young's modulus, which is due to the intermolecular interaction between PVA and CS chains through hydrogen bonds [42,43] but can also be explained by the higher chemical crosslinking of the C2P4 networks compared to C2P2 networks.

When ZnONPs were incorporated into the hydrogels, the tensile strength and elongation at break decreased compared with the sample without metal oxide NPs (Figure 8B,D). A higher effect was observed for the C2P4.15.Z3 sample, but with a further increase of the ZnONPs concentration, the mechanical properties were improved. It is known that the addition of metallic oxide nanofillers in CS films generally improved their mechanical properties [44,45], but in this case, ZnONPs probably interfere with the ionic crosslinking. However, the hydrogels with ZnONPs had a lower modulus of elasticity, compared with C2P4.15. The higher elasticity of the network is attributed to the fact that ZnONPs weakened the intermolecular hydrogen bonds between CS chains or between CS and PVA [45,46].

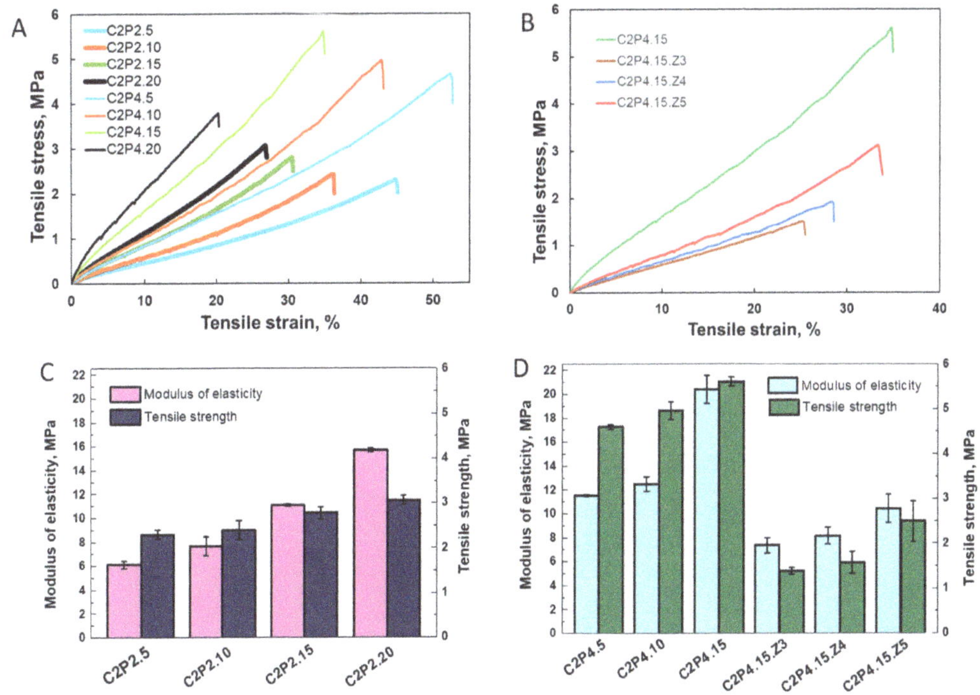

Figure 8. Tensile stress–strain curves for C2P2 and C2P4 series (**A**) and for C2P4.15 sample with and without ZnONPs (**B**). Variation of the modulus of elasticity and tensile strength for the C2P2 series (**C**) and C2P4 series (**D**).

After analyzing the obtained results, two of the samples with the good mechanical properties, C2P4.10 and C2P4.15, were selected for subsequent cytotoxicity and antimicrobial testing.

3.2.5. The In Vitro Cytotoxic Effects

Cytotoxicity tests provide necessary information in order to evaluate the biocompatibility of materials that will be used in biomedical applications. For in vitro cytotoxicity assessment of hydrogels (in the absence and in the presence of ZnONPs), human dermal fibroblast cells (HDFa) were used as model cells. The effect of hydrogels on the viability of fibroblasts after incubation periods of 24 and 48 h was evaluated using the MTT colorimetric assay and the obtained results are shown in Figure 9. The MTT assay was adapted according to the protocol described by Mosmann [47] and calculates cell viability based on mitochondrial function by reducing 3-(4,5-dimethylthiazol-2-yl)-2,5-diphenyltetrazolium bromide (MTT) to a colored insoluble formazane salt.

As can be seen from Figure 9, the viability of cells in contact with hydrogels without ZnONPs was around 89% after 24 h and around 84% after 48 h, which proves that they are non-cytotoxic [48,49]. The addition of an increasing amount of ZnONPs in hydrogels has led to a decrease in cell viability. In the case of hydrogels with ZnONPs, cell viability after 24 h of incubation was between 62% and 78%, demonstrating that these samples show a weak cytotoxicity. After 48 h of incubation, there was a decrease in cell viability, with only the samples with 3% ZnONPs remaining in the weak cytotoxic range, while the others decreased in the moderate cytotoxic range. A small decrease in viability was observed when the amount of covalent crosslinker (GA), used to obtain hydrogels, was increased.

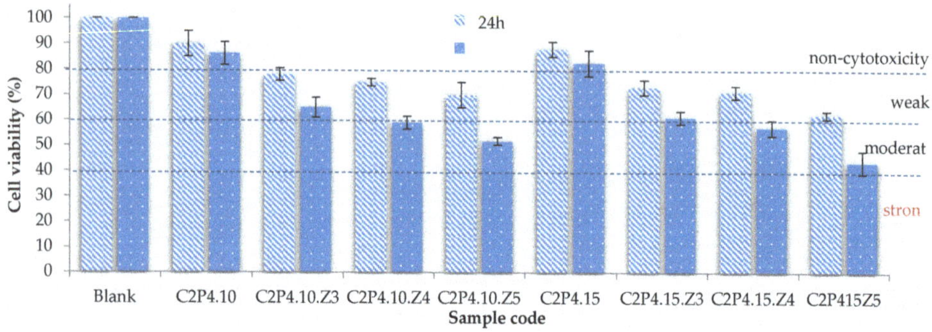

Figure 9. In vitro cell viability of C2P4.10 and C2P4.15 hydrogels obtained in the absence and in the presence of ZnONPs at 24 and 48 h after incubation.

There are studies in the literature that have evaluated the cytotoxic effect of ZnONPs and found that they are toxic to several cell types [50,51]. Therefore, their incorporation into a hydrogel matrix is a solution to reducing their cytotoxic effect, as it was concluded by different studies where hydrogels with ZnONPs based on bacterial cellulose (CS) [52] or sodium alginate [53] were tested.

3.2.6. The Antimicrobial Activities

Agar disk-diffusion testing is the method used in some clinical microbiology laboratories for routine antimicrobial susceptibility testing [54]. This method was also used in this study in order to evaluate the antimicrobial activity for the hydrogels based on CS and PVA, obtained in the absence and in the presence of ZnONPs. Typical reference strains of *S. aureus* (Gram-positive bacteria) and *E. coli*, *K. pneumoniae*, and *P. aeruginosa* (Gram-negative bacteria) were used for these tests (Figure 10). Four samples were placed on each agar plate: the hydrogels without ZnONPs (A-C2P4.10 or B-C2P4.15) and three hydrogel samples with different concentrations of ZnONPs (3% ZnONPs for E-C2P4.10.Z3 or G-C2P4.15.Z3; 4% ZnONPs for F-C2P4.10.Z4 or H-C2P4.15.Z4; 5% ZnONPs for C-C2P4.10.Z5 or D-C2P4.15.Z5).

The diameter of the inhibition zone (d_{iz}), obtained after 24 h of incubation at 37 °C, was transcribed as "−" when no antimicrobial effect was noticed, "+" when the d_{iz} is <15 mm, and "++" when the d_{iz} is between 15 and 25 mm [55]. These results are presented in Figure 10.

As CS and ZnONPs have both antibacterial properties, a synergistic effect might be expected. Antibacterial mechanisms for both materials have been previously reported in the literature [56–59]. Consequently, hydrogel samples containing ZnONPs show significant antibacterial properties, while samples without ZnONPs have no detectable antimicrobial activity (Figure 11). As the concentration of ZnONPs increases, the zone of inhibition also increases. Therefore, the highest antimicrobial activity was observed for samples C2P4.10.Z5 (C) and C2P4.15.Z5 (D) (hydrogels with 5% ZnONPs) against *S. aureus*. These hydrogel samples also showed an antimicrobial effect against *E. coli* and *K. pneumonia*. The diameter of the inhibition zone was smaller in the case of samples C2P4.10.Z4 (F) and C2P4.15.Z4 (H) (hydrogels with 4% ZnONPs), with these having an antimicrobial effect against *E. coli*. Sample H, which was obtained with a higher amount of covalent crosslinker, also has antimicrobial activity against *S. aureus*. However, it appeared that no hydrogel sample showed antimicrobial activity on the *P. aeruginosa* strain. Similar results were observed in studies conducted by other researchers [29].

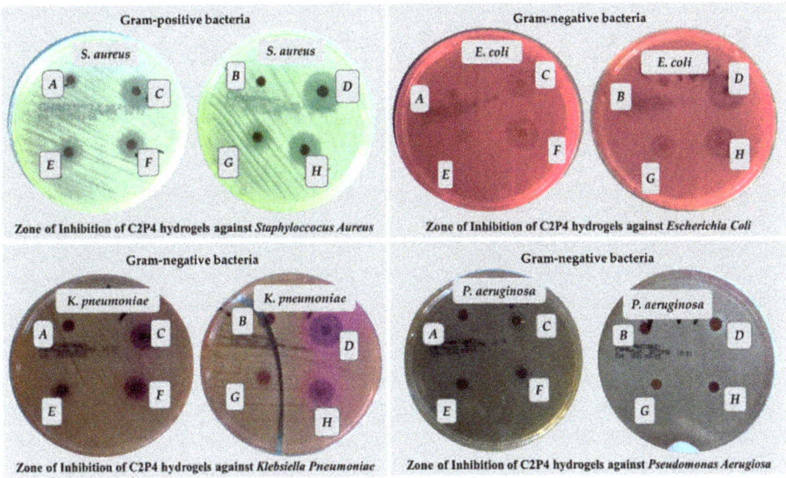

Figure 10. Antibacterial activity of C2P4.10 (A), C2P4.15 (B), C2P4.10.Z5 (C), C2P4.15.Z5 (D), C2P4.10.Z3 (E), C2P4.10.Z4 (F), C2P4.15.Z3 (G), and C2P4.15.Z4 (H) hydrogels against Gram-positive (*S. aureus*) and Gram-negative bacteria (*E. coli, K. pneumonia,* and *P. aeruginosa*).

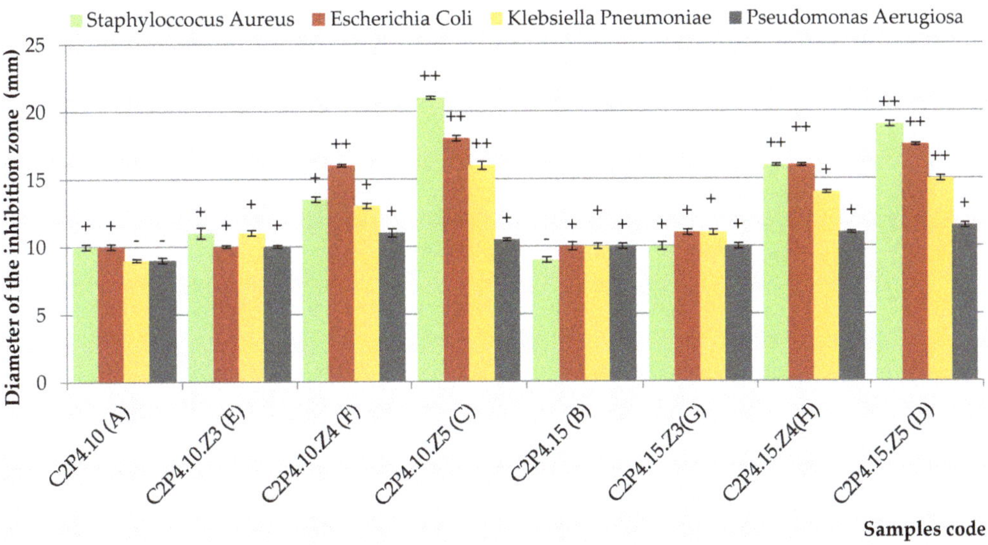

Figure 11. Antimicrobial effect of C2P4-10 (A), C2P4-15 (B), C2P4-10.Z5 (C), C2P4-15.Z5 (D), C2P4-10.Z3 (E), C2P4-10.Z4 (F), C2P4-15.Z3 (G), and C2P4-15.Z4 (H) hydrogels against *S. aureus, E. coli, P. aeruginosa,* and *K. pneumoniae*.

The inhibition zone diameter data from the antibacterial tests for the samples containing 4 and 5% ZnONPs were processed through the one-way ANOVA statistical test in order to evaluate the significance of their antibacterial activity against *S. aureus* (Gram-positive bacteria), *E. coli*, and *K. pneumonia* (Gram-negative bacteria) strains, as compared to the samples without ZnONPs (Table S2).

Since, for all bacteria strains, the significance level is lower than 0.05, it can be stated that the samples containing 4 and 5% ZnONPs have a significant influence on the inhibition of the growth of tested bacteria.

4. Conclusions

Within this study, a new attempt was made in order to obtain double crosslinked hydrogels, based on CS and PVA with embedded ZnONPs, for the potential treatment of bacterial infections. Both the successful crosslinking of the two polymers and the presence of ZnONPs in the biocomposite hydrogel structure were demonstrated by FT-IR spectroscopy. The morphology, swelling, mechanical properties, cytotoxicity, and antimicrobial effect were influenced by the covalent crosslinking degree and the amount of ZnONPs incorporated into hydrogels. The porosity of the hydrogels increased with an increase of the concentration of ZnONPs. On the contrary, the swelling degree decreased when the concentration of the ZnONPs increased. A similar trend was noticed for the cytotoxicity tests after 24 and 48 h. In vitro cytotoxicity tests have shown that only hydrogels without ZnONPs are non-toxic to fibroblast cells. The biocomposite hydrogels proved to be weak (with 3% ZnONPs) and moderately (with 4 and 5% ZnONPs) cytotoxic.

The increased concentration of GA led to increased mechanical properties, but when ZnONPs were incorporated into the hydrogels, the tensile strength and elongation at break decreased compared with the sample without metal oxide NPs. The higher effect was observed for the C2P4.15.Z3 sample, but with a further increase of the ZnONPs concentration, the mechanical properties were improved. However, the hydrogels with ZnONPs had a lower modulus of elasticity, compared with the sample without metal oxide NPs.

In contrast to samples without ZnONPs, hydrogel samples containing 4 and 5% ZnONPs showed significant antibacterial activity against *S. aureus* (Gram-positive bacteria), *E. coli*, and *K. pneumonia* (Gram-negative bacteria) strains with an inhibition zone between 15 and 20 mm. The ANOVA test showed that these results are significant (significant level < 0.05).

The obtained results are promising and therefore the obtained biocomposite hydrogels will be further investigated by in vivo analyses.

Supplementary Materials: The following are available online at https://www.mdpi.com/article/10.3390/pharmaceutics13122079/s1. Table S1: EDAX elemental analysis for biocomposite hydrogels; Table S2: ANOVA statistical test for evaluation of antibacterial activity.

Author Contributions: Conceptualization, D.M.R. and A.N.C.; methodology, D.M.R., A.N.C., L.I.A., O.M.D., I.P., L.E.R. and D.L.I.; software, D.M.R. and A.N.C.; investigation, D.M.R., A.N.C., O.M.D., I.P., L.E.R. and D.L.I.; writing—original draft preparation, D.M.R. and A.N.C.; visualization, D.M.R., A.N.C., L.I.A., O.M.D., I.P., L.E.R. and D.L.I.; supervision, M.P.; project administration, D.M.R. and A.N.C. All authors have read and agreed to the published version of the manuscript.

Funding: This work was supported by a grant of the Romanian Ministry of Education and Research, CCCDI-UEFISCDI, project number PN-III-P2-2.1-PED-2019-1780, within PNCDI III.

Institutional Review Board Statement: Not applicable.

Informed Consent Statement: Not applicable.

Data Availability Statement: The data presented in this study are available on request from the corresponding author.

Conflicts of Interest: The authors declare no conflict of interest.

References

1. Cao, Z.; Yue, L.; Li, Z.; Tan, L.; Liu, X.; Li, C.; Zheng, Y.; Cui, Z.; Yeung, K.W.K.; Liang, Y.; et al. Antibacterial Hybrid Hydrogels. *Macromol. Biosci.* **2021**, *21*, 2000252. [CrossRef]
2. Loo, Y.Y.; Rukayadi, Y.; Nor-Khaizura, M.A.R.; Kuan, C.H.; Chieng, B.W.; Nishibuchi, M.; Radu, S. In Vitro Antimicrobial Activity of Green Synthesized Silver Nanoparticles Against Selected Gram-negative Foodborne Pathogens. *Front. Microbiol.* **2018**, *9*, 1555. [CrossRef] [PubMed]
3. Sánchez-López, E.; Gomes, D.; Esteruelas, G.; Bonilla, L.; Lopez-Machado, A.L.; Galindo, R.; Cano, A.; Espina, M.; Ettcheto, M.; Camins, A.; et al. Metal-Based Nanoparticles as Antimicrobial Agents: An Overview. *Nanomaterials* **2020**, *10*, 292. [CrossRef] [PubMed]
4. Liang, Y.; Chen, B.; Li, M.; He, J.; Yin, Z.; Guo, B. Injectable Antimicrobial Conductive Hydrogels for Wound Disinfection and Infectious Wound Healing. *Biomacromolecules* **2020**, *21*, 1841–1852. [CrossRef] [PubMed]
5. Kim, H.S.; Sun, X.; Lee, J.-H.; Kim, H.-W.; Fu, X.; Leong, K.W. Advanced drug delivery systems and artificial skin grafts for skin wound healing. *Adv. Drug Deliv. Rev.* **2019**, *146*, 209–239. [CrossRef]
6. Radvar, E.; Azevedo, H.S. Supramolecular Peptide/Polymer Hybrid Hydrogels for Biomedical Applications. *Macromol. Biosci.* **2019**, *19*, 1800221. [CrossRef] [PubMed]
7. Chai, Q.; Jiao, Y.; Yu, X. Hydrogels for Biomedical Applications: Their Characteristics and the Mechanisms behind Them. *Gels* **2017**, *3*, 6. [CrossRef]
8. Patil, J.S. Hydrogel System: An Approach for Drug Delivery Modulation. *Adv. Pharmacoepidemiol. Drug. Saf.* **2015**, *4*, e135.
9. Peng, K.T.; Chen, C.F.; Chu, I.M.; Li, Y.M.; Hsu, W.H.; Hsu, R.W.W.; Chang, P.J. Treatment of osteomyelitis with teicoplanin-encapsulated biodegradable thermosensitive hydrogel nanoparticles. *Biomaterials* **2010**, *31*, 5227–5236. [CrossRef] [PubMed]
10. Lai, P.L.; Hong, D.W.; Ku, K.L.; Lai, Z.T.; Chu, I.M. Novel thermosensitive hydrogels based on methoxy polyethylene glycol-co-poly(lactic acidco-aromatic anhydride) for cefazolin delivery. *Nanomedicine* **2014**, *10*, 553–560. [CrossRef] [PubMed]
11. Lequeux, I.; Ducasse, E.; Jouenne, T.; Thebault, P. Addition of antimicrobial properties to hyaluronic acid by grafting of antimicrobial peptide. *Eur. Polym. J.* **2014**, *51*, 182–190. [CrossRef]
12. Glisoni, R.J.; Garcia-Fernandez, M.J.; Pino, M. Beta-Cyclodextrin hydrogels for the ocular release of antibacterial thiosemicarbazones. *Carbohydr. Polym.* **2013**, *93*, 449–457. [CrossRef] [PubMed]
13. Summa, M.; Russo, D.; Penna, I.; Margaroli, N.; Bayer, I.S.; Bandiera, T.; Athanassiou, A.; Bertorelli, R. A biocompatible sodium alginate/povidone iodine film enhances wound healing. *Eur. J. Pharm. Biopharm.* **2018**, *122*, 17–24. [CrossRef] [PubMed]
14. Unal, S.; Oktar, F.N.; Mahirogullari, M.; Gunduz, O. Bone structure and formation: A new perspective. In *Bioceramics*; Elsevier: Amsterdam, The Netherlands, 2021; pp. 175–193.
15. Raţă, D.M.; Chailan, J.-F.; Peptu, C.A.; Costuleanu, M.; Popa, M. Chitosan: Poly(N-vinylpyrrolidone-alt-itaconic anhydride) nanocapsules—a promising alternative for the lung cancer treatment. *J. Nanopart. Res.* **2015**, *17*, 316. [CrossRef]
16. Baghaie, S.; Khorasani, M.T.; Zarrabi, A.; Moshtaghian, J. Wound healing properties of PVA/starch/chitosan hydrogel membranes with nano Zinc oxide as antibacterial wound dressing material. *J. Biomater. Sci. Polym. Ed.* **2017**, *28*, 2220–2241. [CrossRef]
17. Khorasani, M.T.; Joorabloo, A.; Moghaddam, A.; Shamsi, H.; Mansoori Moghadam, Z. Incorporation of ZnO nanoparticles into heparinised polyvinyl alcohol/chitosan hydrogels for wound dressing application. *Int. J. Biol. Macromol.* **2018**, *114*, 1203–1215. [CrossRef]
18. Ahmad Yusof, N.A.; Mat Zain, N. The Effect of ZnOP nanoparticles on the physical, mechanical, and antibacterial properties of Chitosan/Gelatin hydrogel films. *J. Teknol.* **2019**, *81*, 21–26.
19. Raţă, D.M.; Cadinoiu, A.N.; Daraba, O.M.; Mihalache, C.; Mihalache, G.; Burlui, V. Metronidazole-loaded chitosan/ poly(maleic anhydridealt- vinyl acetate) hydrogels for dental treatments. *Int. J. Med. Dent.* **2016**, *20*, 92–97.
20. Ahmadian, Y.; Bakravi, A.; Hashemi, H.; Namazi, H. Synthesis of polyvinyl alcohol/CuO nanocomposite hydrogel and its application as drug delivery agent. *Polym. Bull.* **2019**, *76*, 1967–1983. [CrossRef]
21. Muppalaneni, S.; Omidian, H. Polyvinyl alcohol in medicine and pharmacy: A perspective. *J. Dev. Drugs* **2013**, *2*, 112–117. [CrossRef]
22. Ciobanu, B.C.; Cadinoiu, A.N.; Popa, M.; Desbrieres, J.; Peptu, C.A. Chitosan/Poly (Vinyl Alcohol) Hydrogels for Entrapment of Drug Loaded Liposomes. *Cellul. Chem. Technol.* **2014**, *48*, 485–494.
23. Jătariu, A.N.; Danu, M.; Peptu, C.A.; Ioanid, G.; Ibanescu, C.; Popa, M. Ionically and Covalently Cross-Linked Hydrogels Based on Gelatin and Chitosan. *Soft. Mater.* **2013**, *11*, 45–54. [CrossRef]
24. Jătariu, A.N.; Holban, M.N.; Peptu, C.A.; Sava, A.; Costuleanu, M.; Popa, M. Double crosslinked interpenetrated network in nanoparticle form for drug targeting—Preparation, characterization and biodistribution studies. *Int. J. Pharm.* **2012**, *436*, 66–74. [CrossRef]
25. Cadinoiu, A.N.; Peptu, C.A.; Fache, B.; Chailan, J.F.; Popa, M. Microparticulated systems based on chitosan and poly (vinyl alcohol) with potential ophthalmic applications. *J. Microencapsul.* **2015**, *32*, 381–389. [CrossRef] [PubMed]
26. Jatariu, A.N.; Popa, M.; Curteanu, S.; Peptu, C.A. Covalent and ionic co-cross-linking—An original way to prepare chitosan—Gelatin hydrogels for biomedical applications. *J. Biomed. Mater. Res. Part A* **2011**, *98*, 342–350. [CrossRef]
27. Yang, K.; Han, Q.; Chen, B.; Zheng, Y.; Zhang, K.; Li, Q.; Wang, J. Antimicrobial hydrogels: Promising materials for medical application. *Int. J. Nanomed.* **2018**, *13*, 2217–2263. [CrossRef]
28. Rafieian, S.; Mirzadeh, H.; Mahdavi, H.; Masoumi, M.E. A review on nanocomposite hydrogels and their biomedical applications. *Sci. Eng. Compos. Mater.* **2018**, *26*, 154–174. [CrossRef]
29. Villanueva, M.E.; Cuestas, M.L.; Pérez, C.J.; Dall′ Orto, V.C.; Copello, G.J. Smart release of antimicrobial ZnO nanoplates from a pH-responsive keratin hydrogel. *J. Colloid Interface Sci.* **2019**, *536*, 372–380. [CrossRef]
30. Reddy, L.S.; Nisha, M.M.; Joice, M.; Shilpa, P.N. Antimicrobial activity of zinc oxide (ZnO) nanoparticle against *Klebsiella pneumoniae.* *Pharm. Biol.* **2014**, *52*, 1388–1397. [CrossRef]

31. Daghdari, S.G.; Ahmadi, M.; Saei, H.D.; Tehrani, A.A. The effect of ZnO nanoparticles on infectious wounds. *Nanomed. J.* **2017**, *4*, 232–236.
32. Lallo da Silva, B.; Abuçafy, M.P.; Berbel Manaia, E.; Oshiro Junior, J.A.; Chiari-Andréo, B.G.; Pietro, R.C.R.; Chiavacci, L.A. Relationship between Structure and Antimicrobial Activity of Zinc Oxide Nanoparticles: An Overview. *Int. J. Nanomed.* **2019**, *14*, 9395–9410. [CrossRef] [PubMed]
33. Cadinoiu, A.N.; Rata, D.M.; Atanase, L.I.; Daraba, O.M.; Gherghel, D.; Vochita, G.; Popa, M. Aptamer-Functionalized Liposomes as a Potential Treatment for Basal Cell Carcinoma. *Polymers* **2019**, *11*, 1515. [CrossRef] [PubMed]
34. Hudzicki, J. Kirby-Bauer Disk Diffusion Susceptibility Test Protocol. American Society for Microbiology. 2009. Available online: https://asm.org/Protocols/Kirby-Bauer-Disk-Diffusion-Susceptibility-Test-Pro (accessed on 25 September 2021).
35. Hoffmann, B.; Seitz, D.; Mencke, A.; Kokott, A.; Ziegler, G. Glutaraldehyde and oxidised dextran as crosslinker reagents for chitosan-based scaffolds for cartilage tissue engineering. *J. Mater. Sci Mater. Med.* **2009**, *20*, 1495–1503. [CrossRef] [PubMed]
36. Dellali, K.Z.; Rata, D.M.; Popa, M.; Djennad, M.; Ouagued, A.; Gherghel, D. Antitumoral Drug: Loaded Hybrid Nanocapsules Based on Chitosan with Potential Eects in Breast Cancer Therapy. *Int. J. Mol. Sci.* **2020**, *21*, 5659. [CrossRef]
37. Singh, J.; Kaur, S.; Kaur, G.; Basu, S.; Rawat, M. Biogenic ZnO nanoparticles: A study of blueshift of optical band gap and photocatalytic degradation of reactive yellow 186 dye under direct sunlight. *Green. Process. Synth.* **2019**, *8*, 272–280. [CrossRef]
38. Gadhave, R.V.; Mahanwar, P.A.; Gadekar, P.T. Effect of glutaraldehyde on thermal and mechanical properties of starch and polyvinyl alcohol blends. *Des. Monomers Polym.* **2019**, *1*, 164–170. [CrossRef] [PubMed]
39. Bozdoğan, A.; Aksakal, B.; Denktaş, C.; Salt, Y. Prestretching effect and recovery process of polyvinyl alcohol film crosslinked with tartaric acid. *J. Appl. Polym. Sci.* **2020**, *137*, e49421. [CrossRef]
40. Shi, Y.; Xiong, D.; Li, J.; Wang, K.; Wang, N. In-situ repair of graphene defects and enhancement of its reinforcement effect in polyvinyl alcohol hydrogels. *RSC Adv.* **2017**, *7*, 1045–1055. [CrossRef]
41. Rivero, S.; Garcia, M.A.; Pinotti, A. Crosslinking capacity of tannic acid in plasticized chitosan films. *Carbohydr. Polym.* **2020**, *82*, 270–276.
42. Abraham, A.; Soloman, P.A.; Rejini, V.O. Preparation of chitosan-polyvinyl alcohol blends and studies on thermal and mechanical properties. *Proc. Techno.* **2016**, *24*, 741–748. [CrossRef]
43. Chetouani, A.; Elkolli, M.; Bounekhel, M.; Benachour, D. Chitosan/oxidized pectin/PVA blend film: Mechanical and biological properties. *Polym. Bull.* **2017**, *74*, 4297–4310. [CrossRef]
44. Abdeen, Z.I.; El Faragy, A.F.; Negm, N.A. Nanocomposite framework of chitosan/polyvinyl alcohol/ZnO: Preparation, characterization, swelling and antimicrobial evaluation. *J. Molec. Liq.* **2018**, *250*, 335–343. [CrossRef]
45. Pires, J.; de Paula, C.D.; Souza, V.G.L.; Fernando, A.L.; Coelhoso, I. Understanding the barrier and mechanical behavior of different nanofillers in chitosan films for food packaging. *Polymers* **2021**, *13*, 721. [CrossRef] [PubMed]
46. Li, L.H.; Deng, J.C.; Deng, H.R.; Liu, Z.L.; Xin, L. Synthesis and characterization of chitosan/ZnO nanoparticle composite membranes. *Carbohydr. Res.* **2010**, *345*, 994–998. [CrossRef] [PubMed]
47. Mosmann, T. Rapid colorimetric assay for cellular growth and survival: Application to proliferation and cytotoxicity assays. *J. Immunol. Methods* **1983**, *65*, 55–63. [CrossRef]
48. López-García, J.; Lehocký, M.; Humpolíček, P.; Sáha, P. HaCaT Keratinocytes Response on Antimicrobial Atelocollagen Substrates: Extent of Cytotoxicity, Cell Viability and Proliferation. *J. Funct. Biomater.* **2014**, *5*, 43–57. [CrossRef]
49. International Standard ISO 10993-5: 2009, *Biological Evaluation of Medical Devices—Part 5: Tests for In Vitro Cytotoxicity*; The International Organization for Standardization: Geneva, Switzerland, 2009.
50. Król, A.; Pomastowski, P.; Rafińska, K.; Railean-Plugaru, V.; Buszewski, B. Zinc oxide nanoparticles: Synthesis, antiseptic activity and toxicity mechanism. *Adv. Colloid Interface Sci.* **2017**, *249*, 37–52. [CrossRef]
51. Pinho, A.R.; Martins, F.; Costa, M.E.V.; Senos, A.M.R.; da Cruz e Silva, O.A.B.; Pereira, M.d.L.; Rebelo, S. In Vitro Cytotoxicity Effects of Zinc Oxide Nanoparticles on Spermatogonia Cells. *Cells* **2020**, *9*, 1081. [CrossRef] [PubMed]
52. Kai, J.; Xuesong, Z. Preparation, Characterization, and Cytotoxicity Evaluation of Zinc Oxide–Bacterial Cellulose–Chitosan Hydrogels for Antibacterial Dressing. *Macromol. Chem. Phys.* **2020**, *221*, 2000257. [CrossRef]
53. Wang, T.; Wang, J.; Wang, R.; Yuan, P.; Fan, Z.; Yang, S. Preparation and property of ZnO/sodium alginate bi-layered hydrogel films as novel wound dressings. *New J. Chem.* **2019**, *43*, 8684–8693. [CrossRef]
54. Balouiri, M.; Sadiki, M.; Ibnsouda, S.K. Methods for in vitro evaluating antimicrobial activity: A review. *J. Pharm. Anal.* **2016**, *6*, 71–79. [CrossRef] [PubMed]
55. Suflet, D.M.; Popescu, I.; Pelin, I.M.; Ichim, D.L.; Daraba, O.M.; Constantin, M.; Fundueanu, G. Dual Cross-Linked Chitosan/PVA Hydrogels Containing Silver Nanoparticles with Antimicrobial Properties. *Pharmaceutics* **2021**, *13*, 1461. [CrossRef]
56. Yilmaz Atay, H. Antibacterial Activity of Chitosan-Based Systems. *Funct. Chitosan.* **2020**, *Mar 6*, 457–489.
57. Li, J.; Zhuang, S. Antibacterial activity of chitosan and its derivatives and their interaction mechanism with bacteria: Current state and perspectives. *Eur. Polym. J.* **2020**, *138*, 109984. [CrossRef]
58. Sirelkhatim, A.; Mahmud, S.; Seeni, A.; Kaus, N.H.M.; Ann, L.C.; Bakhori, S.K.M.; Hasan, H.; Mohamad, D. Review on Zinc Oxide Nanoparticles: Antibacterial Activity and Toxicity Mechanism. *Nanomicro Lett.* **2015**, *7*, 219–242. [CrossRef] [PubMed]
59. Abebe, B.; Zereffa, E.A.; Tadesse, A.; Murthy, H.C.A. A Review on Enhancing the Antibacterial Activity of ZnO: Mechanisms and Microscopic Investigation. *Nanoscale Res. Lett.* **2020**, *15*, 190. [CrossRef] [PubMed]

pharmaceutics

MDPI

Article

Preparation of Topical Itraconazole with Enhanced Skin/Nail Permeability and In Vivo Antifungal Efficacy against Superficial Mycosis

Laxman Subedi [1,†], Seung-Yub Song [1,†], Saurav Kumar Jha [1], Sung-Ho Lee [1], Rudra Pangeni [2], Kyo-Tan Koo [3], Beum Joon Kim [4], Seung-Sik Cho [1,2,*] and Jin Woo Park [1,2,*]

1 Department of Biomedicine, Health & Life Convergence Sciences, BK21 Four, Biomedical and Healthcare Research Institute, Mokpo National University, Jeonnam 58554, Korea; laxmansubedi789@gmail.com (L.S.); tgb1007@naver.com (S.-Y.S.); saurav.balhi@gmail.com (S.K.J.); tjdgh0730@naver.com (S.-H.L.)
2 Department of Pharmacy, College of Pharmacy and Natural Medicine Research Institute, Mokpo National University, Jeonnam 58554, Korea; capriconpangeni@gmail.com
3 BioBelief Co., Ltd., Seoul 05841, Korea; kyotankoo@gmail.com
4 Department of Dermatology, Chung-Ang University College of Medicine, Seoul 06973, Korea; beomjoon74@gmail.com
* Correspondence: sjason1@naver.com (S.-S.C.); jwpark@mokpo.ac.kr (J.W.P.); Tel.: +82-61-450-2687 (S.-S.C.); +82-61-450-2704 (J.W.P.)
† These authors contributed equally to this work.

Abstract: In this study, a stable and highly skin-permeable topical delivery system for itraconazole (ITZ) was designed to provide effective treatment against superficial mycosis. Herein, ITZ was incorporated into a solution composed of ethanol, benzyl alcohol, hydrochloric acid, Transcutol P, and cyclomethicone as a delivery vehicle, solubilizer, protonating agent, permeation enhancer, and spreading agent, respectively. At 72 h, the optimal topical ITZ formulation (ITZ–TF#11) exhibited 135% enhanced skin permeability, which led to increases in drug deposition in the stratum corneum, epidermis, and dermis of 479%, 739%, and 2024%, respectively, compared with the deposition of 1% ITZ in ethanol (control). Moreover, on day 7, ITZ–TF#11 demonstrated 2.09- and 2.30-fold enhanced nail flux and drug deposition, compared with the control. At a dose of 40 mg/kg/day, ITZ–TF#11 showed 323% greater lesion recovery, a 165% lower mean erythema severity score, and a 37% lower mean logarithm of viable fungal cells in skin in the treated area, compared with mice that received oral ITZ at the same dose. Overall, the findings imply that ITZ–TF#11 is a superior alternative to oral ITZ for treatment of superficial mycosis.

Keywords: itraconazole; topical solution; superficial mycosis; skin penetration; nail infiltration; antifungal activity

1. Introduction

Superficial mycosis is a fungal disease that is mostly confined to outer layers of skin, nail, or hair but may invade deeper tissue. This disease is common worldwide [1]. Moreover, mycosis has been reported to negatively impact host psychological and psychosocial factors and reduce quality of life [2]. Various types of wide-spectrum antifungal azoles (e.g., efinaconazole [EFN], fluconazole, ketoconazole, voriconazole, and itraconazole [ITZ]) are available for use in mycosis treatment. Importantly, azole groups interfere with the synthesis of ergosterol, an essential fungal membrane component, by blocking 14-α demethylase (a cytochrome P-450 enzyme). In addition, the selectivity of azole groups toward the cytochrome P-450 enzymes of fungi, rather than mammals, enables highly specific antifungal action with comparatively minimal side effects, relative to other antifungal treatments [3]. Among several azole-based antifungal agents, 2-butan-2-yl-4-[4-[4-[4-[[2-(2,4-dichlorophenyl)-2-(1,2,4-triazol-1-ylmethyl)-1,3-dioxolan-4-yl] methoxy] phenyl]-1-piperazinyl] phenyl]-1,2,4-triazol-3-one (ITZ) has strong keratinophilic and lipophilic properties and has therefore emerged as an effective broad-spectrum antifungal treatment for superficial mycosis [4]. Furthermore, ITZ has been used successfully in the treatment and prevention of *Aspergillus* infections, with fewer toxic side effects than amphotericin B, indicating a better therapeutic index [5].

Clinical treatment of cutaneous mycosis with ITZ generally involves oral administration. Despite considerable absolute oral bioavailability (approximately 70%), an absence of blood vessels in the epidermis and stratum corneum can limit the availability of ITZ at the desired site of action [6,7]. In addition, because of its large molecular size (705.64 Da) and enhanced plasma-binding affinity, ITZ may not effectively diffuse from the dermis sublayer to other layers of the epidermis and stratum corneum [8]. Thus, greater concentrations of ITZ have been achieved with typical oral doses of 200–400 mg/day [9,10]. However, the resulting high systemic concentration of ITZ has been reported to cause nephrotoxicity and hepatic damage, as well as various reproductive complications [11]. Therefore, topical administration of ITZ can be advantageous for site-specific delivery, with fewer side effects than oral administration.

Approaches such as creams, ointments, pastes, and liposomal formulations have been used to improve topical delivery of ITZ [12]. Colloidal carrier systems incorporating ITZ can moderately enhance its solubility. However, the semi-solid nature of colloidal carriers minimizes drug release and reduces formulation spreading on the skin surface. It also leads to limited permeability across the skin membrane [13]. Furthermore, residues of particulate carriers on skin after the application of the above formulations may reduce patient compliance. Therefore, incorporation of volatile vehicles may be a suitable alternative approach in topical ITZ solution development. Additionally, the volatile vehicle used in topical formulations has been reported to follow the metamorphism phenomenon, which leads to drug supersaturation within the vehicle system at the applied area over time, which can increase both effective and immediate drug delivery [14]. The incorporation of a permeation enhancer, which interacts with lipid and protein constituents present in the outermost layer of skin, can increase drug penetration and partitioning [15]. Therefore, the generation of a stable ITZ solution with a volatile vehicle, effective permeation enhancer, and wetting agent to promote spreading may enhance the clinical effects of ITZ on superficial mycosis.

This study sought to design a topical delivery system for ITZ with enhanced solubility, robust skin penetration, and longer storage stability and to then compare the in vivo antifungal efficacy of this new system with a commercially available oral formulation. Herein, to prepare a topical ITZ formulation (ITZ–TF), ethanol and benzyl alcohol were used as a volatile vehicle and solubilizing agent, respectively. To enhance the solubility and stability of ITZ in solution, hydrochloric acid was used as a protonating agent and sodium hydroxide was incorporated to maintain the solution pH within the physiological range of skin (pH 4.5–5.5). This pH range was presumed to aid in maintenance of stability after application to skin and to promote effective partitioning of ITZ across the skin with fewer side effects [14]. Furthermore, Transcutol P and cyclomethicone were used as a permeation enhancer and spreading agent, respectively. The permeabilities of the new ITZ formulation across the artificial membrane, human skin, and human nail were examined. Based on the permeability results, the optimal formulation was subjected to comparative in vivo antifungal efficacy testing at a dose identical to that of commercially available oral ITZ, using a BALB/c mouse model of dermal *Candida albicans* infection.

2. Materials and Methods

2.1. Materials

ITZ was obtained from SVR Drugs Pvt Ltd. (Telangana, India). Polyoxyl-40 hydrogenated castor oil (Kolliphor RH 40), polyoxyl-35 castor oil (Cremephor EL), cyclomethicone, butylated hydroxyanisole (BHA), ethylenediamine tetraacetic acid (EDTA), and benzyl alcohol were purchased from Sigma-Aldrich Inc. (St. Louis, MO, USA). Diethylene glycol monoethyl ether (Transcutol P) was obtained from Gattefossé (Saint Priest, France). Ethanol, sodium hydroxide, and hydrochloric acid were purchased from Reagents Duksan (Gyeonggi-do, Korea). Sabouraud dextrose agar was obtained from BD Biosciences (San Jose, CA, USA). ITZ oral capsule (reference control for in vivo study) was provided by Janssen Korea Ltd. (Seoul, Korea). EFN (internal standard for liquid chromatographic mass spectroscopy [LC/MS] analysis) was obtained from Cipla Limited (Mumbai, India). Reagents such as acetonitrile and formic acid, which were required to prepare the mobile phase, were obtained from Sigma-Aldrich Inc. All chemical reagents used in this study were of analytical grade. All solvents were of high-performance liquid chromatography (HPLC) grade.

2.2. Animals

Female BALB/c mice (6 weeks old, 20 g) were purchased from OrientBio (Gwangju, Korea). All mice were housed in a controlled environment in terms of temperature (23 ± 2 °C), relative humidity (55 ± 10%), and light (12/12-h light/dark cycle) throughout the duration of the experiment. In addition, animals were provided free access to a standard laboratory diet (Bestle Purina PetCare Research, St. Louis, MO, USA) and ion-sterilized water. Ethical approval for this study protocol was obtained from the Institutional Animal Care and Use Committee of Mokpo National University (Jeonnam, Korea: approval no. MNU-IACUC-2020-020). All experiments were performed in accordance with the National Institutes of Health guidelines for the Care and Use of Laboratory Animals and the guidelines of the institutional animal care and use committee.

2.3. Solubility of ITZ in Topical Formulations

The major impediment to skin permeation of ITZ is its low aqueous solubility [16]. Therefore, various solvents (e.g., ethanol, benzyl alcohol, Transcutol P, Cremephor EL, Capryol 90, polyethylene glycol 400, oleic acid, isopropyl myristate, tetraethylene glycol, and n-methyl-2-pyrrolidone) were selected and subjected to screening for their ability to enhance ITZ solubility. An excess amount of ITZ was initially added to each individual solvent and mixed using a vortex mixer at room temperature. After 48 h, samples were centrifuged at 10,000 rpm for 10 min and the supernatant was diluted with a mobile phase consisting of acetonitrile and water (pH adjusted to 4.0 with phosphoric acid) (80:20, *v*/*v*). The concentration of ITZ in each diluted sample was measured via HPLC at 263 nm. A Luna C18 column (250 × 4.6 mm, 5-µm particle size, 20-µL injection volume) was used with a mobile phase at a flow rate of 0.7 mL/min.

2.4. Preparation and In Vitro Artificial Membrane Permeability of Topical ITZ Formulation

To formulate 1% ITZ topical formulation (ITZ–TF; 1 g ITZ/100 g of solution), benzyl alcohol and ethanol were selected as a solubilizing agent and delivery vehicle, respectively. To provide additional stability, BHA was incorporated as an antioxidant and EDTA was incorporated as a chelating agent (Table 1). Briefly, 0.05 g of ITZ was weighed in a glass vial and dissolved by using 0.5, 1.0, and 1.5 g of benzyl alcohol, respectively. Then, 0.005 g of BHA and 0.05 g of aqueous solution of EDTA (0.025 mg/mL) were added and vortex mixing was performed. Hydrochloric acid was then added at concentrations of 0.121%, 0.242%, and 0.484% into the topical formulations for the protonation of ITZ, thereby enhancing its solubility and stability. Subsequently, 1.5 g of ethanol was incorporated and mixed for 1 min. To improve the permeability of ITZ across the skin and nails, various amounts of Transcutol P (10%, 15%, and 20%) and cyclomethicone (5.00% and 8.22%) were added to the formulations. Sodium hydroxide was added to adjust the pH to the range 4.5–5.5. Finally, the weight of the formulation was maintained at 100% by addition of ethanol. The effects of multiple concentrations of various excipients on the solubility and miscibility of ITZ, and on other components in the ITZ–TF system, were evaluated. Moreover, changes in drug concentration and physiochemical properties (e.g., color, precipitation, and phase separation) were examined weekly for >3 months.

The in vitro artificial membrane permeability of ITZ–TF was investigated using a synthetic membrane (Strat-M®; EMD Millipore, Temecula, CA, USA) and a Franz diffusion cell system (Labfine, Gyeonggi-do, Korea). In this analysis, each membrane was sandwiched between the donor and receptor compartments with a 0.785-cm^2 diffusion area. The receptor compartment was filled with 5 mL of phosphate-buffered saline (PBS; pH 7.4) containing 15% acetonitrile as a receptor-phase solution to maintain the sink condition [17]. The solution in the receptor cell was stirred using a magnetic stirrer at 600 rpm within a heating system to maintain the membrane surface temperature at 32 °C throughout the experiment. Afterwards, 50 μL of each formulation (636 μg/cm^2) was loaded into the donor compartment. Then, 500-μL aliquots of the receptor phase were withdrawn from the sampling port at pre-determined time intervals (0.5, 1, 1.5, 2, 3, 4, 5, and 6 h) and replaced with equivalent volumes of fresh receptor solution. Collected samples were filtered through a membrane filter (polyvinylidene fluoride [PVDF] with 0.45-μm pore size) and analyzed using an HPLC system at 263 nm as described above.

Table 1. Compositions of topical ITZ formulations.

Formulations (%, *w/w*)	ITZ	BHA	EDTA	Water	Benzyl Alcohol	HCl	1 M NaOH	Transcutol P	Cyclomethicone	Ethanol
ITZ in ethanol	1	0.1	0.0025	0.9975						97.900
ITZ–TF#1	1	0.1	0.0025	0.9975	10					87.900
ITZ–TF#2	1	0.1	0.0025	0.9975	20					77.900
ITZ–TF#3	1	0.1	0.0025	0.9975	30					67.900
ITZ–TF#4	1	0.1	0.0025	0.9975	30	0.121	1.12			66.659
ITZ–TF#5	1	0.1	0.0025	0.9975	30	0.242	2.13			65.528
ITZ–TF#6	1	0.1	0.0025	0.9975	30	0.484	4.12			63.296
ITZ–TF#7	1	0.1	0.0025	0.9975	30	0.484	3.64	10		53.776
ITZ–TF#8	1	0.1	0.0025	0.9975	30	0.484	3.48	15		48.936
ITZ–TF#9	1	0.1	0.0025	0.9975	30	0.484	3.46	20		46.044
ITZ–TF#10	1	0.1	0.0025	0.9975	30	0.484	4.64	15	5.00	47.160
ITZ–TF#11	1	0.1	0.0025	0.9975	30	0.484	4.68	15	8.22	39.156

2.5. In Vitro Permeation and Deposition of ITZ in Full-Thickness Human Skin

To compare the human skin permeabilities of ITZ in each vehicle, excised full-thickness human skin (HansBiomed Corp., Daejeon, Korea) was mounted (with the stratum corneum upward) over the receptor compartments of a Franz diffusion cell filled with 5 mL PBS (pH 7.4): acetonitrile 85:15 (v/v). The solubility of ITZ in this solution was 1.58 ± 0.10 mg/mL, thus 15.8-fold higher than the maximum concentration of ITZ (100 μg/mL) in the receptor compartment; the sink condition was met [17]. Prior to addition of test samples to donor cells, the receptor cells were allowed to equilibrate for at least 1 h and only skin samples exhibiting transepidermal water loss (TEWL) < 10 g/m^2/h and transepithelial electrical resistance (TEER) between 1–2 kΩ were used; this guaranteed the integrity of the epidermal membrane [18]. The receptor compartment was stirred at 600 rpm to maintain the skin surface temperature at 32 °C throughout the experiment. Then, 50 μL amounts of each 1% ITZ formulation (636 μg/cm^2) were loaded into the donor compartments at 24 h intervals. Five hundred-microliter aliquots of the receptor phase were withdrawn from each diffusion cell at 0, 0.5, 1, 1.5, 2, 3, 4, 5, 6, 8, 10, 24, 48, and 72 h and replaced with equivalent volumes of fresh receptor solution. All samples were passed through PVDF membrane filters (0.45-μm in pore diameter) and analyzed using the HPLC method described above.

To evaluate ITZ skin deposition after completion of the full-thickness human skin permeability study, the skin sections were rinsed five times with distilled water and gently wiped with tissue paper to remove residual formulations. The skin sections were weighed to measure initial weight. To isolate the stratum corneum, the skin was stripped 35 times with D-squame® (CuDerm, Dallas, TX, USA) [19]. The stratum corneum weights were calculated by subtracting the weights after stripping from the initial weights. Next, the epidermis and dermis were thermally separated via immersion in distilled water at 60 °C for 1 min. The epidermis was then carefully isolated from the dermis using a pair of forceps. The stripped tape containing stratum corneum, epidermis, and dermis was cut into small pieces and immersed in 5 mL of methanol and then extracted via vortex mixing for 24 h [13]. The resulting suspensions were filtered using membrane filters (0.45-μm pore size) and transferred to glass vials. Internal standard (100 μL of 5 μg/mL EFN) was added and the mixture was dried using a rotary evaporator. Thereafter, the dried residue was reconstituted in 200 μL of a 1:1 (*v*/*v*) mixture of methanol and mobile phase (acetonitrile:0.1% formic acid in water, 60:40 [*v*/*v*]) and subjected to LC/MS analysis. Finally, the concentration of ITZ in the reconstituted sample was determined using an Agilent 6120 quadruple LC/MS system with a Luna C18 column (250 × 4.6 mm, 5 μm) as a stationary phase. Aliquots (20 μL) from each sample were eluted in the mobile phase at a flow rate of 1 mL/min. ITZ and EFN were ionized using an electrospray ionization source in the positive-ion mode under the following conditions: capillary voltage of 4 kV, drying gas flow rate of 10.0 L/min, and drying temperature of 100 °C. Quantitative analyses of protonated molecular ions were performed at ($[M + H]^+ = 706.8$) and ($[M + H]^+ = 349.4$), for ITZ and EFN, respectively.

2.6. In Vitro Human Nail Permeability and Deposition of ITZ

To examine the infiltration of ITZ across the human nail, in vitro nail permeation and deposition analyses were performed using a Franz diffusion cell system. Human nails obtained from Science Care, Inc. (Phoenix, AZ, USA) were stored at −80 °C until use. One day before the experiment, the nails were transferred to room-temperature storage conditions. On the day of the experiment, nail plates were stored in normal saline to ensure sufficient hydration. They were then fixed between two Teflon nail-holder gaskets, which were mounted between the donor and receptor compartments. Receptor solution (5 mL; PBS [pH 7.4]:acetonitrile, 85:15, [*v*/*v*]) was then used to fill the receptor compartment. The solution in the receptor compartment was stirred at 600 rpm to maintain a nail surface temperature of 32 °C throughout the experiment. The respective 1% ITZ–TF with a once-daily dose (20 μL per dose) was then loaded into the donor compartment for 7 days. Aliquots (500 μL) of the receptor phase were withdrawn at intervals of 1, 2, 3, 4, 5, 6, and 7 days and replaced with equivalent volumes of fresh receptor solution. The withdrawn samples were filtered with a membrane filter (0.45-μm PVDF). Afterwards, 100-μL aliquots of filtrate samples were mixed with 100 μL of EFN drug solution (5 μg/mL, internal standard) in methanol and analyzed via LC/MS, as mentioned above.

To monitor ITZ deposition in the nail, diffusion cells were disassembled and nail plates were isolated. All nail plates were washed twice with 5 mL of water. Afterwards, the area effectively exposed to the drug formulation was cut into small pieces and dissolved in 1 M sodium hydroxide:methanol (1:1, *v*/*v*) solution. The nail suspensions were incubated at 60 °C for 4 h and then mixed with 1 mL of methanol and 1 mL of ethanol in conjunction with continuous vortex mixing [20]. After incubation had been completed, the nail suspension was centrifuged at 4000 rpm for 10 min to obtain clear supernatant, which was then mixed with 100 μL of internal standard (5 μg/mL EFN) in a glass vial. The mixture of supernatant and internal standard was evaporated using a rotatory evaporator (Genevac Ltd., Ipswich, UK). Finally, the dried residue was reconstituted in 200 μL of a 1:1 (*v*/*v*) mixture of methanol and mobile phase (acetonitrile:0.1% formic acid in water, 60:40, [*v*/*v*]) and analyzed via LC/MS, as described above.

2.7. In Vitro Antifungal Efficacy

The susceptibility of *C. albicans* to various ITZ samples was evaluated via the disc diffusion method. At first, a *C. albicans* (100 μL, 10^7 colony-forming unit, CFU) was loaded onto potato dextrose agar. Next, paper discs (8 mm, Toyo, Japan) soaked in ITZ sample (40 μL, 100 μg/mL concentration of ITZs) were arranged on a potato dextrose agar plate and then incubated for 24 h at 28 °C. Finally, zones of inhibition were calculated by measuring the mean diameter of the clear area around each ITZ paper disc.

The in vitro minimum inhibitory concentration (MIC, μg/mL) of ITZ was determined via serial dilution in potato dextrose medium. *C. albicans* (10^7 CFU/mL) was exposed to six different concentrations of ITZ (0, 3.125, 6.25, 12.5, 100, and 200 μg/mL) at 28 °C for 24 h. Then, MIC was determined visually and spectrophotometrically at 660 nm. After the MIC test, 100 μL of culture broth containing ITZs was loaded onto Sabouraud dextrose agar plates and incubated at 28 °C for 48 h. The in vitro fungicidal activity (MFC) value was determined using the lowest drug concentration that showed 100% growth inhibition of the fungal strains.

2.8. In Vivo Skin Irritation and Antifungal Efficacy of ITZ–TF

The in vivo skin irritation and antifungal potentials of ITZ–TF were evaluated using BALB/c mice. The mice were randomly divided into six groups (groups 1–6, n = 15 mice per group), as shown in Table 2. Because of the opportunistic nature of *C. albicans* in immunocompromised hosts, mice in groups 3, 4, 5, and 6 underwent immunosuppression using dexamethasone sodium phosphate in water at a dose of 5 mg/kg/day through the intraperitoneal route for 3 days. On the penultimate day of immunosuppression, approximately 2 × 2 cm^2 of dorsal hair was shaved off all mice using an electric clipper. Afterwards, a scratch of approximately 1.75 × 1.75 cm^2 was made on the shaved area in mice from groups 3, 4, 5, and 6 using sterile fine-grit sandpaper (120 mesh). To induce fungal infection, 100 μL of *C. albicans* (10^7 CFU) was continuously applied to the shaved and lightly scarified skin of mice in groups 4, 5, and 6 for 3 days. The induction of fungal infection in groups 4, 5, and 6 was confirmed by comparison of clinical scores (based on the presence of exudate, erythema, and a white substance over the infected area) with the findings in group 3. The symptoms were individually scored as follows: 0 (complete absence), 1 (slightly visible), 2 (clearly visible), 3 (substantially visible in a small area), and 4 (substantially visible in a large area). The mean severity score of each symptom was calculated separately for each group. Moreover, to evaluate antifungal efficacy, the lesion area was measured by means of a digital camera and ImageJ software (LOCI); the healing rate was expressed as a proportion of the remaining wound area. Additionally, skin sections (approximately 2 × 2 cm^2) from the infected sites of 20 randomly selected mice (n = 5 mice per group; groups 3, 4, 5, and 6) were excised for mycological evaluation. The skin samples were punched at three different sites to obtain subsamples 8 mm in diameter. Each subsample was vortexed in Sabouraud dextrose medium (without antibiotics) for 1 min to extract all remaining viable fungi. Then, each extract was loaded onto a Sabouraud dextrose agar plate and incubated at 37 °C. After 48 h, colonies were counted and mean values (CFUs/1 cm^2 skin) calculated. Finally, via comparisons of the severity scores and mycological statuses of mice from groups 4, 5, and 6 with those from group 3, fungal infection was confirmed. Treatment was then initiated for 10 days using ITZ samples in mice from all groups. Treatment of mice with ITZ samples at the dose equivalent to 40 mg/kg/day was performed as follows: groups 1, 3, and 4 [ITZ–TF#11 (vehicle)], groups 2 and 3 [ITZ–TF#11], and group 5 [ITZ (oral)].

Table 2. Group allocations.

Group	Immunosuppression	Scratching	Fungal Infection	Treatment	Administration Route	Dose of ITZ (mg/kg/day)
1				ITZ–TF#11 (vehicle)	Topical	
2				ITZ–TF#11	Topical	40
3	Immuno-suppression	Scratched		ITZ–TF#11 (vehicle)	Topical	
4	Immuno-suppression	Scratched	Infected	ITZ–TF#11 (vehicle)	Topical	
5	Immuno-suppression	Scratched	Infected	ITZ oral capsule	Oral	40
6	Immuno-suppression	Scratched	Infected	ITZ–TF#11	Topical	40

To examine the skin-irritation and antifungal-efficacy characteristics of each ITZ sample, continuous severity evaluation was performed throughout treatment. On the 11th day of treatment, all mice were sacrificed and their skins were isolated for further analysis. Mycological evaluation was performed as described above to directly access the antifungal activities of the ITZ samples. Antifungal efficacy at the site of infection was evaluated via histopathological assay via hematoxylin and eosin (H&E) and periodic acid–Schiff (PAS) staining. For H&E staining, the skin was blocked using paraffin and stained with H&E dye to enable microscopic examination and characterization of inflammatory patterns, blood vessels, and exudate. For PAS staining, skin specimens were rinsed three times with PBS and fixed overnight in Carnoy's solution. Skin specimens were then immersed in sterile PBS, washed and processed for dehydration with ethanol, vitrified with xylene, and immersed and embedded in paraffin. Microscopic examination of skin was performed to visualize bright reddish-pink color indicating carbohydrate analogs present in the fungal cell wall or cell capsule [21].

2.9. Statistical Analyses

All data are expressed as means ± standard deviations. Unpaired data were evaluated using Student's *t*-test for comparisons between two mean values, and one-way analysis of variance followed by Tukey's multiple-comparison test was performed for comparisons among more than two mean values. In all analyses, $p < 0.05$ was considered statistically significant.

3. Results and Discussion

3.1. Drug Stability and In Vitro Artificial Membrane Permeability of ITZ–TF

To develop an effective topical ITZ formulation (ITZ–TF) with enhanced solubility–permeability, a biphasic system composed of volatile and non-volatile components was designed. First, based on its safety profile, ethanol was selected as a delivery vehicle. BHA, EDTA, and 1 M sodium hydroxide were used as an antioxidant, chelating agent, and pH-adjusting agent, respectively, for additional stability. Next, based on the solubility of ITZ, benzyl alcohol was used as a co-solvent. After the addition of benzyl alcohol at 10%, 20%, and 30% concentrations in ITZ–TF#1, ITZ–TF#2, and ITZ–TF#3, respectively, approximately 29-fold greater drug concentrations were achieved in the solubilized form than in 1% ITZ in ethanol (control solution) on day 0 (Table 3). In addition, the artificial membrane permeability of ITZ–TF#3 was enhanced by 12.9-fold, compared with the control (Table 3). However, on day 30, ITZ–TF#3 demonstrated 69.7% drug content, which suggested precipitation of the drug through an anti-solvent crystallization phenomenon due to the presence of polar solvents (e.g., ethanol and water) that enable low ITZ solubility [22]. Thus, to prevent drug precipitation, hydrochloric acid at concentrations of 0.121%, 0.242%, and 0.484% was added to ITZ–TF#4, ITZ–TF#5, and ITZ–TF#6, respectively. The solubility and stability in the vehicle with a polar system were thus achieved via drug protonation and via the formation of a stable complex between ITZ and hydrochloric acid [23]. The results showed that ITZ–TF#6 exhibited a 99.5% drug content until day 30 (Table 3). However, the drug permeability of ITZ–TF#6 was not enhanced, compared with that of ITZ–TF#3. To enhance the permeability, we incorporated 10%, 15%, and 20% concentrations of Transcutol P in ITZ–TF#7, ITZ–TF#8, and ITZ–TF#9, respectively. The artificial-membrane drug permeability of ITZ–TF#7, ITZ–TF#8, and ITZ–TF#9 was enhanced by 2.89-, 4.08-, and 4.11-fold, respectively, compared with that of ITZ–TF#6. The drug contents of those formulations were maintained at approximately 99% until day 30 (Table 3). However, the almost identical drug penetration properties of ITZ–TF#8 and 9 suggested that 15% Transcutol P provided maximal enhancement of drug permeability, presumably due to the saturation of Transcutol P to maintain the necessary thermodynamic driving force for effective flux across the barrier membrane [24]. Furthermore, 5% and 8.22% concentrations of cyclomethicone as a wetting agent were added to ITZ–TF#10 and 11, respectively. ITZ–TF#11 then showed 2.35-fold greater drug permeability, compared with ITZ–TF#8, which may be due to the enhancing effect of cyclomethicone on drug infiltration by lowering surface tension and increasing the spreadability of topical formulations (Table 3) [20]. Moreover, ITZ–TF#11 was able to maintain the drug content within the range of 95.0–105.0% for more than 90 days, and no changes in physical appearance (e.g., precipitation, phase separation, or color change) were observed.

Table 3. Initial drug concentration, drug content after 1 month, and cumulative permeated drug from ITZ–TF through the artificial skin membrane.

Formulation	Initial Drug Concentration (mg/mL)	Drug Content after 1 Month (%) [1]	Cumulative Permeated Drug (μg/mL) [2]
ITZ in ethanol	0.32 ± 0.01	12.7 ± 0.69	0.244 ± 0.033
ITZ–TF#1	9.25 ± 0.01	38.9 ± 0.88	1.244 ± 0.120
ITZ–TF#2	9.33 ± 0.07	51.1 ± 0.30	2.560 ± 0.170
ITZ–TF#3	9.32 ± 0.95	58.4 ± 0.46	3.156 ± 0.012
ITZ–TF#4	9.24 ± 0.47	69.7 ± 0.70	3.250 ± 0.150
ITZ–TF#5	9.25 ± 0.20	80.0 ± 0.46	3.420 ± 0.105
ITZ–TF#6	9.38 ± 1.65	99.5 ± 0.54	3.814 ± 0.003
ITZ–TF#7	9.41 ± 0.21	99.8 ± 1.80	11.03 ± 1.465
ITZ–TF#8	9.36 ± 0.90	99.6 ± 0.15	15.59 ± 0.250
ITZ–TF#9	9.42 ± 0.11	99.8 ± 0.23	15.68 ± 0.941
ITZ–TF#10	9.40 ± 0.27	99.8 ± 1.21	24.75 ± 1.384
ITZ–TF#11	9.39 ± 0.21	99.9 ± 0.60	36.65 ± 1.045

[1] Drug content after 1 month was calculated by defining the initial drug concentration as 100%. [2] Cumulative permeated drug concentrations from 1% ITZ dispersed in ethanol and ITZ–TFs were measured using an artificial skin membrane after 6 h of permeation. Each value represents the mean ± standard deviation (n = 6).

3.2. *In Vitro Permeation and Deposition of ITZ in Full-thickness Human Skin*

The skin permeations and deposition levels of ITZ from 1% ITZ in ethanol, ITZ–TF#6, ITZ–TF#8, and ITZ–TF#11 were determined using the Franz diffusion system. ITZ from ITZ–TF#6, ITZ–TF#8, and ITZ–TF#11 was detected in the receptor phase after 0.5 h of permeation. However, skin permeation of ITZ from 1% ITZ in ethanol commenced at only 1.5 h (Figure 1a). The cumulative permeated ITZ level from ITZ–TF#6 was 1.45 ± 0.051 μg/cm^2 at 24 h, thus 67% greater than the value from 1% ITZ in ethanol. Such permeability enhancement may be attributable to the 29.8-fold greater drug concentration of ITZ–TF#6 compared to that of 1% ITZ in ethanol on day 0. In addition, skin drug penetration from ITZ–TF#8 was more rapid than that from ITZ–TF#6; the 24 h cumulative permeated drug level from ITZ–TF#8 was 29.3% higher than that from ITZ–TF#6. This may reflect an effect of Transcutol P, which creates stratum corneum channels via lipid extraction from the bilayer matrix, fluidization of the lipid bilayer, changes in stratum corneum protein conformations, and creation of a concentration gradient that drives drug permeability. Furthermore, amphiphilic Transcutol P enhances drug solubility in lipids of the stratum corneum, and prevents precipitation of lipophilic drugs in water during skin penetration [24]. The permeated drug levels from ITZ–TF#11 were 7.00% and 28.0% greater than those from ITZ–TF#8 at 24 and 72 h, respectively (Figure 1b). This may reflect the presence of cyclomethicone in ITZ–TF#11, unlike ITZ–TF#8. Cyclomethicone enhances skin moisture retention and, in synergy with Transcutol P, promotes intercellular lipid fluidization in the stratum corneum.

The ITZ deposition levels in various skin layers after application of 1% ITZ in ethanol, ITZ–TF#6, ITZ–TF#8, and ITZ–TF#11 over 3 days are shown in Figure 1c. The deposition levels were highest in the stratum corneum, followed by the epidermis and dermis, presumably because of the lipophilic nature of ITZ, which should bind readily to lipids of the stratum corneum [25,26]. ITZ–TF#6 dermal deposition was 2.43-fold greater than dermal deposition from the 1% ITZ in ethanol. The ITZ–TF#8 drug depositions in the stratum corneum, epidermis, and dermis were 210%, 169%, and 399% those of ITZ–TF#6. The 2.11-fold greater dermal drug deposition of ITZ–TF#11, compared to ITZ–TF#8, may reflect enhanced drug partitioning of ITZ–TF#11 across human skin.

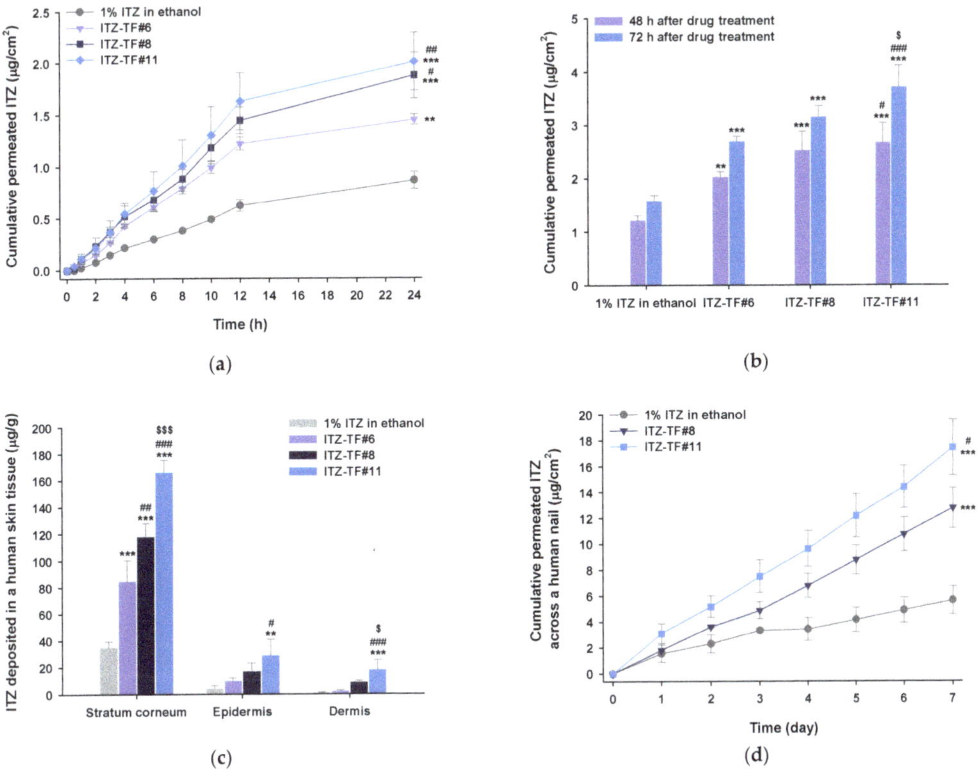

Figure 1. The in vitro full-thickness human skin and nail permeabilities of itraconazole (ITZ) from 1% ITZ in ethanol, ITZ–TF#6, ITZ–TF#8, and ITZ–TF#11. (**a**) The cumulative ITZ levels that permeated human skin over 24 h. (**b**) Cumulative human skin ITZ infiltrations at 48 and 72 h. (**c**) ITZ concentrations/g of stratum corneum, epidermal, and dermal tissues. (**d**) The time course of cumulative, partitioned drug concentrations across human nails from day 1 to 7. Each value is a mean ± standard deviation (n = 4). $^{**}\ p < 0.01$, $^{***}\ p < 0.001$ compared to 1% ITZ in ethanol. $^{\#}\ p < 0.05$, $^{\#\#}\ p < 0.01$, $^{\#\#\#}\ p < 0.001$ compared to ITZ–TF#6. $^{\$}\ p < 0.05$, $^{\$\$\$}\ p < 0.001$ compared to ITZ–TF#8.

3.3. In Vitro Human Nail Permeability and Deposition

The permeation potential across the nail plates of ITZ in 1% ITZ in ethanol, ITZ–TF#8, and ITZ–TF#11 was evaluated as shown in Figure 1d. After 24 h of drug loading, the permeability of ITZ from ITZ–TF#8 was 1.80 ± 0.093 μg/cm^2, which was 117% greater than the permeability of ITZ from 1% ITZ in ethanol (Figure 1d). In addition, ITZ–TF#8 demonstrated 226% greater cumulative permeated drug concentration, compared with 1% ITZ in ethanol, on day 7. These enhanced permeabilities of ITZ–TF#8 may have been due to the greater drug concentration in solubilized form in ITZ–TF#8, as well as the presence of bipolar Transcutol P in ITZ–TF#8, which can interact with both hydrophilic and lipophilic components present in the nail matrix [27,28]. Moreover, on day 7, ITZ–TF#11 produced a 36.7% enhancement of drug penetration across the nail, compared with ITZ–TF#8. This enhancement may have been due to the presence of cyclomethicone, which lowers the surface tension of ITZ–TF#11 and accelerates drug penetration from the small channels generated after conformational changes in the nail keratin structure caused by Transcutol P [29,30]. Furthermore, after continuous application for 7 days, 1% ITZ in ethanol, ITZ–TF#8, and ITZ–TF#11 showed 31.4 ± 5.68, 42.9 ± 4.23, and 72.2 ± 14.9 μg/g deposition of the drug in the nail plate, respectively. Thus, based on the comparative enhanced nail penetration and deposition of the drug in nail plates, ITZ–TF#11 was identified as the optimal formulation, compared with 1% ITZ in ethanol and ITZ–TF#8.

3.4. In Vitro Antifungal Activity

The potential of a formulation to inhibit the growth of fungal cells effectively is known as its fungicidal activity. This fungicidal effect is mediated by its solubility and permeability across the fungal cellular membrane. Although ITZ has considerable antifungal capability, its poor solubility and stability are important obstacles that restrict permeability across the fungi cell membrane and substantially reduce its in vitro antifungal effect. Based on its solubility, stability, and enhanced in vitro permeability across human skin and nail, we selected ITZ–TF#11 to evaluate its in vitro antifungal effects against *C. albicans*. In addition, we used ITZ–TF#11 (vehicle) and 1% ITZ in benzyl alcohol for comparison. As shown in Figure 2, the mean zones of inhibition of ITZ–TF#11 were 244% and 163% greater than the mean zones of inhibition of ITZ–TF#11 (vehicle) and 1% ITZ in benzyl alcohol, respectively. Moreover, MFCs of both ITZ–TF#11 and 1% ITZ in benzyl alcohol were 4-fold lower than the MFC of ITZ–TF#11 (vehicle). The MIC of ITZ–TF#11 was 4- and 10.6-fold lower than the MICs of 1% ITZ in benzyl alcohol and ITZ–TF#11 (vehicle), respectively (Table 4). Such enhanced fungicidal activity of ITZ–TF#11 against *C. albicans* may reflect increased drug permeation mediated by Transcutol P and rapid drug penetration facilitated by the wetting agent (cyclomethicone). They also enhance drug cellular uptake via endocytosis and passive drug diffusion into the cytoplasm; Transcutol P and cyclomethicone increase lipid extraction and fluidization of the lipid bilayer and improve the partition coefficients of drugs relative to those of cell membranes [31–33].

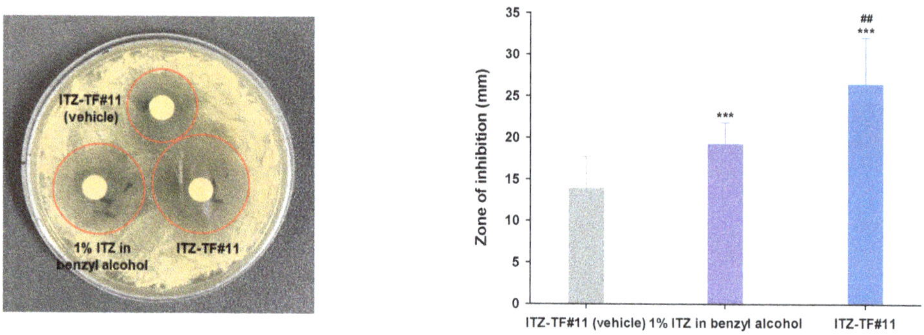

(a) (b)

Figure 2. In vitro antifungal efficacies of ITZ–TF#11 (vehicle), 1% ITZ in benzyl alcohol, and ITZ–TF#11. (**a**) Image depicting zones of inhibition of *C. albicans* in a Petri dish. (**b**) Zone of inhibition (mm). Each value represents the mean ± standard deviation (n = 4). *** $p < 0.001$ compared with ITZ–TF#11 (vehicle). ## $p < 0.01$ compared with 1% ITZ in benzyl alcohol.

Table 4. Minimum inhibitory concentration (MIC) and minimum fungicidal concentration (MFC) of ITZ–TF#11 (vehicle), 1% ITZ in benzyl alcohol, and ITZ–TF#11.

Sample	MIC (µg/mL)	MFC (µg/mL)
ITZ–TF#11 (vehicle)	50.0	100
1% ITZ in benzyl alcohol	12.5	25
ITZ–TF#11	4.68	25

3.5. In Vivo Antifungal Activity of ITZ–TF

3.5.1. Evaluation of Symptoms Related to Dermal Fungal Infection

Responses during in vivo skin irritation and antifungal studies were evaluated based on lesion size, as well as the presence or absence of exudate, skin redness (erythema), and dry scaly skin (Figure 3a). Mice in groups 1 [normal skin treated with ITZ–TF#11 (vehicle)] and 2 [normal skin treated with ITZ–TF#11] did not show any obvious symptoms related to skin irritation, indicating suitable safety profiles of ITZ–TF#11 (vehicle) and ITZ–TF#11 in normal mouse skin. In addition, on day 0, the exudate severity scores for mice in groups 4 [fungal-infected skin treated with ITZ–TF#11 (vehicle)], 5 [fungal-infected skin treated with ITZ (oral)], and 6 [fungal-infected skin treated with ITZ–TF#11] were greater than the exudate severity score for mice in group 3 [scratched skin treated with ITZ–TF#11 (vehicle)]. This finding demonstrates successful fungal infection in the skin of mice in groups 4, 5, and 6 (Figure 3b). Similar severity scores of both exudate and erythema in groups 4, 5, and 6 on day 0 indicate comparable extents of fungal infection (Figure 3b,c). After 10 days of continuous treatment, mice in group 4 showed no obvious improvement in the infected area. In contrast, mice in groups 5 and 6 demonstrated 77.5% and 93.1% reductions in lesion size in the infected area (Figure 3d,e). Exudate symptoms were absent after 7 days of treatment among mice in group 6 (Figure 3b). Additionally, mice in group 6 showed a 165% lower mean erythema severity score, compared with mice in group 5 (Figure 3c). This enhanced treatment efficacy of ITZ–TF#11, with respect to symptoms related to superficial mycosis and lesion recovery at the infected site, may have been due to the immediate drug availability that permitted robust local action.

3.5.2. Mycological Assay

Next, we performed a mycological assay to investigate fungal viability in mouse skin during infection and after treatment with ITZ–TF#11 (vehicle) (group 4), ITZ (oral) (group 5), and ITZ–TF#11 (group 6). Groups 4, 5, and 6 exhibited similar mean viable fungal counts (log CFUs/skin section) on day 0, implying that all mice had been infected with a comparable number of fungi (Figure 4a). To directly assess antifungal efficacy, the CFUs of groups 4, 5, and 6 were measured after 10 days of continuous treatment. Figure 4b reveals the absence of fungal growth in groups 1 and 2, indicating the absence of microbial contamination during the mycological assay. The number of colonies did not significantly decrease in the mice of group 4. However, groups 5 and 6 exhibited 2.25- and 4.99-fold reductions in the log CFUs/skin section area, compared to the day 0 data, indicating complete removal of fungal cells from infected skin (Figure 4). Therefore, based on the antifungal response and the mycological assay findings, ITZ–TF#11 effectively treated superficial mycosis.

3.5.3. Histological Evaluation of Skin

To confirm the structural alteration and inflammation of skin from mice in groups 1 [normal skin treated with ITZ–TF#11 (vehicle)], 2 [normal skin treated with ITZ–TF#11], 3 [scratched skin treated with ITZ–TF#11(vehicle)], 4 [fungal-infected skin treated with ITZ–TF#11 (vehicle)], 5 [fungal-infected skin treated with ITZ (oral)], and 6 [fungal-infected skin treated with ITZ–TF#11] after 10 days of the indicated treatment, we performed histological examinations by staining with H&E. Groups 1, 2, and 3 did not exhibit any variations in skin structure, which supported the safety profiles of ITZ–TF#11 (vehicle) and ITZ–TF#11 (Figure 5a). However, group 4 demonstrated epidermal rupture and irregular granulation of fibroblast cells in the epidermis and dermis. In comparison with groups 4 and 5, skin from mice in group 6 demonstrated well-organized epidermis and dermis, together with thinner granulation tissue and denser mature collagen fiber bundles. Next, to confirm the cell viability among layers of skin, mouse skin sections from each group were examined using PAS staining. Skin from mice in groups 1, 2, and 3 did not demonstrate any obvious PAS markers related to fungi (Figure 5b). However, skin from mice in group 4 showed red-stained viable fungal cells (indicated by black arrows) in various layers of the skin. Conversely, skin from mice in group 6 demonstrated very few viable fungi cells, compared with skin from mice in groups 4 and 5. Overall, the enhanced efficacy of ITZ–TF#11 was presumably due to immediate drug availability at application sites that achieved maximal lysis of fungi, thereby promoting natural tissue proliferation and tissue remodeling, resulting in well-organized skin layers [34,35].

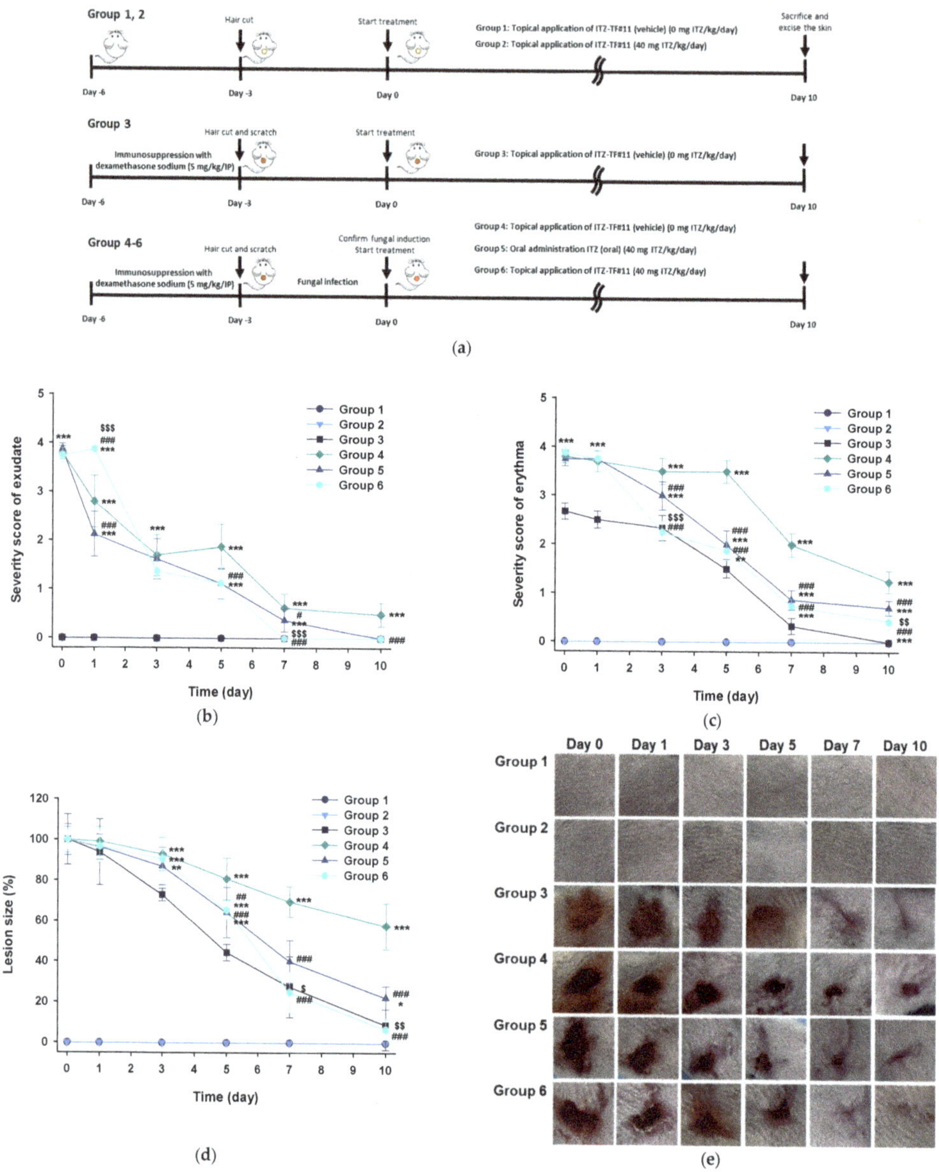

Figure 3. In vivo antifungal efficacies of once-daily oral or topical ITZ treatments in a mouse model of dermal *Candida albicans* infection. (**a**) Fungal infections and treatment schedules for once-daily oral and topical administrations of 40 mg/kg ITZ and ITZ–TF#11 (vehicle) or ITZ–TF#11 (equivalent to 40 mg/kg ITZ), respectively. (**b**) Exudates and (**c**) erythema from/on mouse skin lesions infected or not with *C. albicans* and treated with topical ITZ–TF#11 (vehicle) or ITZ–TF#11 or oral ITZ for 10 days. (**d**) Lesion sizes and (**e**) representative photographs of normal skin treated with ITZ–TF#11 (vehicle) (group 1) or ITZ–TF#11 (group 2); scratched skin treated with ITZ–TF#11 (group 3); and scratched and fungus-infected skin treated with ITZ–TF#11 (vehicle) (group 4), oral ITZ (group 5), and ITZ–TF#11 (group 6) for 10 days. Each value represents a mean ± standard deviation (n = 10 for each group). * $p < 0.05$, ** $p < 0.01$, *** $p < 0.001$ compared to scratched skin treated with ITZ–TF#11 (group 3). ## $p < 0.01$, ### $p < 0.001$ compared to scratched and fungus-infected skin treated with ITZ–TF#11 (vehicle) (group 4). \$ $p < 0.05$, \$\$ $p < 0.01$, \$\$\$ $p < 0.001$ compared to scratched and fungus-infected skin treated with oral ITZ (group 5).

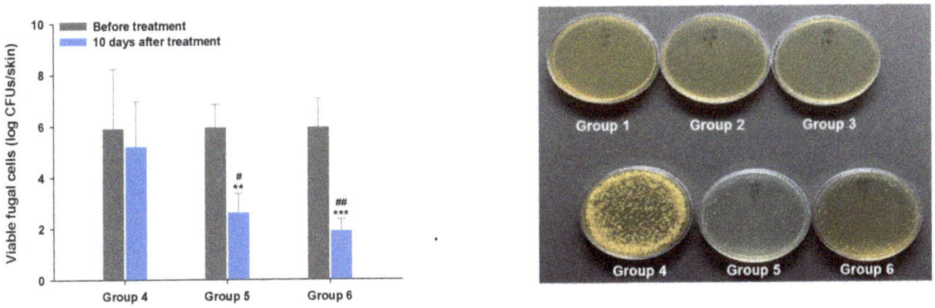

(a) (b)

Figure 4. Mycological assay in full-thickness excisional normal, scratched, and scratched and fungal-infected mouse skins after once-daily treatment with ITZ–TF #11 (vehicle) (groups 1, 3, and 4) and ITZ–TF#11 (groups 2 and 6) based on 40 mg/kg ITZ or once-daily oral administration of 40 mg/kg ITZ (group 5) for 10 days. (**a**) Viable fungal cell counts in *C. albicans*-infected mouse skin lesions in groups 4, 5, and 6. Each value represents the mean ± standard deviation (n = 5). ** $p < 0.01$, *** $p < 0.001$ compared with initial log CFUs/skin section in each group. # $p < 0.05$, ## $p < 0.01$ compared with log CFUs/skin section in group 4 after 10 days of treatment. (**b**) Representative photographs of viable fungal cells after 24 h incubation of excised normal, scratched, or scratched and fungal-infected mouse skins, following 10 days of treatment.

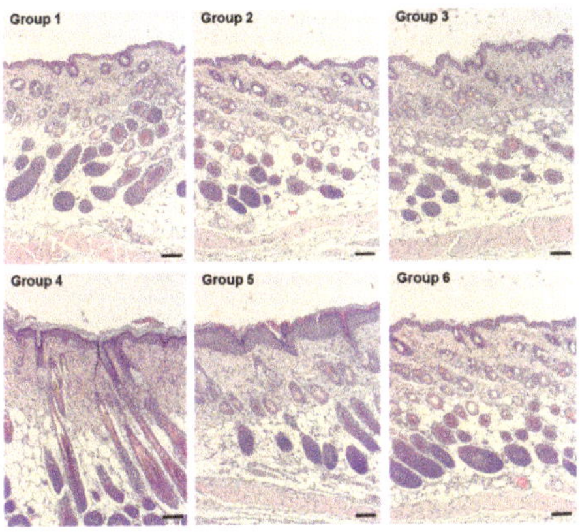

(a)

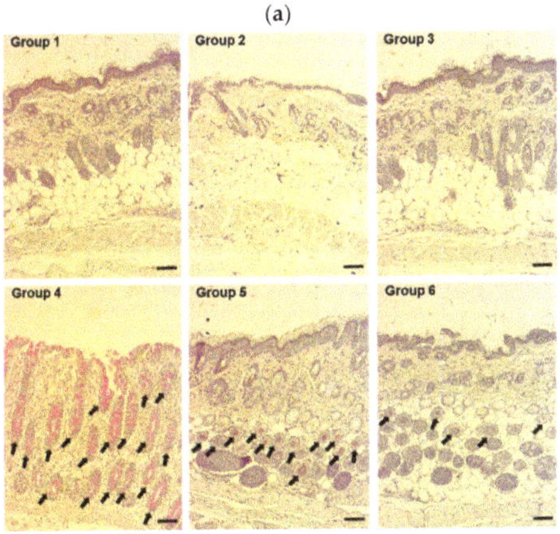

(b)

Figure 5. Representative light micrographs of tissue sections from normal, scratched, and scratched and fungal-infected mouse skin stained with (**a**) hematoxylin and eosin (H&E) and (**b**) periodic acid-Schiff (PAS) after 10 days of oral or topical ITZ treatment. PAS-positive reactions (red) were observed in the basement membrane of the epidermis, hair follicles, and sebaceous glands (black arrows). Scale bar = 100 μm.

4. Conclusions

This study demonstrated that topical ITZ formulations prepared via the incorporation of ITZ into a biphasic system composed of volatile and non-volatile components substantially promoted drug solubility and stability during long-term storage. The optimal formulation (ITZ–TF#11) showed 132% greater skin permeability at 24 h, as well as 209% greater nail infiltration on day 7, compared with 1% ITZ in ethanol. In addition, permeability analysis showed that concentrations of drug deposition in skin and nail from ITZ–TF#11 were 539% and 230% greater than concentrations of drug deposition from 1% ITZ in ethanol, respectively. Moreover, ITZ–TF#11 achieved 323% greater fungal-infected skin lesion recovery, a 165% lower mean erythema severity score, and a 37% lower mean logarithm of viable fungal cells in skin at the treatment area, compared with skin treated with oral ITZ at the same dose. Thus, the long-term stability and improved permeability of ITZ–TF#11 resulted in the immediate availability of locally applied ITZ at the site of application, in comparison with ITZ (oral). The findings indicate that ITZ–TF#11 offers an effective alternative formulation against topical fungal infections.

Author Contributions: Conceptualization, B.J.K., S.-S.C. and J.W.P.; methodology, L.S., S.-Y.S. and R.P.; validation, L.S., S.-Y.S. and S.K.J.; formal analysis, S.K.J., S.-H.L. and R.P.; investigation, S.K.J., S.-H.L. and K.-T.K.; resources, K.-T.K., S.-S.C. and J.W.P.; data curation, S.K.J., S.-H.L. and K.-T.K.; writing—original draft preparation, L.S. and S.-Y.S.; writing—review and editing, S.-S.C. and J.W.P.; visualization, S.-S.C.; supervision, B.J.K. and J.W.P.; project administration, B.J.K., S.-S.C. and J.W.P.; funding acquisition, K.-T.K., B.J.K. and J.W.P. All authors have read and agreed to the published version of the manuscript.

Funding: This research was supported by the Basic Research Program through the National Research Foundation of Korea (NRF), funded by the Korean Government Ministry of Science and ICT (MSIT), grant number NRF-2020R1F1A1069889.

Institutional Review Board Statement: The study was conducted according to the guidelines of the Declaration of Helsinki and approved by the Institutional Review Board (or Ethics Committee) of the Institutional Animal Care and Use Committee of Mokpo National University (Jeonnam, Republic of Korea). (Protocol code no. MNU-IACUC-2020-020, 12 October 2020).

Informed Consent Statement: Not applicable.

Data Availability Statement: Not applicable.

Acknowledgments: We greatly appreciated using the Convergence Research Laboratory (established by the MNU Innovation Support Project in 2020) to conduct this research.

Conflicts of Interest: The authors declare no conflict of interest.

References

1. Dorr, P. 7.14—Fungi and Fungal Disease. In *Comprehensive Medicinal Chemistry II*; Taylor, J.B., Triggle, D.J., Eds.; Elsevier: Oxford, UK, 2007; pp. 419–443.
2. Narang, T.; Bhattacharjee, R.; Singh, S.; Jha, K.; Mahajan, R. Dogra Quality of life and psychological morbidity in patients with superficial cutaneous dermatophytosis. *Mycoses* **2019**, *62*, 680–685. [CrossRef] [PubMed]
3. Sheehan, D.J.; Hitchcock, C.A.; Sibley, C.M. Current and emerging azole antifungal agents. *Clin. Microbiol. Rev.* **1999**, *12*, 40–79. [CrossRef] [PubMed]
4. Vlaminck, K.M.J.A.; Engelen, M.A.C.M. Itraconazole: A treatment with pharmacokinetic foundations. *Vet. Dermatol.* **2004**, *15*, 8. [CrossRef]
5. Como, J.A.; Dismukes, W.E. Oral azole drugs as systemic antifungal therapy. *N. Engl. J. Med.* **1994**, *330*, 263–272.

6. Kolarsick, P.A.J.; Kolarsick, M.A.; Goodwin, C. Anatomy and Physiology of the Skin. *J. Dermatol. Nurses Assoc.* **2011**, *3*, 203–213. [CrossRef]
7. Shin, J.H.; Choi, K.Y.; Kim, Y.C.; Lee, M.G. Dose-dependent pharmacokinetics of itraconazole after intravenous or oral administration to rats: Intestinal first-pass effect. *Antimicrob. Agents Chemother.* **2004**, *48*, 1756–1762. [CrossRef] [PubMed]
8. Zheng, W.-S.; Fang, X.-Q.; Wang, L.-L.; Zhang, Y.-J. Preparation and quality assessment of itraconazole transfersomes. *Int. J. Pharm.* **2012**, *436*, 291–298. [CrossRef]
9. Doncker, P.D.; Pande, S.; Richarz, U.; Garodia, N. Itraconazole: What clinicians should know? *Indian J. Drugs Dermatol.* **2017**, *3*, 4–10.
10. Gupta, A.K.; Nolting, S.; de Prost, Y.; Delescluse, J.; Degreef, H.; Theissen, U.; Wallace, R.; Marynissen, G.; De Doncker, P. The use of itraconazole to treat cutaneous fungal infections in children. *Dermatology* **1999**, *199*, 248–252. [CrossRef]
11. Debruyne, D.; Coquerel, A. Pharmacokinetics of antifungal agents in onychomycoses. *Clin. Pharmacokinet.* **2001**, *40*, 441–472. [CrossRef] [PubMed]
12. Garg, T.; Rath, G.; Goyal, A.K. Comprehensive review on additives of topical dosage forms for drug delivery. *Drug Deliv.* **2015**, *22*, 969–987. [CrossRef] [PubMed]
13. Kim, H.; Jung, S.; Yeo, S.; Kim, D.; Na, Y.C.; Yun, G.; Lee, J. Characteristics of Skin Deposition of Itraconazole Solubilized in Cream Formulation. *Pharmaceutics* **2019**, *11*, 195. [CrossRef] [PubMed]
14. Surber, C.; Knie, U. Metamorphosis of Vehicles: Mechanisms and Opportunities. *pH Skin Issues Chall.* **2018**, *54*, 152–165.
15. Gupta, R.; Dwadasi, B.S.; Rai, B.; Mitragotri, S. Effect of Chemical Permeation Enhancers on Skin Permeability: In silico screening using Molecular Dynamics simulations. *Sci. Rep.* **2019**, *9*, 1456. [CrossRef]
16. Domínguez-Gil Hurlé, A.; Sánchez Navarro, A.; García Sánchez, M.J. Therapeutic drug monitoring of itraconazole and the relevance of pharmacokinetic interactions. *Clin. Microbiol. Infect.* **2006**, *12*, 97–106. [CrossRef]
17. Alam, S.; Iqbal, Z.; Ali, A.; Khar, R.K.; Ahmad, F.J.; Akhter, S.; Talegaonkar, S. Microemulsion as a Potential Transdermal Carrier for Poorly Water Soluble Antifungal Drug Itraconazole. *J. Dispers. Sci. Technol.* **2009**, *31*, 84–94. [CrossRef]
18. Zhang, Q.; Murawsky, M.; LaCount, T.; Kasting, G.B.; Li, S.K. Transepidermal water loss and skin conductance as barrier integrity tests. *Toxicol. In Vitro* **2018**, *51*, 129–135. [CrossRef] [PubMed]
19. Clausen, M.L.; Slotved, H.C.; Krogfelt, K.A.; Agner, T. Tape Stripping Technique for Stratum Corneum Protein Analysis. *Sci. Rep.* **2016**, *6*, 19918. [CrossRef] [PubMed]
20. Lee, B.C.; Pangeni, R.; Na, J.; Koo, K.-T.; Park, J.W. Preparation and in vivo evaluation of a highly skin- and nail-permeable efinaconazole topical formulation for enhanced treatment of onychomycosis. *Drug Deliv.* **2019**, *26*, 1167–1177. [CrossRef]
21. Song, X.; Wei, Y.X.; Lai, K.M.; He, Z.D.; Zhang, H.J. In vivo antifungal activity of dipyrithione against Trichophyton rubrum on guinea pig dermatophytosis models. *Biomed. Pharmacother.* **2018**, *108*, 558–564. [CrossRef]
22. Sriamornsak, P.; Burapapadh, K. Characterization of recrystallized itraconazole prepared by cooling and anti-solvent crystallization. *Asian J. Pharm. Sci.* **2015**, *10*, 230–238. [CrossRef]
23. Bagavatula, H.; Lankalapalli, S.; Vinai, V.; Tenneti, K.; Beeraka, N.M.R.; Bulusu, B.T. Comparative Studies on Solubility and Dissolution Enhancement of Different Itraconazole Salts and Their Complexes. *Adv. Pharmacol. Pharm.* **2014**, *2*, 85–95. [CrossRef]
24. Osborne, D.W.; Musakhanian, J. Skin Penetration and Permeation Properties of Transcutol®-Neat or Diluted Mixtures. *AAPS PharmSciTech* **2018**, *19*, 3512–3533. [CrossRef]
25. Takahashi, H. Action of topical anti-trichophyton agents. *Jpn. J. Dermatol.* **1972**, *82*, 421–437.
26. Hashiguchi, T.; Kodama, A.; Ryu, A.; Otagiri, M. Retention capacity of topical imidazole antifungal agents in the skin. *Int. J. Pharm.* **1998**, *161*, 195–204. [CrossRef]
27. Mura, P.; Faucci, M.T.; Bramanti, G.; Corti, P. Evaluation of transcutol as a clonazepam transdermal permeation enhancer from hydrophilic gel formulations. *Eur. J. Pharm. Sci.* **2000**, *9*, 365–372. [CrossRef]
28. Matsuda, Y.; Sugiura, K.; Hashimoto, T.; Ueda, A.; Konno, Y.; Tatsumi, Y. Efficacy Coefficients Determined Using Nail Permeability and Antifungal Activity in Keratin-Containing Media Are Useful for Predicting Clinical Efficacies of Topical Drugs for Onychomycosis. *PLoS ONE* **2016**, *11*, e0159661. [CrossRef]
29. Bseiso, E.A.; Nasr, M.; Sammour, O.A.; Abd El Gawad, N.A. Novel nail penetration enhancer containing vesicles "nPEVs" for treatment of onychomycosis. *Drug Deliv.* **2016**, *23*, 2813–2819. [CrossRef]
30. Bhatt, V.; Pillai, R. Efinaconazole topical solution, 10%: Formulation development program of a new topical treatment of toenail onychomycosis. *J. Pharm. Sci.* **2015**, *104*, 2177–2182. [CrossRef]
31. Varis, T.; Kivistö, K.T.; Backman, J.T.; Neuvonen, P.J. The cytochrome P450 3A4 inhibitor itraconazole markedly increases the plasma concentrations of dexamethasone and enhances its adrenal-suppressant effect. *Clin. Pharmacol. Ther.* **2000**, *68*, 487–494. [CrossRef]
32. Sha, X.; Yan, G.; Wu, Y.; Li, J.; Fang, X. Effect of self-microemulsifying drug delivery systems containing Labrasol on tight junctions in Caco-2 cells. *Eur. J. Pharm. Sci.* **2005**, *24*, 477–486. [CrossRef] [PubMed]
33. Spósito, P.Á.; Mazzeti, A.L.; de Oliveira Faria, C.; Aurbina, J.; Pound-Lana, G.; Bahia, M.T.; Mosqueira, V.F. Ravuconazole self-emulsifying delivery system: In vitro activity against Trypanosoma cruzi amastigotes and in vivo toxicity. *Int. J. Nanomed.* **2017**, *12*, 3785–3799. [CrossRef]

34. Chellan, G.; Neethu, K.; Varma, A.K.; Mangalanandan, T.S.; Shashikala, S.; Dinesh, K.R.; Sundaram, K.R.; Varma, N.; Jayakumar, R.V.; Bal, A.; et al. Targeted treatment of invasive fungal infections accelerates healing of foot wounds in patients with Type 2 diabetes. *Diabet. Med.* **2012**, *29*, 255–262. [CrossRef] [PubMed]
35. Choi, J.U.; Lee, S.W.; Pangeni, R.; Byun, Y.; Yoon, I.S.; Park, J.W. Preparation and in vivo evaluation of cationic elastic liposomes comprising highly skin-permeable growth factors combined with hyaluronic acid for enhanced diabetic wound-healing therapy. *Acta Biomater.* **2017**, *57*, 197–215. [CrossRef] [PubMed]

pharmaceutics

MDPI

Article

Implant Surfaces Containing Bioglasses and Ciprofloxacin as Platforms for Bone Repair and Improved Resistance to Microbial Colonization

Irina Negut [1], Carmen Ristoscu [1,*], Tatiana Tozar [1,2], Mihaela Dinu [3], Anca Constantina Parau [3], Valentina Grumezescu [1], Claudiu Hapenciuc [1], Marcela Popa [4], Miruna Silvia Stan [5], Luminita Marutescu [4], Ion N. Mihailescu [1] and Mariana Carmen Chifiriuc [5,6,7,8]

1 National Institute for Lasers, Plasma and Radiation Physics, 409 Atomistilor Street, 077125 Magurele, Romania; negut.irina@inflpr.ro (I.N.); tatiana.alexandru@inflpr.ro (T.T.); valentina.grumezescu@inflpr.ro (V.G.); hapenciuc.claudiu@inflpr.ro (C.H.); ion.mihailescu@inflpr.ro (I.N.M.)
2 Extreme Light Infrastructure-Nuclear Physics, Horia Hulubei National Institute for R&D in Physics and Nuclear Engineering, 077125 Magurele, Romania
3 National Institute of Research and Development for Optoelectronics-INOE2000, 409 Atomistilor St., 077125 Magurele, Romania; mihaela.dinu@inoe.ro (M.D.); anca.parau@inoe.ro (A.C.P.)
4 Microbiology Immunology Department, Faculty of Biology, University of Bucharest, 077206 Bucharest, Romania; marcela.popa@bio.unibuc.ro (M.P.); luminita.marutescu@bio.unibuc.ro (L.M.)
5 Department of Biochemistry and Molecular Biology, Faculty of Biology, University of Bucharest, Splaiul Independentei 91-95, 050095 Bucharest, Romania; miruna.stan@bio.unibuc.ro (M.S.S.); carmen.chifiriuc@bio.unibuc.ro (M.C.C.)
6 Department of Microbiology, Faculty of Biology, University of Bucharest, Aleea Portocalelor Str. 1-3, District 5, 060101 Bucharest, Romania
7 Romanian Academy of Scientists, 3 Ilfov Str., District 5, 050044 Bucharest, Romania
8 The Romanian Academy, Calea Victoriei 25, District 1, 010071 Bucharest, Romania
* Correspondence: carmen.ristoscu@inflpr.ro

Abstract: Coatings are an attractive and challenging selection for improving the bioperformance of metallic devices. Composite materials based on bioglass/antibiotic/polymer are herein proposed as multifunctional thin films for hard tissue implants. We deposited a thin layer of the polymeric material by matrix-assisted pulsed laser evaporation—MAPLE onto Ti substrates. A second layer consisting of bioglass + antibiotic was applied by MAPLE onto the initial thin film. The antimicrobial activity of MAPLE-deposited thin films was evaluated on *Staphylococcus aureus*, *Enterococcus faecalis*, *Escherichia coli*, and *Pseudomonas aeruginosa* standard strains. The biocompatibility of obtained thin films was assessed on mouse osteoblast-like cells. The results of our study revealed that the laser-deposited coatings are biocompatible and resistant to microbial colonization and biofilm formation. Accordingly, they can be considered viable candidates for biomedical devices and contact surfaces that would otherwise be amenable to contact transmission.

Keywords: thin film; antibiotic delivery; biocompatibility; bioglass; laser deposition

1. Introduction

Accidents, falls or direct strikes to the body, deficiency of particular nutrients, persistent bone and skeletal illnesses, or injuries while conducting daily tasks are all possible causes of bone fractures. As bones protect humans' vital organs like the heart, lungs, and brain, fractures must be treated as quickly as possible [1]. Metallic implants such as bone plates, rods, nails, and other compression plates offer artificial support when the fracture is severe. These devices provide mechanical stability, allowing for optimal bone alignment and function during physiologic loads. As a result, implants alleviate pain, restore mobility, and easily resume the human body's functions [2].

Implants and bone healing are intimately linked, therefore the stabilized bone should "self-heal" [2]. Major advances have been made in recent years in the understanding and design of implantable devices, notably in the areas of biocompatibility, bioavailability, and efficacy [3]. Several synthetic biomaterials have been designed as substitutes for natural bones. Among them, bioactive glasses (BG) are the most used owing to their biocompatibility and their capacity to form a bone-like mineral phase when in contact with the native hard tissue [4].

Any device implanted into the human body is prone to infections caused by bacteria (e.g., *Staphylococcus aureus*, *Enterococcus faecalis*, *Escherichia coli*, *Pseudomonas aeruginosa*), which further affect bone healing or ingrowth, nonunion of fractured bones, and even implant loosening [5]. These infections are difficult to treat, and they frequently necessitate further surgical procedures such as implant replacement. In the worst-case scenario, they can also result in amputation or death, posing a significant societal and economic burden [6].

For an implant-associated infection to occur, bacteria must first adhere to the implant surface. These infections anchored to the surface are known as biofilms. Biofilms are formed by complex groups of microbial cells of both Gram-positive and Gram-negative bacteria, generating a protective extracellular polysaccharide matrix [7]. Removal of these communities is challenging since biofilms are resilient to the host's immune system and antibiotic treatment [8].

The dose of systemically administered antibiotics does not often reach the site of infection; bones are moderately perfused organs and the blood flow in infected bone tissues is usually reduced [5]. Locally administered antibiotics for the treatment of bone infections have several advantages over systemic administration, such as the reduction of adverse effects and the risk of overdose while boosting the drug's bioavailability effectively reaching the target site. Therefore, an implant surface must meet the requirement of a multifunctional system that responds to microbiological signals by releasing antimicrobials or other alleviating compounds [1]. As a result, antimicrobial surfaces are considered as "contact killing" and antimicrobial agent eluting surfaces. Many studies have been reported on antimicrobial coatings developed for the treatment of implantable medical device-associated infections [9–11].

Matrix-assisted pulsed laser evaporation (MAPLE) has been applied for the growth of high-quality thin films for an assortment of medical devices, with different biological and antimicrobial functionalities [12–14].

Our previous BG-related work demonstrated that MAPLE is a suitable technique for fabricating biocompatible and bone-healing thin films containing combinations of BG and antimicrobial compounds, including doxycycline [15], Neem [16], turmeric and holy basil [17]. These films showed a high antibacterial efficiency against Gram-positive and Gram-negative bacterial strains, and this activity was mediated by a contact-killing effect.

We report herewith composite coatings based on a polymer (Poly(methyl methacrylate)), BG and carboxyfluoroquinoline antibiotic (ciprofloxacin) obtained by MAPLE in order to: (i) reduce corrosion of implant materials; (ii) increase the biocompatibility of implants by promoting bone regeneration and growth, without toxic effects, and (iii) inhibit microbial biofilm formation.

2. Materials and Methods

2.1. Materials

Chemicals, namely chloroform ($CHCl_3$), analytical grade acetone (C_6H_6O), ethanol (C_2H_5OH), and reagents required for the preparation of the simulated body fluid (SBF), such as NaCl, $NaHCO_3$, KCl, $K_2HPO_4{\cdot}3H_2O$, $MgCl_2{\cdot}6H_2O$, HCl, $CaCl_2$, Na_2SO_4, and $(CH_2OH)_3CNH_2$ were acquired from Sigma-Aldrich Chemie GmbH (Steinheim, Germany). SBF with an ionic composition identical to blood plasma was prepared according to the corrected Kokubo's SBF similar to that of human blood plasma [18] by mixing the reagents, following the precise order and quantity.

Grade 4 titanium (Ti) of (10 × 10) and (5 × 5) mm^2 disks were used as deposition substrates. Preceding the deposition, the disks were cleaned with acetone, ethanol, and deionized water in an Elma X-Tra 30 H ultrasonic bath (Elma Schmidbauer GmbH, Singen, Germany). The choice of Ti is determined by the well-known compatibility of this metal with bone which justifies its intensive use as a biomaterial for designing implantable devices [19].

Since the report on its bioactivity, the 45S5 Bioglass® (45S5) has been the most extensively used in biomedical applications, ranging from scaffolds for tissue engineering, bone grafts, periodontal applications, to coatings for bioinert implants ([20] p. 5). Its composition can be considered as the predecessor of many BG, obtained by removing or adding ions, with the aim of producing materials with novel properties, suitable for specific clinical requirements [21]. Our chosen BG contains 56.5% SiO_2, 15% CaO, 11% Na_2O, 8.5% MgO, 6% P_2O_5, and 3% K_2O (in wt%). It is fabricated following the protocol described in references [22,23].

Poly(methyl methacrylate)—PMMA, $(C_5H_8O_2)n$, is a synthetic biopolymer demonstrating good mechanical stability and chemical (acid and base) resistivity [24]. Due to its inexpensiveness, lightweight features, superior toughness, and cost-effective fabrication, it is widely used in medical applications such as bone cement preparation for drug delivery and release [25], optical systems such as contact lenses [26], or the flat panel industry [27].

Ciprofloxacin (CIPRO), a small molecular drug, is a broad-spectrum antimicrobial carboxyfluoroquinoline agent with demonstrated activity against most strains of Gram-negative bacteria and against certain Gram-positive bacteria [28]. CIPRO proved a powerful antibiotic to be applied for bone infections [29]. CIPRO hydrochloride powder of 99.5% purity was purchased from Gen Hunter (Nashville, TN, USA).

2.2. MAPLE Target Preparation and Experimental Conditions

MAPLE represents an adjustable laser-based processing method that implies a vacuum deposition chamber where a pulsed laser beam hits and evaporates a rotating cryogenic target. The target is made from a diluted mixture of the complex organic material of interest; the material to be deposited is dissolved in a high vapor-pressure light-absorbent solvent. The volatile solvent is pumped away by the vacuum system. The laser beam transfers softly and without heat and/or biological deterioration and harms diverse compounds, including big molecular weight species, such as polymeric or organic molecules. MAPLE experimental set-up can be seen in Figure 1.

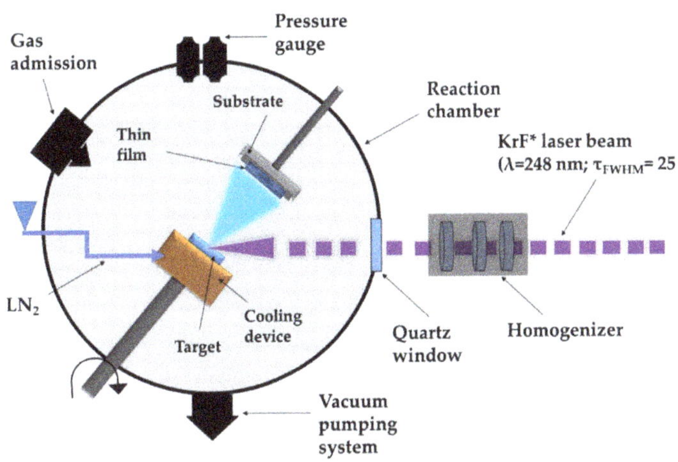

Figure 1. Schematic representation of MAPLE experimental set-up.

In the first experiment we deposited a thin layer of PMMA by MAPLE onto Ti substrates. A second layer consisting of BG and CIPRO was applied also by MAPLE onto the PMMA initial thin film. The resulting samples are further denoted as BG+CIPRO/PMMA.

For the first MAPLE thin film, 0.6 g PMMA was dissolved into 19.3 mL chloroform to obtain a frozen target in liquid nitrogen. For the second layer, 0.05 g BG and 0.125 g ciprofloxacin were dissolved into 12 mL deionized water and frozen.

The experiments were carried out with a KrF* (λ = 248 nm, $\tau_{FWHM} \leq 25$ ns) excimer laser source operated at 7 Hz repetition rate and a fluence of 0.55 J/cm^2. For the deposition of each thin film, 10,000 subsequent laser pulses were applied. In order to obtain a uniform layer and to avoid target drilling, both the substrate and target were simultaneously rotated with 50 rpm, while the background pressure inside the deposition chamber was maintained at 2×10^{-2} mbar. Throughout the deposition experiments, the targets were kept at liquid nitrogen temperature using a cryogenic setting. The target to substrate separation distance was set at 5 cm.

2.3. *Thin Films Characterization*

2.3.1. Morphological and Structural Characterization

The surface morphology of samples was inspected by scanning electron microscopy (SEM) using a FEI Inspect S electron microscope (Hillsboro, OR, USA). The investigations were carried out in top-view and cross-section modes, at 20 kV acceleration voltages, in high vacuum, using secondary electron mode. For the reduction of electrical charging through analysis, all samples were capped with a thin gold film.

The surface topography of samples was obtained by atomic force microscopy (AFM), by using an AFM TT-Workshop apparatus (Signal Hill, CA, USA), with a scan area of (15 × 15) μm^2 and (10 × 10) μm^2, in non-contact operating mode. Because of the samples' high roughness, the scanning speed was reduced to 0.5 Hz at 650 mV.

The wettability of the deposited coating was analyzed by the sessile drop method, using an Attension Theta Lite (TL) 101 optical tensiometer (version 1.0.3, Biolin Scientific, Vastra Frolunda, Sweden), in atmospheric conditions (22 °C, 30% relative humidity). The contact angle between the test liquid and surface of the specimens was measured.

The stoichiometry and chemical functions integrity of thin films were studied by means of Fourier-transform infrared spectroscopy (FTIR). The FTIR study was performed using a Nicolet ™ FT-IR iS ™ 50 spectrometer (Thermo Fisher Scientific, Waltham, MA, USA) equipped with an attenuated total reflection (ATR) module, within 3700–700 cm^{-1} range at 4 cm^{-1} resolutions. For the ATR, a ZnSe crystal was used, with the following characteristics: 1.5 mm diameter, one internal reflection at 42° incidence angle, penetration depth of 2.03 μm at 1000 cm^{-1}, and 2.4 refractive index. The spectra were taken in absorbance mode, between 1800–650 cm^{-1} with a resolution of 4 cm^{-1}. The recorded spectra were averages of 16 scans. FTIR ATR spectra of single components were performed for CIPRO, BG, and PMMA in powder form. The IR spectra of BG+CIPRO/PMMA structures after different immersion times in SBF were recoded.

2.3.2. Drug Release Behavior Evaluation

To simulate the insertion of implants into the human body and the phenomena occurring at the tissue-implant interface as a result of interaction with physiological fluids, BG+CIPRO/PMMA samples were immersed in 1 mL of SBF in glass flasks, at 37 °C, in a Binder microbiological incubator. They were investigated by FTIR.

The volume of the SBF used for drug release evaluation tests was calculated from Equation (1):

$$Vs = Sa/10 \tag{1}$$

where Vs is the volume of SBF (mL) and Sa is the apparent surface area of the sample (mm^2).

The SBF solutions containing the released products from thin films were analyzed by UV-Vis after different immersion times. The release profiles of CIPRO over time were recorded by means of an Evolution 220 spectrophotometer (ThermoFisher Scientific, Darmstadt, Germany) within the range 190–1200 nm, in absorbance mode. All measurements were carried out in triplicate, in accordance with ISO/FDIS 23317: 2007 (E). The absorption peaks amplitude, which is proportional to the drug concentration delivered by coatings to the adjacent tissue, was determined.

2.3.3. Electrochemical Investigation

The electrochemical behavior of the investigated samples was analyzed by potentiodynamic polarization and electrochemical impedance spectroscopy (EIS), using a VersaSTAT 3 Potentiostat/Galvanostat system (Princeton Applied Research, Oak Ridge, USA). The measurements were performed in SBF at 37 °C, using a typical three electrode setup with a Pt grid counter electrode (CE) and an Ag/AgCl (saturated KCl) (0.197 V) as the reference electrode (RE). The working electrode consisted of uncoated and BG+CIPRO/PMMA-coated Ti, respectively.

The open circuit potential (EOC) was monitored for 36 h, starting immediately after the sample's immersion, whereas EIS data was recorded after 1, 12, 24, and 36 h of immersion. The EIS measurements were performed over a range of frequencies (0.2 ÷ 103 Hz), by applying a sinusoidal signal of 10 mV RMS vs. EOC. The data were recorded by VersaStudio software (version 2.60.6, Princeton Applied Research, Oak Ridge, TN, USA) and the EIS fitting procedure was performed using ZView software (version 12136-4, Scribner Associates Inc., Southern Pines, NC, USA).

2.4. Biological Evaluation

2.4.1. Biocompatibility Assay

Cell Culture

Mouse osteoblast-like MC3T3-E1 cells were grown in Dulbecco Modified Eagle's Medium (Invitrogen, Carlsbad, CA, USA) with 10% fetal bovine serum (Gibco, Grand Island, NY, USA) at 37 °C in a humidified atmosphere with 5% CO_2. The cells were seeded at a cell density of 5 × 104 cells/ cm^2 on the tissue culture plastic surface (TCPS) which served as a control, or on the top of the tested samples (Ti and BG+CIPRO/PMMA) which were previously sterilized under UV light. After 24 h of incubation in standard conditions, the biocompatibility tests were performed.

MTT Assay

Cellular viability was measured using the 3-(4,5-dimethylthiazol-2-yl)-2,5-diphenyltetrazolium bromide (MTT; Sigma-Aldrich, St. Louis, MO, USA) assay. At the end of the incubation period, the culture medium was removed and the cells were incubated with 1 mg/mL MTT for 2 h at 37 °C. The purple formazan crystals formed in the viable cells were dissolved with 2-propanol (Sigma-Aldrich, St. Louis, MO, USA) and the absorbance was measured at 595 nm using a plate multireader (GENios Tecan, Ramsey, MN, USA).

Griess Assay

Using the Griess reagent, a stoichiometric solution (v/v) of 0.1% naphthyl ethylenediamine dihydrochloride and 1% sulphanilamide, the level of nitric oxide (NO) released in the culture medium during the 24 h of incubation was measured. Equal volumes of culture supernatants and Griess reagent were mixed and their absorbance was read at 550 nm using the GENios Tecan multireader.

F-Actin Staining

At the end of incubation, the cells were fixed with 4% paraformaldehyde for 20 min and permeabilized with 0.1% Triton X-100—2% bovine serum albumin for 45 min. The actin filaments were stained with 10 µg/mL phalloidin-FITC (fluorescein isothiocyanate) and the nuclei were counterstained with 2 µg/mL DAPI (4′,6-diamino-2-phenylindole). The images were obtained with an Olympus IX71 inverted fluorescence microscope.

2.4.2. Antimicrobial Activity Evaluation

The antimicrobial activity of the materials was evaluated on standard strains *Staphylococcus aureus* ATCC 25923, *Enterococcus faecalis* ATCC 29212, *Escherichia coli* ATCC 25922, and *Pseudomonas aeruginosa* ATCC 27853. Following the UV sterilization, the materials were immersed in a 10^5 CFU (colony forming unit)/mL bacterial suspension and incubated at 37 °C in a shaker. After different periods of time (T0—after initial contact, 15, 30, and 45 min, and 1, 2, 4, and 24 h), the microbial viability was evaluated through serial dilutions plating on nutrient medium (Plate Count Agar, Oxoid, Sigma-Aldrich Chemie GmbH, Taufkirchen, Germany). Controls were represented by bacterial strains cultivated in the same conditions and in presence of Ti in order to evaluate the material's influence on microbial viability.

3. Results and Discussion

3.1. Surface Investigation of As-Deposited Structures

The micro-topography of the thin films is observable from typical SEM images presented in Figure 2. One can note a uniform morphology with spherical, ovoid, or elongated formations of variable dimensions, representative for pulsed laser deposition methods [30]. The micrometric droplets are arbitrarily scattered throughout the surface (Figure 2a). They have a favorable outcome on cell adhesion and growth/proliferation [31]. Considering the magnification and the software of the SEM microscope, estimated thickness for BG+CIPRO/PMMA films was ~2.052 (±5%) µm (Figure 2c).

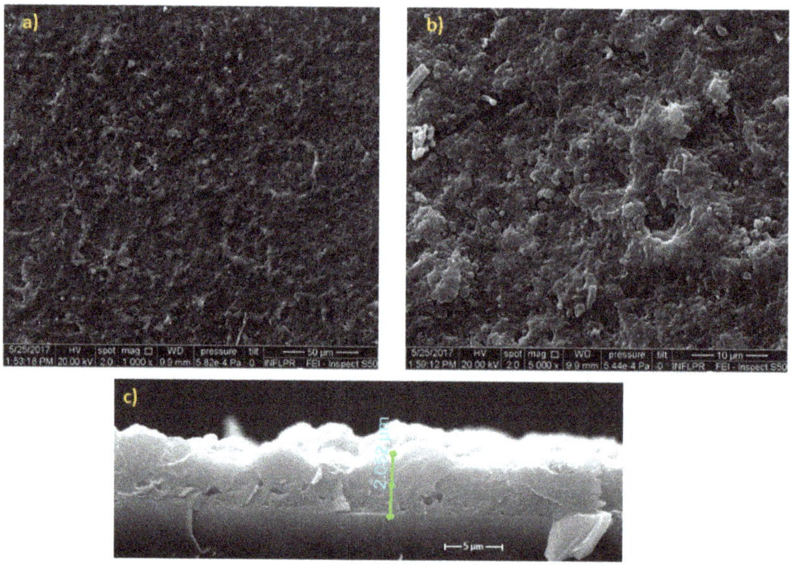

Figure 2. SEM images of BG+CIPRO/PMMA MAPLE films on Si substrates covered with PMMA: (**a**) overview, (**b**) detail, and (**c**) cross-section.

The MAPLE films preserve the composition of raw materials as demonstrated by corresponding FTIR spectra in Figure 3. Peaks belonging to BG, PMMA, and CIPRO in powdery and thin film forms were identified. For the BG+CIPRO/PMMA thin film, FTIR analysis revealed representative peaks for PMMA and CIPRO. PMMA was identified with bands at 1723 cm^{-1} attributed to stretching vibrations acrylate carboxyl groups, at 1447 cm^{-1} and 1431 cm^{-1} attributed to the bending vibration of the C–H bonds of the –CH_3 groups, and at 1385 cm^{-1} and 751 cm^{-1} responsible for α-methyl group vibrations [32]. The presence of PMMA bands in the IR spectrum suggests that the polymer did not suffer structural modifications during the deposition process. The peak at 1726 cm^{-1} represented by C=O stretching vibration of carboxyl group is characteristic for both CIPRO and PMMA [33,34]. In the case of CIPRO, two typical peaks were identified: at 1629 cm^{-1} attributed to C–O stretching vibration (quinoline) and 1041 cm^{-1} attributed to $-CH_2$ deformation vibration [35]. The other CIPRO specific bands between 1500 cm^{-1} and 1000 cm^{-1} overlap the PMMA ones and are not observable. It should be noted that the thin film IR spectrum fingerprint is determined by the weight ratio of PMMA and BG. Thus, for ratios greater than 7.5, PMMA IR peaks predominate, whereas for ratios less than 7.5, higher BG peaks predominate [36]. In our case, the PMMA/BG weight ratio was 12, therefore the IR spectrum of BG+CIPRO/PMMA lacked the characteristic IR bands of BG.

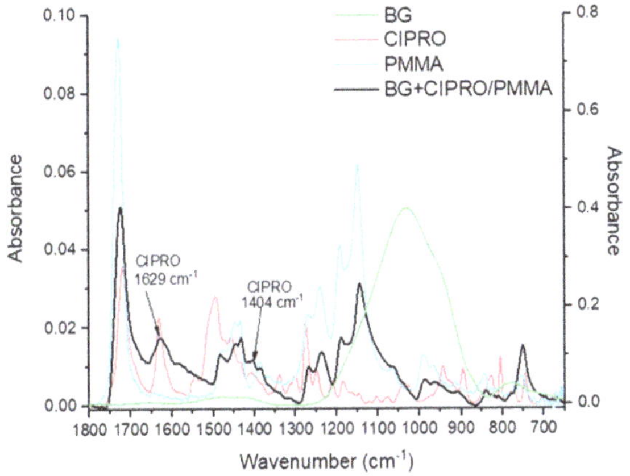

Figure 3. FTIR spectra of BG+CIPRO/PMMA thin film deposited by MAPLE and of single components, namely PMMA, CIPRO, and BG. Note: OY axis (left)—absorbance of CIPRO, PMMA, and BG+CIPRO/PMMA thin film; OY axis (right)—absorbance of BG.

The surface topography of the BG+CIPRO/PMMA sample was characterized by a high roughness, which did not allow the scanning of large 15 × 15 μm^2 areas. In order to avoid contacting the higher peaks present in the topography that could unfavorably affect the image quality, images with an area of 10 × 10 μm^2 were acquired at two different locations on the sample (Figure 4a,b).

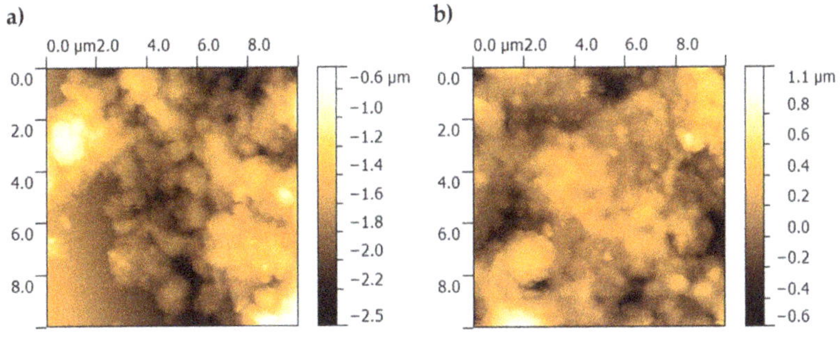

Figure 4. AFM images of as-deposited film surface, at two different locations on the sample. (**a**) BG+CIPRO/PMMA_1 and (**b**) BG+CIPRO/PMMA_2.

From AFM analyses, one can observe that the BG+CIPRO/PMMA sample (Figure 4a,b) shows high local roughness, reaching peaks of 1 μm. The clusters on BG+CIPRO/PMMA_1 surface contain particles of ~300 nm in size, while the BG+CIPRO/PMMA_2 sample presents large clusters incorporating particles with a diameter of ~200 nm.

AFM data (Table 1) are in good accordance with SEM observations regarding the surface roughness and the presence and dimension of particles.

Table 1. Roughness values of thin films before immersion in SBF.

Sample	RMS (μm)	Ra (μm)
BG+CIPRO/PMMA_1	0.283	0.222
BG+CIPRO/PMMA_2	0.211	0.160

The surface wettability investigation comprises the contact established between a liquid (in our case SBF) in form of a droplet and a solid (uncoated or MAPLE-coated samples). The contact angle (CA) represents a reliable parameter for biomaterials characterization as tailoring the hydrophobicity/hydrophilicity of surfaces is of great interest for cell adhesion and proliferation.

Figure 5 shows representative images of SBF CA as measured on samples. The uncoated Ti substrate was the least hydrophilic and presented a contact angle of 115.79°. The sample coated with BG+CIPRO/PMMA achieved a remarkable improvement in hydrophilicity, with a contact angle of 25.76°. The high wetting of an implant surface increases the bone cells' contact with an implant, which leads to a rapid and a long-term osseointegration and improves wound healing of hard and soft tissues [37]. As known, hydrophilic properties are influenced by the chemical composition and the surface roughness [38]. As shown in Figure 5, after applying BG+CIPRO/PMMA coating on the substrate, the CA was decreased significantly to 25.76°. One of the possible explanation reasons for the high hydrophilicity of this sample is the presence of the antibiotic. The oxygen functional groups are exposed on the surface of the composite coating, and they act as key sites for water adsorption and ultimately increase the surface wettability. Another factor that influences the wetting properties is the presence of droplets (Figure 2) and particles (Figure 3) on the surface. Roughening the surface has previously been shown to improve the wettability and therefore reduce the contact angle [39–42].

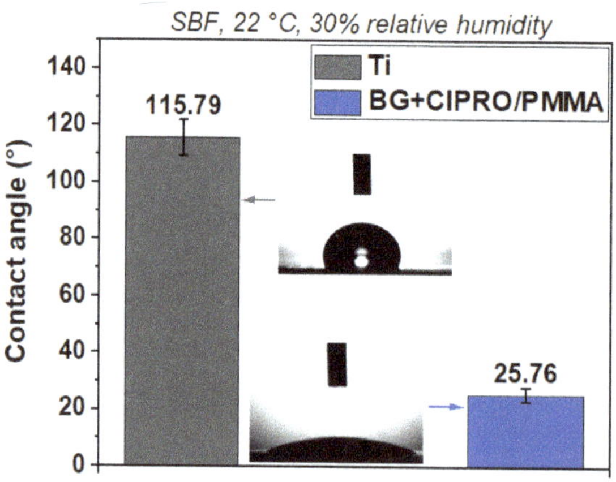

Figure 5. SBF CA measurements on bare Ti and Ti substrate coated with BG+CIPRO/PMMA.

3.2. *Characterization of Structures after the Immersion in SBF*

As observed by SEM images in Figure 6, BG+CIPRO/PMMA thin films deposited on Ti substrates, after 21 days of immersion in SBF, consist of a rather smooth matrix covered with arbitrarily scattered, micronic, isolated, or agglomerated spheroidal particles, probably from the biological apatite.

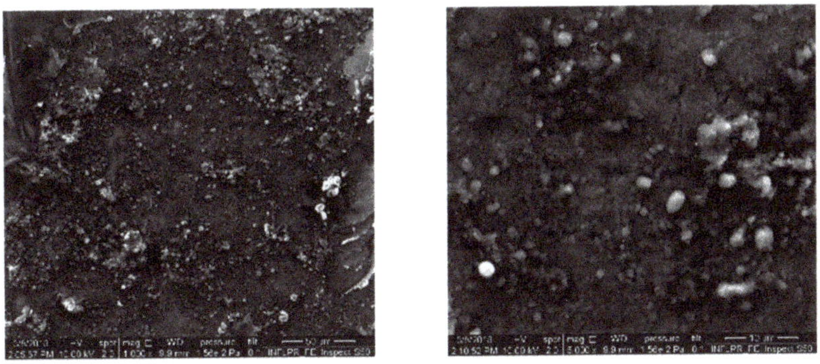

Figure 6. SEM images of BG+CIPRO/PMMA thin films after 21 days of immersion in SBF, at different magnifications.

In parallel to SEM, AFM presents 15 × 15 μm^2 areas with unevenly distributed particles on surfaces (Figure 7). The topology is characterized by an RMS in the range 0.2–0.3 μm (Table 2). The registered values are due to the non-equilibrium character of the depositing technique only, but also to the formation of salts during the immersion in SBF.

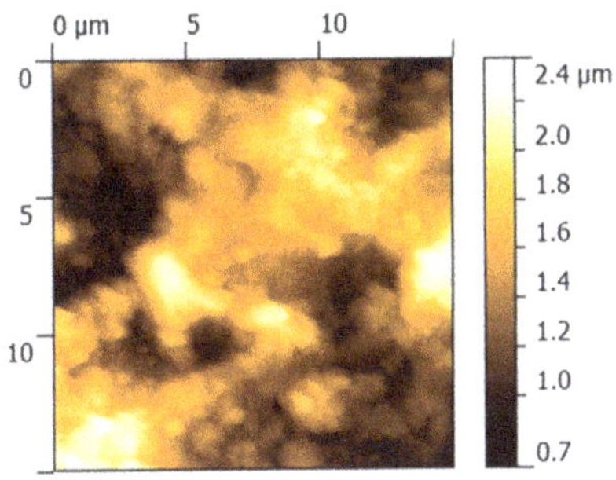

Figure 7. AFM image on 15 × 15 μm^2 surface areas of BG+CIPRO/PMMA thin films after 21 days of immersion in SBF.

Table 2. Roughness values of thin films after 21 days of immersion in SBF.

Sample	RMS (μm)	Ra (μm)
BG+CIPRO/PMMA	0.330	0.277

SBF containing products released from samples was analyzed by UV-Vis. The drug release kinetics was studied for a prolonged period (21 days) and is displayed in Figure 8. The maximum wavelength (λmax), obtained by scanning all samples from 200 to 400 nm, was found to be 267 nm. As it can be observed, the antibiotic is released at a constant rate throughout the 21 days immersion interval. This process suggests the gradual dissolution of BG in SBF, simultaneously with the release of the antibiotic. There is a demand to design drug release systems that allow longer release periods, e.g., the treatment time for most patients with bone infection caused by bacteria is usually at least 4–6 weeks [43]. This represents an important reason in favor of local treatment compared to oral ones, as a result of prolonged storage of the drug in the MAPLE-deposited coating.

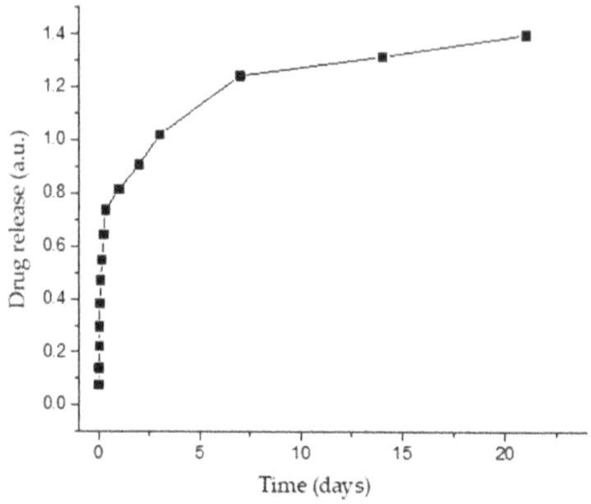

Figure 8. Drug release as function of time from BG+CIPRO/PMMA coatings.

The electrochemical performance of the investigated systems during 36 h of immersion in SBF and their behavior after 1, 12, 24, and 36 h was monitored by EIS. Nyquist impedance plots of Ti and BG+CIPRO/PMMA-coated Ti (Figure 9a,b) showed a different AC polarization response with higher semicircles in the case of Ti, ascribed to higher charge transfer resistance. Moreover, a slight variation as a function of immersion time was observed when BG+CIPRO/PMMA structure was analyzed. Note the increase in semicircle diameter corresponding to the spectrum recorded after 36 h of immersion as compared with the initial EIS data, proving the higher corrosion resistance at the end of the immersion time.

As observed in Bode amplitude and phase plots, the impedance modulus varies significantly over the applied frequency range, more evident for BG+CIPRO/PMMA, revealing the dielectric properties of the investigated systems, observed at high-frequency range (HF). Data in the literature reported that the time constant present in the phase plot at medium frequencies (MF) provide information related to coating–solution interface, whereas the second time constant present in the low-frequency range (LF) is indicative of a corrosion process occurring at the substrate–solution interface [44]. The BG+CIPRO/PMMA-coated Ti is characterized by two well-defined time constants (indicated by arrows in Figure 9f). In the case of Ti, literature data reported the formation of a bilayer structure of oxide film, with thickness and characteristics as a function of the testing solution [45] and mainly formed of TiO_2, Ti_3O_5, Ti_2O_3, and TiO oxides [46,47]. A two-time constant representative equivalent circuit was selected taking into consideration the results of XPS analysis showing the presence of the mentioned oxides and suboxides [46,48]. In the current study, the Bode phase plot characteristic to Ti sample shows a convolution for the two phases over the selected immersion time (indicated by the arrow in Figure 9e). Therefore, the recorded impedance data were fitted with the electrical equivalent circuit (EEC) presented in the inset of Figure 9a,b). In order to investigate the electrolyte–coating interface and also electrolyte–substrate interface, the following electrical components were considered: R_s (solution resistance), CPE_{coat} (coating capacitance), R_{pore} (resistance associated to the current flow through the pores generated by the coating defects), CPE_{dl} (double layer capacitance), and R_{ct} (charge transfer resistance). A constant phase element (CPE) was used to take into consideration the deviations from the ideal behavior [49]. As pointed out in reference [50], these deviations can be due to surface disorder or inhomogeneity, geometric irregularities, working electrode roughness, or porosity.

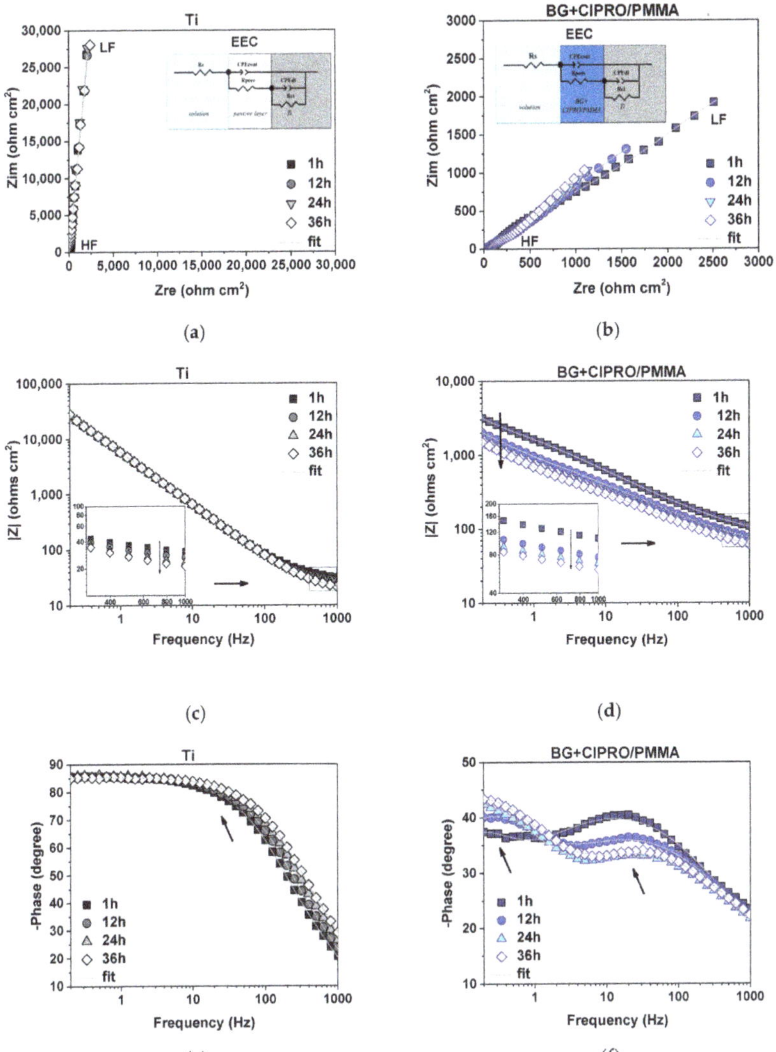

Figure 9. Nyquist and Bode amplitude and phase plots of Ti (**a**,**c**,**e**) and BG+CIPRO/PMMA-coated (**b**,**d**,**f**) Table 1; 12, 24, and 36 h of immersion in SBF at 37 °C (HF = high frequency range, HL = low frequency range, EEC = equivalent electrical circuit used for fitting the impedance data).

The electrochemical parameters obtained after the fitting procedure of impedance data with the mentioned EEC are given in Table 3. The associated relative error of each parameter and χ^2 used as an indication of the goodness of fit is also presented. The values of relative error were lower than ~7% (calculated for R_{ct} in the case of BG+CIPRO/PMMA), whereas in the case of Ti the mentioned parameter could not be determined in the selected frequency range. Moreover, the low values obtained for χ^2 parameter ($7 \times 10^{-5} \div 2 \times 10^{-4}$) represent a validation of the fitting procedure.

Table 3. The open circuit potential (EOC) and EIS fitted parameters with the associated relative errors of Ti and BG+CIPRO/PMMA-coated Ti after 1, 12, 24, and 36 h of immersion in SBF at 37 °C.

Sample	Ti				BG+CIPRO/PMMA			
	1 h	12 h	24 h	36 h	1 h	12 h	24 h	36 h
E_{OC} (mV)	64.72	−29.75	8.28	−3.07	−63.79	−82.81	−62.57	−60.42
Rs (Ω cm^2)	27.03 (0.28%)	22.51 (0.28%)	19.01 (0.34%)	16.51 (0.37%)	66.45 (0.68%)	44.99 (0.39%)	39.04 (0.36%)	34.19 (0.37%)
Qcoat (μF s$^{(\alpha-1)}$ cm^{-2})	19.89 (0.15%)	19.95 (0.14%)	20.25 (0.15%)	19.05 (0.16 %)	161.86 (0.26%)	257.06 (0.15 %)	318.27 (0.15%)	343.91 (0.14%)
α_{coat}	0.96 (0.03%)	0.96 (0.03%)	0.96 (0.03%)	0.97 (0.03%)	0.55 (0.09%)	0.53 (0.05%)	0.52 (0.05%)	0.51 (0.05%)
R_{pore} (Ω cm^2)	87.23 (2.01%)	74.98 (1.98%)	70.11 (2.32%)	58.34 (2.22%)	4337 (1.04%)	1735 (0.61%)	988 (0.52%)	964 (0.55%)
Q_{dl} (μF s$^{(\alpha-1)}$ cm^{-2})	9.90 (0.36%)	9.68 (0.35%)	9.13 (0.41%)	10.45 (0.37%)	191.39 (2.99%)	189.38 (0.97%)	284.68 (0.65%)	257.68 (0.66%)
α_{dl}	0.93 (0.09%)	0.93 (0.09%)	0.93 (0.10%)	0.92 (0.09%)	0.89 (1.81%)	0.79 (0.58%)	0.73 (0.41 %)	0.73 (0.41%)
R_{ct} (Ω cm^2)	-	-	-	-	7631 (7.43%)	7717 (2.74%)	11,401 (3.95%)	12,054 (3.97%)
χ^2	2×10^{-4}	2×10^{-4}	2×10^{-4}	2×10^{-4}	3×10^{-4}	8×10^{-5}	7×10^{-5}	7×10^{-5}

The open-circuit potential (EOC) was recorded before each EIS measurement as a function of immersion time. A higher variation range was obtained for the Ti sample, showing the instability of the oxide layer formed when in contact with body fluids. As showed in reference [51], the mentioned layer is destroyed by Cl^- ions from SBF. The layer's breakdown leads to Q_{coat} increase up to 36 h, when the higher value was recorded, as compared to the initial capacitance (Q_{coat_Ti} (1 h) = 19.89 μF s$^{(\alpha-1)}$ cm^{-2} and to Q_{coat_Ti} (36 h) = 19.05 μF s$^{(\alpha-1)}$ cm^{-2}). Even though the spectra displayed by the Bode amplitude plot indicated a stable behavior in the middle-frequency range specific to the formed protective layer, the fitted parameters showed low values of R_{pore} with decreasing tendency over time, generating pathways for the electrolyte ingress. However, no corrosion process was detected. Moreover, a 5% increase (calculated at 0.2 Hz for 36 h immersion spectra as compared with the initial spectra) of EIS recorded data in LF range of Bode amplitude plot was observed. All the above results indicate the formation of new outer oxides that continuously rearrange after breakdown, having slightly higher protective properties against corrosion attack at the end of immersion.

On the other hand, BG+CIPRO/PMMA showed a more stable EOC (approximately $-60 \div -80$ mV) with less variation and a higher Rpore value as compared with Ti sample at the end of the test ($R_{pore\_BG+CIPRO/PMMA}$ (36 h) ~964 Ω cm^2, whereas R_{pore_Ti} (36 h) ~58 Ω cm^2). For both tested systems, solution resistance gradually decreased as the immersion time increased (as indicated in the inset of Figure 9c,d), with higher values obtained in the case of coated sample. A degradation over 36 h immersion of BG+CIPRO/PMMA coating was occurring, pointed out by the increase in Q_{coat} and associated αcoat parameter, which started at 0.55 and recorded a 0.51 value at the end of the test. A similar decreasing tendency was shown by the double-layer capacitance and αdl, respectively. Moreover, a 52% decrease of impedance modulus calculated at 0.2 Hz for 36 h immersion spectra as compared with the initial spectra was obtained. However, the charge transfer resistance recorded large values reaching 12,054 Ω cm^2 after 36 h, indicating an improvement with time of protecting properties, thus delaying the contact between the substrate and corrosive ions. The decrease of maximum phase angle in MF range and the increase of phase angle in LF range suggest the BG dissolution in time, revealing the capacitive behavior of PMMA layer, which is in accordance with previous studies [17,51].

Figure 10 shows the IR spectra of the BG+CIPRO/PMMA thin films after 2 and 8 h, and 1, 3, 7, and 21 days of immersion in SBF, respectively. Transformations on the surface of the thin film were observed after the first day of immersion in SBF, indicating BG dissolution and the formation of a hydroxyl carbonate apatite layer on the film surface. The hydroxyl carbonate apatite peaks were found at 1553, 1300, and 1038 cm^{-1}, respectively. The band at 1553 cm^{-1} was assigned to carbonate group absorption [52]. The peak around 1300 cm^{-1} was attributed to phosphate absorptions. The band at 1038 cm^{-1} corresponds to the asymmetric stretching of the P–O bond in $(PO_4)^{3-}$. This behavior is consistent with the apatite formation mechanism described in reference [53].

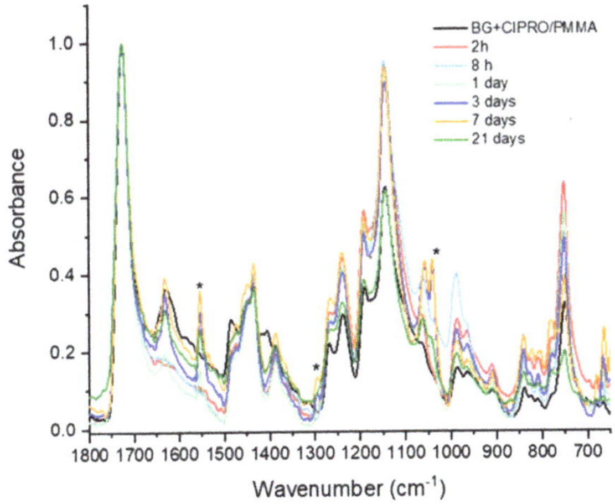

Figure 10. FTIR spectra of samples after different immersion times. * hydroxyl carbonate apatite peaks.

3.3. Biological Assays

As it can be observed in Figure 11, the fluorescence staining of actin filaments in MC3TE-E1 osteoblasts revealed almost similar cell densities for both uncoated and coated Ti samples compared to TCPS control after 24 h of incubation. The images revealed cell-to-cell junctions on tested samples, established by numerous lamellipodia and filopodia. These structures confirmed the good cell adhesion and migration on the coated surface, the osteoblasts having a structured actin cytoskeleton.

The MTT assay results (Figure 12) were in good agreement with the fluorescence microscopy, the level of viable cells being almost similar for both control and samples. Their biocompatibility was also validated by Griess assay, as the level of nitric oxide did not increase, as compared to TCPS.

Gram-positive bacteria, such as *S. aureus* and *E. faecalis*, are dangerous and challenging infectious agents due to their continuously increasing resistance to antibiotics [54,55]. Among these, *S. aureus* is the most frequently isolated bacterial population from implant-associated infections [56]. Enterococci, including *E. faecalis*, are the second most frequently isolated Gram-positive bacteria [57].

Bacteria growth and adhesion on surfaces take place in two stages: (a) initial and reversible physical phase, and (b) irreversible chemical and cellular phase. In the first stage, many surface-related factors such as surface roughness and surface wettability affect the process. This stage usually takes 6 h and if bacteria attach to the surface during this stage, they can grow rapidly to form a biofilm [58].

After 24 h of direct contact with bacterial strains in liquid media, Ti and BG+CIPRO/PMMA samples did not inhibit the growth of *S. aureus* (Figure 13) and *E. faecalis* (Figure 14).

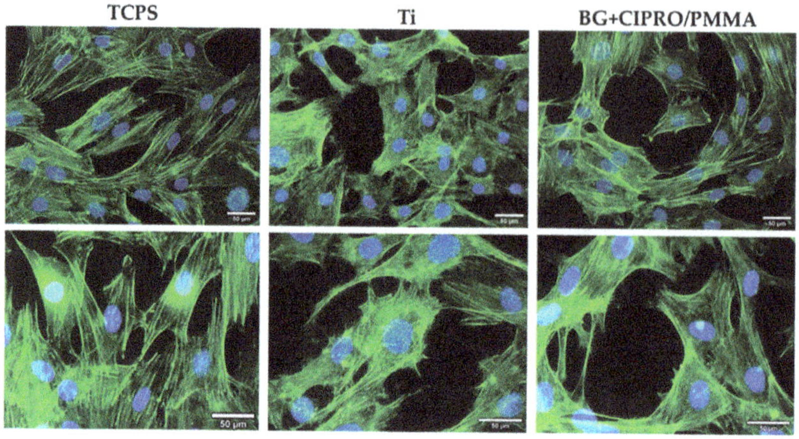

Figure 11. Fluorescence microscopy images of MC3T3-E1 osteoblasts (green: F-actin labelled with phalloidin-FITC; blue: nuclei labelled with DAPI; scale bar: 50 μm) after 24 h of incubation on TCPS control, bare Ti, and BG+CIPRO/PMMA samples.

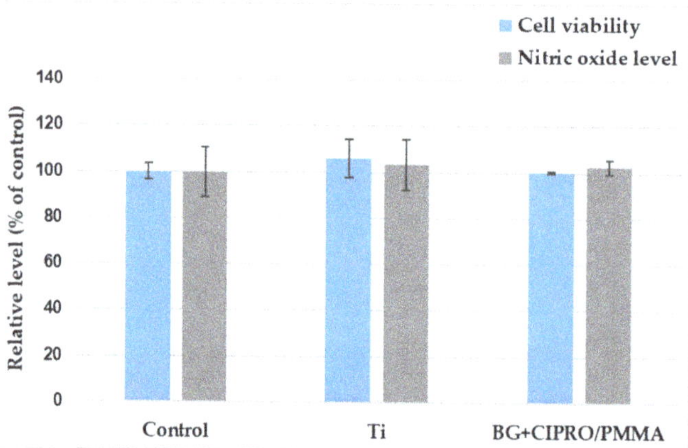

Figure 12. MTT (cell viability) and Griess (NO level). Results are presented as mean ± standard deviation of three independent experiments and represented as percentages of control.

E. coli represents one of the most common Gram-negative species implicated in the etiology of device-associated infections of endogenous origin [59], while *P. aeruginosa* is responsible for a great variety of medical device- and tissue biofilm-associated infections [60].

The bare Ti had no inhibitory effect on Gram-negative or Gram-positive bacteria, while the MAPLE-obtained coating interfered with the Gram-negative bacterial strain growth, a decrease of viable cell counts being noticed after the first 45 min of incubation. A 2-log reduction of the total viable cells was observed for BG+CIPRO/PMMA samples as compared to the control or Ti (Figures 15 and 16).

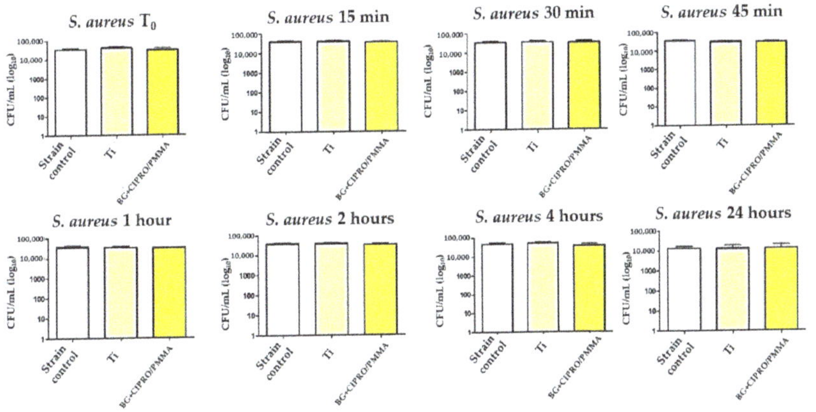

Figure 13. Evaluation of *S. aureus* viability in the presence of bare Ti and BG+CIPRO/PMMA samples.

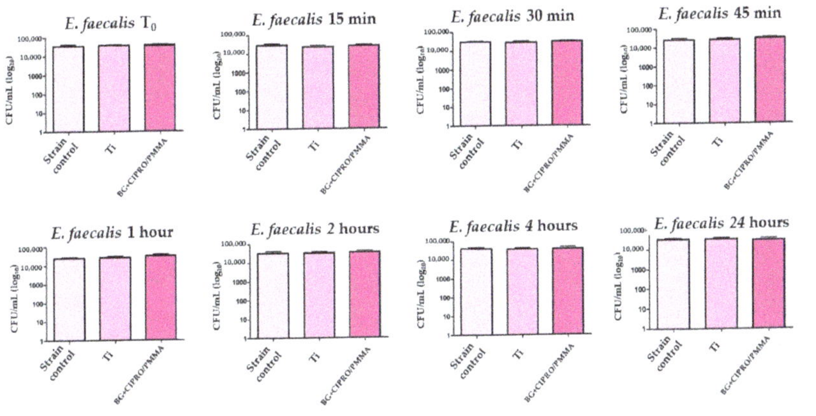

Figure 14. Evaluation of *E. faecalis* viability in the presence of bare Ti and BG+CIPRO/PMMA samples.

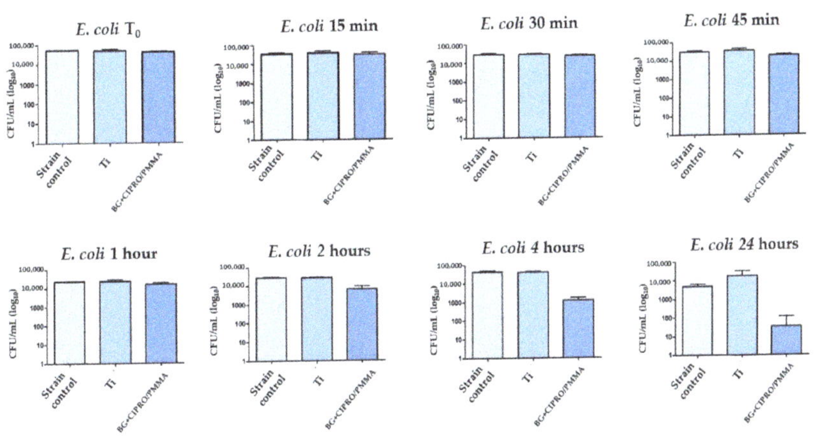

Figure 15. Evaluation of *E. coli* viability in the presence of bare Ti and BG+CIPRO/PMMA samples.

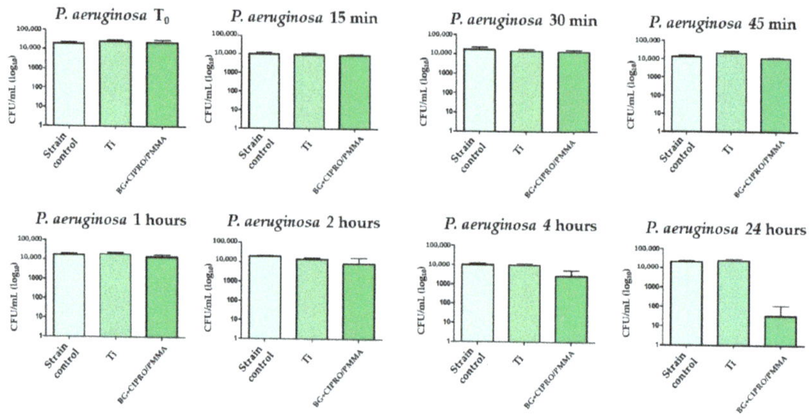

Figure 16. Evaluation of *P. aeruginosa* viability in the presence of bare Ti and BG+CIPRO/PMMA samples.

Our results demonstrate that the antibiotic is released from MAPLE layers and is effective against Gram-negative bacteria starting with 45 min after the initial contact, while for Gram-positive strains we did not obtain an inhibition. Although ciprofloxacin is a second-generation quinolone with an extended spectrum including Gram-negative bacteria and some Gram-positive strains, the lack of an inhibitory effect against *S. aureus* and *E. faecalis* could be explained by the different structure of the cellular wall (e.g., the thick peptidoglycan of the Gram-positive strains delaying the intracellular accumulation of active concentrations). In the Gram-negative strains, the influx of fluoroquinolones is favored by porins expressed in their outer membrane, such as OmpF and OmpC in the case of *E. coli* [61,62].

4. Conclusions

The potential of BG+CIPRO/PMMA to bond to bone tissue is demonstrated by the presence of the carbonated hydroxyapatite layer.

The MAPLE-processed coating proved to be a good candidate for the design of efficient implant surfaces.

BG+CIPRO/PMMA coatings successfully inhibited Gram-negative bacterial growth with no harmful effects on MC3T3-E1 osteoblasts.

It can be concluded that our coating is promising in terms of its bioactivity and may find potential applications for bone repair and tissue engineering. In bone surgery, it offers more possibilities to adapt it with the bone metabolism activities when a rapid bone biomaterial interface is necessary.

Author Contributions: Conceptualization, I.N., C.R., I.N.M.; methodology, I.N., C.R.; investigation, I.N., T.T., V.G., C.H., M.D., A.C.P., M.P., M.S.S., and L.M. writing—original draft preparation, I.N., C.R., writing—review and editing, C.R., M.C.C.; supervision, I.N.M., M.C.C. All authors have read and agreed to the published version of the manuscript.

Funding: This research was funded by a grant of the Ministry of Research, Innovation and Digitization, CNCS—UEFISCDI, project number PN-III-P1-1.1-PD-2021-0598, within PNCDI III.

Institutional Review Board Statement: Not applicable.

Informed Consent Statement: Not applicable.

Acknowledgments: This work was supported by the Romanian Ministry of Education and Research, under Romanian National Nucleu Program LAPLAS VI—contract no 16N/2019. I.N. acknowledges the support by a grant of the Ministry of Research, Innovation and Digitization, CNCS—UEFISCDI, project number PN-III-P1-1.1-PD-2021-0598, within PNCDI III. M.D. and, A.C.P. acknowledges the support of the Romanian National Core Program 18N/2019. T.T. acknowledges the support by a grant of the Romanian Ministry of Education and Research, CNCS—UEFISCDI, project number PN-III-P1-1.1-PD-2019-1117, within PNCDI III. C.R and I.N.M. acknowledge the support of the Ministry of Research, Innovation and Digitization, CNCS—UEFISCDI, contract PCE 113/2021.

Conflicts of Interest: The authors declare no conflict of interest.

References

1. Tiwari, A.; Sharma, P.; Vishwamitra, B.; Singh, G. Review on Surface Treatment for Implant Infection via Gentamicin and Antibiotic Releasing Coatings. *Coatings* **2021**, *11*, 1006. [CrossRef]
2. Filipović, U.; Dahmane, R.G.; Ghannouchi, S.; Zore, A.; Bohinc, K. Bacterial adhesion on orthopedic implants. *Adv. Colloid Interface Sci.* **2020**, *283*, 102228. [CrossRef] [PubMed]
3. Utomo, E.; Stewart, S.A.; Picco, C.J.; Domínguez-Robles, J.; Larrañeta, E. Classification, material types, and design approaches of long-acting and implantable drug delivery systems. In *Long-Acting Drug Delivery Systems*; Pharmaceutical, Clinical, and Regulatory Aspects; Woodhead Publishing Series in Biomaterials; Woodhead Publishing: Sawston, UK, 2022; pp. 17–59. ISBN 978-0-12-821749-8. [CrossRef]
4. Kaur, G.; Pandey, O.; Singh, K.; Homa, D.; Scott, B.; Pickrell, G. A review of bioactive glasses: Their structure, properties, fabrication and apatite formation. *J. Biomed. Mater. Res. Part A* **2014**, *102*, 254–274. [CrossRef] [PubMed]
5. Boulila, S.; Oudadesse, H.; Badraoui, R.; Lefeuvre, B.; Mabrouk, M.; Chaabouni, K.; Mostafa, A.; Makni-Ayedi, F.; Barroug, A.; Rebai, T.; et al. Antioxidative/oxidative effects and retarding osteoconductivity of ciprofloxacin-loaded porous polyvinyl alcohol/bioactive glass hybrid. *Med Biol. Eng. Comput.* **2017**, *55*, 17–32. [CrossRef] [PubMed]
6. Haider, S.; Haider, A.; Bano, B.; Khan, R.; Bukhari, N.; Alrahlah, A. Microbial Biofilm Infections in Tissue Implant: A Review. *J. Biomed. Res. Environ. Sci.* **2021**, *2*, 1163–1167. [CrossRef]
7. Caldara, M.; Belgiovine, C.; Secchi, E.; Rusconi, R. Environmental, Microbiological, and Immunological Features of Bacterial Biofilms Associated with Implanted Medical Devices. *Clin. Microbiol. Rev.* **2022**, *35*, e00221-20. [CrossRef]
8. Li, X.; Sun, L.; Zhang, P.; Wang, Y. Novel Approaches to Combat Medical Device-Associated BioFilms. *Coatings* **2021**, *11*, 294. [CrossRef]
9. Racero, F.J.; Bueno, S.; Gallego, M.D. Predicting Students' Behavioral Intention to Use Open Source Software: A Combined View of the Technology Acceptance Model and Self-Determination Theory. *Appl. Sci.* **2020**, *10*, 2711. [CrossRef]
10. González, A.; Riego, A.; Vega, V.; García, J.; Galié, S.; del Río, I.G.; de Yuso, M.M.; Villar, C.; Lombó, F.; De la Prida, V. Functional Antimicrobial Surface Coatings Deposited onto Nanostructured 316L Food-Grade Stainless Steel. *Nanomaterials* **2021**, *11*, 1055. [CrossRef]
11. Villegas, M.; Alonso-Cantu, C.; Rahmani, S.; Wilson, D.; Hosseinidoust, Z.; Didar, T.F. Antibiotic-Impregnated Liquid-Infused Coatings Suppress the Formation of Methicillin-Resistant *Staphylococcus aureus* Biofilms. *ACS Appl. Mater. Interfaces* **2021**, *13*, 27774–27783. [CrossRef]
12. Gherasim, O.; Grumezescu, A.; Grumezescu, V.; Negut, I.; Dumitrescu, M.; Stan, M.; Nica, I.; Holban, A.; Socol, G.; Andronescu, E. Bioactive Coatings Based on Hydroxyapatite, Kanamycin, and Growth Factor for Biofilm Modulation. *Antibiotics* **2022**, *11*, 160. [CrossRef] [PubMed]
13. Caciandone, M.; Niculescu, A.-G.; Roșu, A.R.; Grumezescu, V.; Negut, I.; Holban, A.M.; Oprea, O.; Vasile, B.; Bîrcă, A.C.; Grumezescu, A.M.; et al. PEG-Functionalized Magnetite Nanoparticles for Modulation of Microbial Biofilms on Voice Prosthesis. *Antibiotics* **2022**, *11*, 39. [CrossRef] [PubMed]
14. Negut, I.; Visan, A.I.; Popescu, C.; Cristescu, R.; Ficai, A.; Grumezescu, A.M.; Chifiriuc, M.C.; Boehm, R.D.; Yamaleyeva, D.; Taylor, M.; et al. Successful Release of Voriconazole and Flavonoids from MAPLE Deposited Bioactive Surfaces. *Appl. Sci.* **2019**, *9*, 786. [CrossRef]
15. Floroian, L.; Ristoscu, C.; Mihailescu, N.; Negut, I.; Badea, M.; Ursutiu, D.; Chifiriuc, M.C.; Urzica, I.; Dyia, H.M.; Bleotu, C.; et al. Functionalized Antimicrobial Composite Thin Films Printing for Stainless Steel Implant Coatings. *Molecules* **2016**, *21*, 740. [CrossRef] [PubMed]
16. Negut, I.; Floroian, L.; Ristoscu, C.; Mihailescu, C.N.; Mirza Rosca, J.C.; Tozar, T.; Badea, M.; Grumezescu, V.; Hapenciuc, C.; Mihailescu, I.N. Functional Bioglass—Biopolymer Double Nanostructure for Natural Antimicrobial Drug Extracts Delivery. *Nanomaterials* **2020**, *10*, 385. [CrossRef]
17. Floroian, L.; Ristoscu, C.; Candiani, G.; Pastori, N.; Moscatelli, M.; Mihailescu, N.; Negut, I.; Badea, M.; Gilca, M.; Chiesa, R. Antimicrobial thin films based on ayurvedic plants extracts embedded in a bioactive glass matrix. *Appl. Surf. Sci.* **2017**, *417*, 224–233. [CrossRef]
18. Bohner, M.; Lemaitre, J. Can bioactivity be tested in vitro with SBF solution? *Biomaterials* **2009**, *30*, 2175–2179. [CrossRef]
19. Anene, F.; Jaafar, C.A.; Zainol, I.; Hanim, M.A.; Suraya, M. Biomedical materials: A review of titanium based alloys. *Proc. Inst. Mech. Eng. Part C: J. Mech. Eng. Sci.* **2021**, *235*, 3792–3805. [CrossRef]
20. Bhaskar, P.; Kumar, R.; Maurya, Y.; Ravinder, R.; Allu, A.R.; Das, S.; Gosvami, N.N.; Youngman, R.E.; Bødker, M.S.; Mascaraque, N.; et al. Cooling rate effects on the structure of 45S5 bioglass: Insights from experiments and simulations. *J. Non-Crystalline Solids* **2020**, *534*, 119952. [CrossRef]

21. Gupta, S.; Majumdar, S.; Krishnamurthy, S. Bioactive glass: A multifunctional delivery system. *J. Control. Release* **2021**, *335*, 481–497. [CrossRef]
22. Tanaskovic, D.; Jokic, B.; Socol, G.; Popescu, A.; Mihailescu, I.; Petrovic, R.; Janackovic, D. Synthesis of functionally graded bioactive glass-apatite multistructures on Ti substrates by pulsed laser deposition. *Appl. Surf. Sci.* **2007**, *254*, 1279–1282. [CrossRef]
23. Gyorgy, E.; Grigorescu, S.; Socol, G.; Mihailescu, I.; Janackovic, D.; Dindune, A.; Kanepe, Z.; Palcevskis, E.; Zdrentu, E.; Petrescu, S. Bioactive glass and hydroxyapatite thin films obtained by pulsed laser deposition. *Appl. Surf. Sci.* **2007**, *253*, 7981–7986. [CrossRef]
24. Scientific Reports. A Simple and Low-Cost Approach for Irreversible Bonding of Polymethylmethacrylate and Polydimethylsiloxane at Room Temperature for High-Pressure Hybrid Microfluidics. Available online: https://www.nature.com/articles/s41598-021-83011-8 (accessed on 4 May 2022).
25. Saxena, P.; Shukla, P. A comparative analysis of the basic properties and applications of poly (vinylidene fluoride) (PVDF) and poly (methyl methacrylate) (PMMA). *Polym. Bull.* **2021**. [CrossRef]
26. Ma, X.; Ahadian, S.; Liu, S.; Zhang, J.; Liu, S.; Cao, T.; Lin, W.; Wu, D.; de Barros, N.R.; Zare, M.R.; et al. Smart Contact Lenses for Biosensing Applications. *Adv. Intell. Syst.* **2021**, *3*, 2000263. [CrossRef]
27. Butoi, B.; Staicu, D. Cured pmma thin films for greatly improving the electro- optical response time in nematic liquid crystal pixels. *Dig. J. Nanomater. Biostructures* **2019**, *14*, 175–182.
28. Vardanyan, R.S.; Hruby, V.J. 33–Antimicrobial Drugs. In *Synthesis of Essential Drugs*; Elsevier: Amsterdam, The Netherlands, 2006; pp. 499–523. [CrossRef]
29. Yayehrad, A.T.; Wondie, G.B.; Marew, T. Different Nanotechnology Approaches for Ciprofloxacin Delivery Against Multidrug-Resistant Microbes. *Infect. Drug Resist.* **2022**, *15*, 413–426. [CrossRef]
30. Floroian, L.; Sima, F.; Florescu, M.; Badea, M.; Popescu, A.; Serban, N.; Mihailescu, I. Double layered nanostructured composite coatings with bioactive silicate glass and polymethylmetacrylate for biomimetic implant applications. *J. Electroanal. Chem.* **2010**, *648*, 111–118. [CrossRef]
31. Brunette, D.M.; Tengvall, P.; Textor, M.; Thomsen, P. Titanium in Medicine. In *Material Science, Surface Science, Engineering, Biological Responses and Medical Applications*; Springer: Berlin, Germany, 2001; pp. 145–162. Available online: https://link.springer.com/book/10.1007/978-3-642-56486-4 (accessed on 4 May 2022).
32. Duan, G.; Zhang, C.; Li, A.; Yang, X.; Lu, L.; Wang, X. Preparation and Characterization of Mesoporous Zirconia Made by Using a Poly (methyl methacrylate) Template. *Nanoscale Res. Lett.* **2008**, *3*, 118–122. [CrossRef]
33. Sugumaran, D.; Karim, K.J.A. Removal of copper (II) ion using chitosan-graft-poly(methyl methacrylate) as adsorbent. *Eproc. Chem.* **2017**, *2*, 1–11. [CrossRef]
34. Tommasini, F.J.; Ferreira, L.D.C.; Tienne, L.G.P.; Aguiar, V.D.O.; Da Silva, M.H.P.; Rocha, L.F.D.M.; Marques, M.D.F.V. Poly (Methyl Methacrylate)-SiC Nanocomposites Prepared Through in Situ Polymerization. *Mater. Res.* **2018**, *6*, 21. [CrossRef]
35. Tozar, T.; Boni, M.; Staicu, A.; Pascu, M. Optical Characterization of Ciprofloxacin Photolytic Degradation by UV-Pulsed Laser Radiation. *Molecules* **2021**, *26*, 2324. [CrossRef] [PubMed]
36. Cuppoletti, J. *Metal, Ceramic and Polymeric Composites for Various Uses*; BoD–Books on Demand: Norderstedt, Germany, 2011.
37. Nadimi, M.; Dehghanian, C. Incorporation of ZnO–ZrO_2 nanoparticles into TiO_2 coatings obtained by PEO on Ti–6Al–4V substrate and evaluation of its corrosion behavior, microstructural and antibacterial effects exposed to SBF solution. *Ceram. Int.* **2021**, *47*, 33413–33425. [CrossRef]
38. ScienceDirect. Antibacterial and Bioactive Properties of Stabilized Silver on Titanium with a Nanostructured Surface for Dental Applications. Available online: https://www.sciencedirect.com/science/article/pii/S0169433218312674 (accessed on 4 May 2022).
39. Ryan, B.J.; Poduska, K.M. Roughness effects on contact angle measurements. *Am. J. Phys.* **2008**, *76*, 1074–1077. [CrossRef]
40. Giljean, S.; Bigerelle, M.; Anselme, K.; Haidara, H. New insights on contact angle/roughness dependence on high surface energy materials. *Appl. Surf. Sci.* **2011**, *257*, 9631–9638. [CrossRef]
41. Shi, L.B.; Wang, C.; Ding, H.H.; Kvarda, D.; Galas, R.; Omasta, M.; Hartl, M. Influence of surface texture directionality and roughness on wettability, sliding angle, contact angle hysteresis, and lubricant entrapment capability. *Tribol. Int.* **2021**, *158*, 106932. [CrossRef]
42. Sajid, H.U.; Kiran, R. Influence of corrosion and surface roughness on wettability of ASTM A36 steels. *J. Constr. Steel Res.* **2018**, *144*, 310–326. [CrossRef]
43. Li, W.; Ding, Y.; Rai, R.; Roether, J.A.; Schubert, D.W.; Boccaccini, A.R. Preparation and characterization of PHBV microsphere/45S5 bioactive glass composite scaffolds with vancomycin releasing function. *Mater. Sci. Eng. C* **2014**, *41*, 320–328. [CrossRef]
44. Flores, J.F.; Olaya, J.J.; Colás, R.; Rodil, S.E.; Valdez-Salas, B.; Fuente, I.G. Corrosion behaviour of TaN thin PVD films on steels. *Corros. Eng. Sci. Technol.* **2006**, *41*, 168–176. [CrossRef]
45. Qin, P.; Chen, L.; Liu, Y.; Jia, Z.; Liang, S.; Zhao, C.; Sun, H.; Zhang, L. Corrosion and passivation behavior of laser powder bed fusion produced Ti-6Al-4V in static/dynamic NaCl solutions with different concentrations. *Corros. Sci.* **2021**, *191*, 109728. [CrossRef]
46. Milošev, I.; Metikoš-Huković, M.; Strehblow, H.-H. Passive film on orthopaedic TiAlV alloy formed in physiological solution investigated by X-ray photoelectron spectroscopy. *Biomaterials* **2000**, *21*, 2103–2113. [CrossRef]

47. Bignon, Q.; Martin, F.; Auzoux, Q.; Miserque, F.; Tabarant, M.; Latu-Romain, L.; Wouters, Y. Oxide formation on titanium alloys in primary water of nuclear pressurised water reactor. *Corros. Sci.* **2019**, *150*, 32–41. [CrossRef]
48. Milošev, I.; Kosec, T.; Strehblow, H.-H. XPS and EIS study of the passive film formed on orthopaedic Ti–6Al–7Nb alloy in Hank's physiological solution. *Electrochim. Acta* **2008**, *53*, 3547–3558. [CrossRef]
49. Abouzari, M.S.; Berkemeier, F.; Schmitz, G.; Wilmer, D. On the physical interpretation of constant phase elements. *Solid State Ionics* **2009**, *180*, 922–927. [CrossRef]
50. Córdoba-Torres, P. Relationship between constant-phase element (CPE) parameters and physical properties of films with a distributed resistivity. *Electrochim. Acta* **2017**, *225*, 592–604. [CrossRef]
51. Floroian, L.; Samoila, C.; Badea, M.; Munteanu, D.; Ristoscu, C.; Sima, F.; Negut, I.; Chifiriuc, M.C.; Mihailescu, I.N. Stainless steel surface biofunctionalization with PMMA-bioglass coatings: Compositional, electrochemical corrosion studies and microbiological assay. *J. Mater. Sci. Mater. Electron.* **2015**, *26*, 195. [CrossRef]
52. Ślósarczyk, A.; Paszkiewicz, Z.; Paluszkiewicz, C. FTIR and XRD evaluation of carbonated hydroxyapatite powders synthesized by wet methods. *J. Mol. Struct.* **2005**, *744–747*, 657–661. [CrossRef]
53. Hench, L.L. Bioceramics: From Concept to Clinic. *J. Am. Ceram. Soc.* **1991**, *74*, 1487–1510. [CrossRef]
54. Idrees, M.; Sawant, S.; Karodia, N.; Rahman, A. *Staphylococcus aureus* Biofilm: Morphology, Genetics, Pathogenesis and Treatment Strategies. *Int. J. Environ. Res. Public Health* **2021**, *18*, 7602. [CrossRef]
55. Bernardi, S.; Anderson, A.; Macchiarelli, G.; Hellwig, E.; Cieplik, F.; Vach, K.; Al-Ahmad, A. Subinhibitory Antibiotic Concentrations Enhance Biofilm Formation of Clinical *Enterococcus faecalis* Isolates. *Antibiotics* **2021**, *10*, 874. [CrossRef]
56. Masters, E.A.; Ricciardi, B.F.; Bentley, K.L.; Moriarty, T.F.; Schwarz, E.M.; Muthukrishnan, G. Skeletal infections: Microbial pathogenesis, immunity and clinical management. *Nat. Rev. Microbiol.* **2022**. [CrossRef]
57. Nawaz, A.; Ali, S.M.; Rana, N.F.; Tanweer, T.; Batool, A.; Webster, T.J.; Menaa, F.; Riaz, S.; Rehman, Z.; Batool, F.; et al. Ciprofloxacin-Loaded Gold Nanoparticles against Antimicrobial Resistance: An In Vivo Assessment. *Nanomaterials* **2021**, *11*, 3152. [CrossRef]
58. Shanaghi, A.; Mehrjou, B.; Ahmadian, Z.; Souri, A.R.; Chu, P.K. Enhanced corrosion resistance, antibacterial properties, and biocompatibility by hierarchical hydroxyapatite/ciprofloxacin-calcium phosphate coating on nitrided NiTi alloy. *Mater. Sci. Eng. C* **2021**, *118*, 111524. [CrossRef] [PubMed]
59. Crémet, L.; Broquet, A.; Brulin, B.; Jacqueline, C.; Dauvergne, S.; Brion, R.; Asehnoune, K.; Corvec, S.; Heymann, D.; Caroff, N. Pathogenic potential of *Escherichia coli* clinical strains from orthopedic implant infections towards human osteoblastic cells. *Pathog. Dis.* **2015**, *73*, ftv065. [CrossRef]
60. Jamal, M.; Ahmad, W.; Andleeb, S.; Jalil, F.; Imran, M.; Nawaz, M.A.; Hussain, T.; Ali, M.; Rafiq, M.; Kamil, M.A. Bacterial biofilm and associated infections. *J. Chin. Med. Assoc.* **2018**, *81*, 7–11. [CrossRef] [PubMed]
61. Sharma, D.; Patel, R.P.; Zaidi, S.T.R.; Sarker, M.R.; Lean, Q.Y.; Ming, L.C. Interplay of the Quality of Ciprofloxacin and Antibiotic Resistance in Developing Countries. *Front. Pharmacol.* **2017**, *8*, 546. [CrossRef] [PubMed]
62. Mahendran, K.R.; Kreir, M.; Weingart, H.; Fertig, N.; Winterhalter, M. Permeation of Antibiotics through Escherichia coli OmpF and OmpC Porins: Screening for Influx on a Single-Molecule Level. *J. Biomol. Screen.* **2010**, *15*, 302–307. [CrossRef] [PubMed]

pharmaceutics

MDPI

Article

Liposomal Encapsulation Increases the Efficacy of Azithromycin against *Chlamydia trachomatis*

Anita Bogdanov [1], László Janovák [2], Jasmina Vraneš [3,4], Tomislav Meštrović [5,6], Sunčanica Ljubin-Sternak [3,4], Zsuzsanna Cseh [1], Valéria Endrész [1], Katalin Burián [1], Željka Vanić [7] and Dezső P. Virok [1,*]

1 Department of Medical Microbiology, Albert Szent-Györgyi Health Center and Albert Szent-Györgyi Medical School, University of Szeged, Semmelweis Str. 6, 6725 Szeged, Hungary; varga-bogdanov.anita@med.u-szeged.hu (A.B.); cseh.zsuzs@gmail.com (Z.C.); endresz.valeria@med.u-szeged.hu (V.E.); burian.katalin@med.u-szeged.hu (K.B.)
2 Interdisciplinary Excellence Centre, Department of Physical Chemistry and Materials Science, University of Szeged, Rerrich Béla Sqr. 1, 6720 Szeged, Hungary; laszlo.janovak@gmail.com
3 Molecular Microbiology Department, Dr. Andrija Štampar Teaching Institute of Public Health, Mirogojska cesta 16, 10000 Zagreb, Croatia; jasmina.vranes@stampar.hr (J.V.); sljsternak@stampar.hr (S.L.-S.)
4 Medical Microbiology Department, University of Zagreb School of Medicine, Šalata 3, 10000 Zagreb, Croatia
5 Clinical Microbiology and Parasitology Unit, Dr. Zora Profozić Polyclinic, Bosutska 19, 10000 Zagreb, Croatia; tomislav.mestrovic@gmail.com
6 University Centre Varaždin, University North, Ul. 104. Brigade 3, 42000 Varaždin, Croatia
7 Department of Pharmaceutical Technology, Faculty of Pharmacy and Biochemistry, University of Zagreb, A. Kovačića 1, 10000 Zagreb, Croatia; zvanic@pharma.hr
* Correspondence: virok.dezso.peter@med.u-szeged.hu

Abstract: *Chlamydia trachomatis* (*C. trachomatis*) is an obligate intracellular bacterium linked to ocular and urogenital infections with potentially serious sequelae, including blindness and infertility. First-line antibiotics, such as azithromycin (AZT) and doxycycline, are effective, but treatment failures have also been reported. Encapsulation of antibiotics in liposomes is considered an effective approach for improving their local effects, bioavailability, biocompatibility and antimicrobial activity. To test whether liposomes could enhance the antichlamydial action of AZT, we encapsulated AZT in different surface-charged elastic liposomes (neutral, cationic and anionic elastic liposomes) and assessed their antibacterial potential against the *C. trachomatis* serovar D laboratory strain as well as the clinical isolate *C. trachomatis* serovar F. A direct quantitative polymerase chain reaction (qPCR) method was used to measure chlamydial genome content 48 h post infection and to determine the recoverable chlamydial growth. All the liposomes efficiently delivered AZT to HeLa 229 cells infected with the laboratory *Chlamydia* strain, exhibiting the minimal inhibitory concentrations (MIC) and the minimal bactericidal concentrations (MBC) of AZT even 4–8-fold lower than those achieved with the free AZT. The tested AZT-liposomes were also effective against the clinical *Chlamydia* strain by decreasing MIC values by 2-fold relative to the free AZT. Interestingly, the neutral AZT-liposomes had no effect on the MBC against the clinical strain, while cationic and anionic AZT-liposomes decreased the MBC 2-fold, hence proving the potential of the surface-charged elastic liposomes to improve the effectiveness of AZT against *C. trachomatis*.

Keywords: *Chlamydia*; *Chlamydia trachomatis*; elastic liposomes; azithromycin; Tween 80; surface charge; PCR; qPCR; MIC; MBC

1. Introduction

C. trachomatis ocular and urogenital serovars are responsible for the acute and chronic inflammations of ocular and urogenital mucosal surfaces. The chronic inflammation could lead to serious complications, such as blindness and infertility. According to a World Health Organization report, 44 countries are affected by trachoma, 137 million people are at risk of the infection and 1.9 million people suffer from infection-related impaired vision or blindness [1]. *C. trachomatis*-caused sexually transmitted infections are estimated to be acquired by 131 million people annually, where mostly young (16–24 years-old) people are affected. For instance, 1,758,668 and 406,406 cases of chlamydia infection were reported in 2018 in the USA and in 26 European countries, respectively [2].

The macrolide antibiotic AZT inhibits protein synthesis by irreversibly binding to the 50S subunit of the bacterial ribosome. AZT is able to accumulate in the host cells reaching a high intracellular concentration [3]; therefore, it can be used to inhibit the obligate intracellular *Chlamydia* species. *C. trachomatis* infections can be treated effectively with a single dose of AZT or a one-week course of another bacterial protein synthesis inhibitor doxycycline. Due to the lesser need of compliance, a single-dose AZT regimen may be preferable over to doxycycline. While both antibiotic regimens are effective, treatment failures have also been reported. Systemic reviews and meta-analyses of clinical trials assessed the AZT effectivity of 96–97%, but the relative risk of treatment failure was 2.45 compared to that for doxycycline [4]. Multiple factors could lead to *C. trachomatis* treatment failures including a higher bacterial load [5], homotypic resistance due to chromosomal mutations, such as the mutations of the 23S rRNA [6], and heterotypic resistance, where a certain sub-population of a *Chlamydia* pool displays higher antimicrobial resistance, but this phenotypic feature is not encoded genetically [7].

Liposomal encapsulation is a promising tool for improving antibiotic efficacy and overcoming microbial resistance, thus reducing treatment failures [8]. Liposomal antibiotics have been proven to enable prolonged and/or controlled release of the encapsulated drug permitting a higher local drug concentrations and a greater efficacy against extracellular and certain facultative intracellular bacteria, such as *Listeria monocytogenes*, *Brucella melitensis*, *Francisella tularensis*, *Salmonella enterica* serovar Typhimurium, *Mycobacterium avium* complex and *Mycobacterium tuberculosis* [9]. Liposomes are additionally attractive as antimicrobial carriers due to their targeting ability and physiological acceptability [10]. Appropriate adjustment of their physicochemical properties (size, bilayer elasticity/rigidity and surface modification) enables enhanced delivery of the entrapped drug to the targeted/infected cells [11] or bacteria, either by facilitating fusion with bacteria or allowing release of the drug in the close vicinity of the microorganisms [12–14]. Improved intracellular delivery into the targeted cells can be achieved by several different mechanisms including passive diffusion of the released drug across the cell membrane, adsorption of liposomes on the cell surface, followed by membrane lipid exchange and release of the drug. Specially designed membrane active liposomes containing fusogenic lipids are proposed to fuse with the cell membrane delivering its cargo into the cytoplasm, while receptor-mediated endocytosis is characteristic for ligand-targeted liposomes [11].

Among the different types of liposomes, elastic (deformable) liposomes have already been proven to improve delivery of entrapped drugs deeper in the skin [15,16] and vaginal tissue [17]. Compared to conventional liposomes, consisting of phospholipids with or without addition of cholesterol, elastic liposomes contain phospholipids and certain lower portions of the edge activator (single chain surfactant) and/or solvent (ethanol, propylene glycol, glycerol), making their bilayers fluid and more squeezable [16,18]. Encapsulation of AZT in anionic elastic liposomes have demonstrated stronger activity against biofilm-forming methicillin-resistant *Staphylococcus aureus* strains (MRSA) than conventional liposomes of the same surface charge [18]. We have previously found that elastic liposomes were also more effective in delivering AZT to the *C. trachomatis* infected HeLa cells than conventional liposomes with rigid bilayers, although their activity was lower than the free drug. It was speculated that such results were due to the faster AZT release from elastic liposomes compared to slower release profile achieved by conventional liposomes and the fact that the released lipophilic drug is more readily taken up via the HeLa cells than liposomes. This was also supported by ex vivo permeation assessment demonstrating increased penetration of the free AZT through the vaginal tissue [19].

Continuing previous research [19], and to improve anti-chlamydial activity of liposomal AZT for local therapy, we have prepared and tested several novel preparations of elastic AZT-liposomes, differing in the surface charge (cationic, neutral and anionic), to select the ones with improved antichlamydial action. Our hypothesis was that by facilitating interaction of the liposomes with the *C. trachomatis* infected HeLa cells increased intracellular AZT delivery could be obtained. Namely, cationic liposomes have been demonstrated to contribute interaction with the negatively surface charged cell membranes [20] as well as bacteria [13,14], and significantly increase the activity of encapsulated drugs against planktonic [21] and biofilm-forming bacteria [10].

2. Materials and Methods

2.1. Materials

Lipoid E PC S (from egg; phosphatidylcholine content $\geq$ 96%) (EPC) and egg phosphatidylglycerol (EPG) were generous gifts from Lipoid GmbH (Ludwigshafen, Germany), while AZT in the form of a dihydrate was kindly donated by PLIVA Croatia Ltd. (Zagreb, Croatia). Dimethyldioctadecylammonium bromide (DODAB) and polyoxyethylene(20) sorbitanmonooleate (Tween 80) were provided by Sigma–Aldrich (St. Louis, MO, USA). Ethanol and KH_2PO_4 were purchased from Kemika (Zagreb, Croatia). For the preparation of liposomes, 0.01 M phosphate buffer (PB) was prepared by dissolving 1.3609 g KH_2PO_4 in demineralized water up to 1000 mL, and the pH was adjusted to 7.5 by the addition of 10 M KOH [19].

2.2. *C. trachomatis* Strains

The laboratory reference strain *C. trachomatis* serovar D (UW-3/CX, ATCC) and a *C. trachomatis* serovar F clinical isolate were used in this study. *C. trachomatis* clinical isolate was obtained from a male urethra sample during routine testing at the Referral Center for Diagnostics of Sexually Transmitted Infections in the Dr. Andrija Štampar Teaching Institute for Public Health, Zagreb, Croatia. The specimen was collected using the collection kit comprised of a Dacron swab and MicroTest M4RT transport medium (Remel Inc., Lenexa, KS, USA) and stored at −80 °C until further propagation. Serovar characterization of the clinical isolate was performed by sequencing the *omp1* gene as described previously [22]. *Chlamydia* strains were propagated and partially purified as described previously [23].

2.3. Preparation of AZT-Liposomes

AZT-liposomes were prepared by the modified proliposome method [19,24]. Briefly, the (phospho)lipids, AZT and Tween 80 (Table 1) were dissolved in ethanol (200 mg) during magnetic stirring (600 rpm, 50 °C). Then, 0.2 mL PB, pH 7.5 (pre-heated to the same temperature) was added, and the mixture was stirred to form an initial proliposome mixture. After cooling to room temperature, the AZT-proliposome mixture was transformed to AZT-liposomal dispersion by the drop-wise addition of PB, pH 7.5 to obtain the final volume of 10 mL. The dispersion was stirred (600 rpm) for 40 min at room temperature. All the liposomal dispersions were homogenized by ultrasonication for 10 s at 60 µW, using the Cole-Parmer Ultrasonic Homogenizer 4710 Series (Vernon Hills, IL, USA), and stored in the refrigerator at 4 °C. Prior to their use, the liposomes were incubated at room temperature and well mixed.

Table 1. Composition of AZT-liposomes.

Liposomes (Code)	EPC (mg)	EPG (mg)	DODAB (mg)	AZT (mg)	Tween 80 (mg)	PB, pH 7.5 (mL)
+EL-AZT	190	-	10	10	50	10
−EL-AZT	190	10	-	10	50	10
EL-AZT	200	-	-	10	50	10
+EL	190	-	10	-	50	10
−EL	190	10	-	-	50	10
EL	200	-	-	-	50	10

AZT, azithromycin dihydrate; DODAB, dimethyldioctadecylammonium bromide; EL, elastic liposomes composed of only neutral phospholipid (neutral liposomes); +EL, elastic liposomes embedding cationic lipid (cationic liposomes); −EL, elastic liposomes embedding anionic phospholipid (anionic liposomes); EPC, egg phosphatidylcholine; EPG, egg phosphatidylglycerol; PB, phosphate buffer; Tween 80, polyoxyethylene (20) sorbitan monooleate.

2.4. Size Measurements of AZT-Liposomes

The average diameters, polydispersity indexes (PDIs) and size distributions of the different AZT-liposomes were determined by dynamic light scattering, using a Horiba SZ-100 Nanoparticle Analyzer (Kyoto, Japan). Prior to the measurements, the liposomal dispersions were thermostated at room temperature and appropriately diluted with PBS, pH 7.4 to 0.02 mg/L (based on the AZT concentration in liposomes). All the measurements were carried out at 25 °C at a scattering angle of 90°, where the acquisition lag time of each measurement was 600. The results are expressed as the mean ± S.D (n = 5).

2.5. Zeta Potential Measurements of AZT-Liposomes

The zeta potentials of the different AZT-liposomes were determined on Zetasizer Ultra Red (Malvern Panalytical Ltd., Malvern, UK) using a folded capillary cell. The liposomal samples were previously diluted in PB, pH 7.5 to obtain a satisfactory quality factor (approximately to 0.02 mg/L). The measurements were performed at 25 °C. The results are expressed as the mean ± S.D (n = 5).

2.6. Liposomal Bilayer Elasticity Determination

The membrane elasticity of the different surface-charged liposomes was evaluated by previously reported method [25]. Briefly, the liposomal dispersions were continuously extruded (LiposoFast, Avestin, Ottawa, ON, Canada) at room temperature through a 100 nm pore-size membrane (r_p) for 5 min by applying an external pressure of 2.5 bar, and the average diameter of the liposomes following extrusion (r_v) as along with the mass of the extruded liposomes (J) were determined. For the assessment of AZT-liposomes, the non-encapsulated drug was pre-separated from the liposomally-encapsulated AZT by ultracentrifugation (2.7). A calculation of the degree of bilayer elasticity (E) was performed using the following equation:

$$E = J\left(\frac{r_v}{r_p}\right)^2$$

r_v of the different liposomes was determined by dynamic light scattering on Zetasizer Ultra Red (Malvern Panalytical Ltd., Malvern, UK) at scattering angle of 90°. The liposomal dispersions were properly diluted by PB, pH 7.5 and the measurements were performed at 25 °C using disposable DTS0012 cuvettes. The results are expressed as the mean ± S.D (n = 3).

2.7. AZT Encapsulation Efficiency (EE) Determination

Encapsulation of AZT in the different surface-charged elastic liposomes was determined after separation of the non-encapsulated drug by ultracentrifugation. For this purpose, AZT-liposomes (1 mL) were diluted with 5 mL PB and ultracentrifuged (Beckman

Optima LE-80 K Ultracentrifuge, Beckman Coulter Inc., Fullerton, CA, USA) for 1 h at 150,000× *g* (20 °C). The supernatant containing non-encapsulated AZT was separated from the pellet (liposomes). The pellet was washed with PB and ultracentrifuged under the same conditions. The amount of non-entrapped AZT was assessed in supernatants spectrophotometrically using a Varian Cary 50 UV/Vis spectrophotometer (Varian Australia Pty Ltd., Belrose, Australia) at 210 nm. The encapsulated AZT was calculated using the following equation:

$$EE\ (\%) = 100 - \left(\frac{non - encapsulated\ AZT}{total\ AZT}\right) \times 100$$

2.8. Cytotoxicity Measurement by 3-(4,5-Dimethyl-2-thiazolyl)-2,5-diphenyl-2H-tetrazolium Bromide (MTT) Assay

MTT assay was performed to find the maximum non-toxic concentration of the AZT-liposomes. HeLa 229 (ATCC) cells were placed into 96-well plates (Sarstedt) at a density of 4×10^4 cells/well in 100 µL of minimal essential medium (MEM) with Earle salts supplemented with 10% heat-inactivated FBS (Gibco, Germany), 2 mmol/L L-glutamine, 1× MEM vitamins, 1× non-essential amino acids, 0.005% Na-pyruvate, 25 µg/mL gentamycin, 1 µg/mL Fungisone. The plates were incubated for 1 h at room temperature (RT) and then overnight at 37 °C, 5% CO_2. The next day, when the cells reached a ~90% confluence, the medium was supplemented with serial 2-fold dilutions of the AZT-liposomes in three parallel wells for each concentration. Concentration ranges of 1–0.0156 µg/mL AZT for each type of AZT-liposomes, the corresponding concentrations of empty liposomes and free AZT were tested. The MTT (Sigma) labeling reagent (10 µL, final concentration 0.5 mg/mL) was added to each well, after 48 h incubation. The plate was incubated for 4 h at 37 °C, 5% CO_2, then 100 µL of the solubilization solution (10% SDS in 1 N HCl) was added into each well and the plate was incubated overnight at 37 °C, 5% CO_2. Labsystems Multiskan Ex 355 microtiter plate reader (Thermo Fisher Scientific, Waltham, MA, USA) was used to measure the optical density of the wells. The absorbance of the formazan product was measured at 540 nm. Three parallel measurements were performed.

2.9. Culture of HeLa 229 Cells and Direct qPCR Measurement of the Impact of AZT-Liposomes on C. trachomatis Growth

HeLa 229 cells (ATCC) were cultured in 96-well plates (Sarstedt, Nümbrecht, Germany) at a density of 4×10^4 cells/well in 100 µL of minimal essential medium (MEM) with Earle salts supplemented with 10% heat-inactivated fetal bovine serum (FBS) (Gibco, Waltham, MA, USA), 2 mmol/L L-glutamine, 1× non-essential amino acids, 8 mmol/L HEPES, 25 µg/mL gentamycin and 1 µg/mL Fungisone. HeLa 229 cells were incubated overnight at 37 °C, 5% CO_2 to get a 90% confluent cell layer. All the reagents were purchased from Sigma–Aldrich, unless otherwise indicated. HeLa cells were infected with *C. trachomatis* (multiplicity of infection (MOI) 1) for 1 h in 0.5% glucose containing medium without centrifugation. After infection, the cells were washed twice with PBS. The culture medium with cycloheximide was supplemented with the serial 2-fold dilutions of the AZT-liposomes and was added to triplicate wells. The cationic (+EL-AZT), anionic (−EL-AZT) and neutral (EL-AZT) AZT-liposomes as well as free AZT (6/4, *v*/*v*, ethanol/water solution) were diluted in culture medium with cycloheximide. Concentration ranges of 0.5–0.0002 µg/mL AZT for each AZT-liposomes and free AZT, with 2-fold dilutions were tested. The plates were incubated for 48 h at 37 °C, 5% CO_2. Measurement of recoverable chlamydial growth was performed on McCoy (ATCC) cells. McCoy cells were transferred into wells of the 96-well plate at a density of 4×10^4 cells/well in 100 µL of MEM, and were incubated overnight at 37 °C, 5% CO_2 to obtain a 90% confluent cell layer. Before the infection, the wells were washed twice with 100 µL/well of PBS (pH 7.2). After the washing steps, 90 µL of culture medium with glucose was added to each well. For the determination of the recoverable inclusion forming units, 10 µL of the treated and *C. trachomatis* infected HeLa 229 cells were transferred onto the McCoy cells. The cells were centrifuged for 1 h at 800× *g* and were incubated for 48 h in cycloheximide-containing (1 µg/mL) growth medium. For direct qPCR, the supernatants of the infected HeLa 229 cells and McCoy cells were removed, and the cells were washed with 100 µL/well PBS twice. After the second wash, 100 µL Milli-Q (MQ) (Millipore, Billerica, MA, USA) water was added to each well and the samples were exposed to two freeze-thaw cycles with a quick freezing (−80 °C, 15 min) and a quick thawing on a plate shaker at room temperature. After the lysis, the cell lysates were thoroughly mixed and the mixed lysates were used as templates in the qPCR to measure the chlamydial genomic DNA. The qPCR was performed in a Bio-Rad CFX96 real time system (Bio-Rad, Hercules, CA, USA). The 5× HOT FIREPol® EvaGreen® qPCR Supermix (Solis BioDyne, Tartu, Estonia) master mix and *C. trachomatis* pykF gene-specific primer pairs were used for the amplification. Direct qPCR for chlamydial genome concentration measurements were performed as described previously [26]. Briefly, the composition of the reaction included 2 µL 5× HOT FIREPol® EvaGreen® qPCR Supermix, 1–1 µL forward and reverse primers (10 pmol each), 1 µL template and 5 µL MQ water. After a 10 min at 95 °C polymerase activation step, 40 PCR cycles of 20 s at 95 °C and 1 min at 64 °C were completed. Melting curve analysis was used to validate the specificity of amplification. For each sample the cycle threshold (Ct) value corresponding to the qPCR cycle where the amplification curve crossed the threshold was determined. Ct values were used to calculate the MIC and MBC values as described before [26].

2.10. Statistical Analysis

Statistical data analyses were completed using the GraphPad 9.2.0 Prism program (GraphPad Software, Inc., San Diego, CA, USA). For comparisons between two groups, Student's *t*-test was used, while for comparisons of three or more groups one-way ANOVA and Tukey post-hoc test was used. Threshold for statistical significance was $p < 0.05$.

3. Results

3.1. Physico-Chemical Properties of the AZT-Liposomes

Neutral AZT-liposomes were prepared from egg lecithin containing at least 96% phosphatidylcholine (PC) with the addition of the edge activator, i.e., Tween 80 (neutral liposomes), while for the preparation of anionic or cationic liposomes negatively charged phospholipid (EPG) or cationic lipid (DODAB,) were additionally used (Table 1).

Assessment of the physicochemical characteristics of the liposomes, such as the size, zeta potential and bilayer elasticity/rigidity, is of great importance as these properties influence the interactions of the liposomes with the microorganisms, as well as the cells and tissues in the biological environment, subsequently determining the therapeutic outcome. Elastic AZT-liposomes were prepared by a simple and reproducible method acceptable for scale up, followed by the short cycle of ultrasonication to homogenize the original AZT-liposomal dispersions. As a result, mean diameters of the originally prepared liposomes decreased and all the AZT-liposomes displayed moderate polydispersity (Figure 1A–C), characteristic for deformable (elastic) liposomes [27]. Those comprising the negatively charged EPG were the smallest (164 ± 10 nm). Liposomes incorporating positively charged DODAB were somewhat larger (175 ± 34 nm), but not significantly (ANOVA, $p > 0.05$), while liposomes, consisting of only neutral phospholipid, were the largest (187 ± 20 nm). The size distributions of AZT-liposomes (Figure 1) also indicate the presence of a smaller population of larger vesicles, typical for the selected preparation method [28], which is consistent with the previous studies [29,30] and could be the result of a short cycle of ultrasonication applied, too.

Although phospholipid composition slightly affected the mean diameters of AZT-liposomes, it defined their surface charge (Figure 2). Embedding only 5% (*w/w*) EPG or DODAB in the liposomal bilayers significantly decreased or increased the zeta potentials of anionic or cationic AZT-liposomes (ANOVA, $p < 0.05$), respectively. These results are in accordance with the previous findings for deformable liposomes [25], and are considered beneficial regarding physical stability of the AZT-liposomes. Bilayers of all the liposomes consisted of 20% non-ionic single chain surfactant, i.e., polyoxyethylene (20) sorbitan monooleate (Tween 80), which has already been reported as the edge activator in deformable (elastic) liposomes [31]. The edge activator destabilizes the tight packing of phospholipids in the liposomal bilayers, making them deformable (flexible) and squeezable, thus subsequently enhancing permeation of the vesicles and/or encapsulated drug into/through the skin and cervicovaginal tissue [18,19,32].

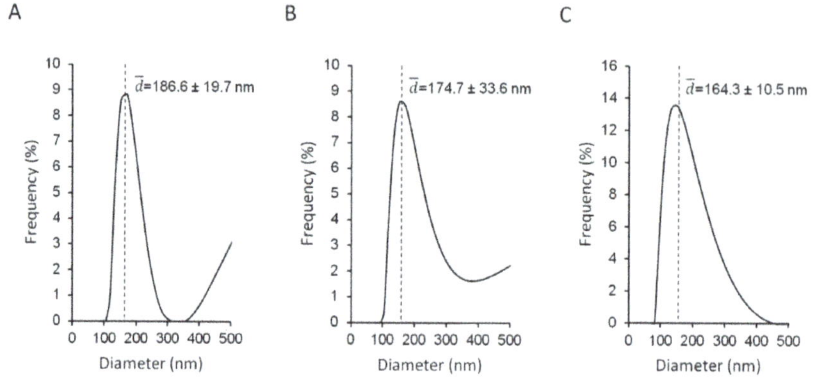

Figure 1. Size distributions and mean diameters of the different surface-charged AZT-liposomes (25 °C). (**A**), neutral (EL-AZT), (**B**), cationic (+EL-AZT) and (**C**), anionic (−EL-AZT) liposomes with azithromycin (AZT). The corresponding PDI values were: 0.48 ± 0.02 (**A**), 0.45 ± 0.04 (**B**) and 0.36 ± 0.05 (**C**), respectively. Data are mean ± S.D (n = 5).

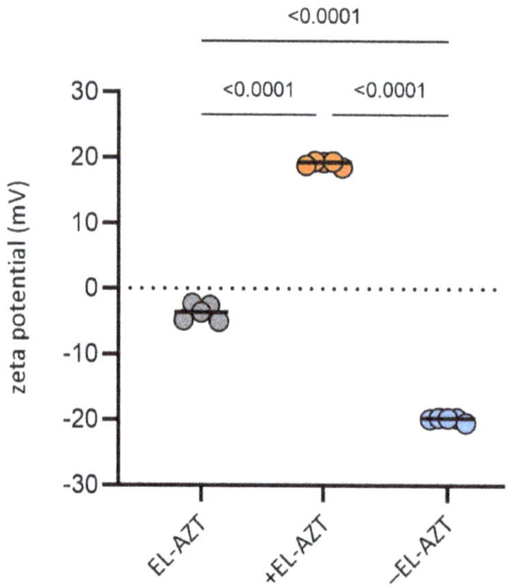

Figure 2. Zeta potentials of the different surface-charged AZT-liposomes. Data are mean and individual values (n = 5). Statistically significant differences with p values are shown. Statistical analysis was performed by one-way ANOVA with post-hoc Tukey test. EL-AZT, neutral liposomes with azithromycin; +EL-AZT, cationic liposomes with azithromycin; −EL-AZT, anionic liposomes with azithromycin.

Evaluation of the liposomes' deformability has proved that the liposomes prepared without AZT (empty liposomes) were highly elastic vesicles with a degree of bilayer elasticity (E) between 14 and 18 (Figure 3). However, the incorporation of AZT in cationic and neutral liposomes strengthened their membranes relative to the bilayers of the empty liposomes. This was confirmed by significantly decreased E values (3-4) (ANOVA, $p < 0.05$), while the membranes of anionic AZT-liposomes retained better their bilayer flexibility (E > 12).

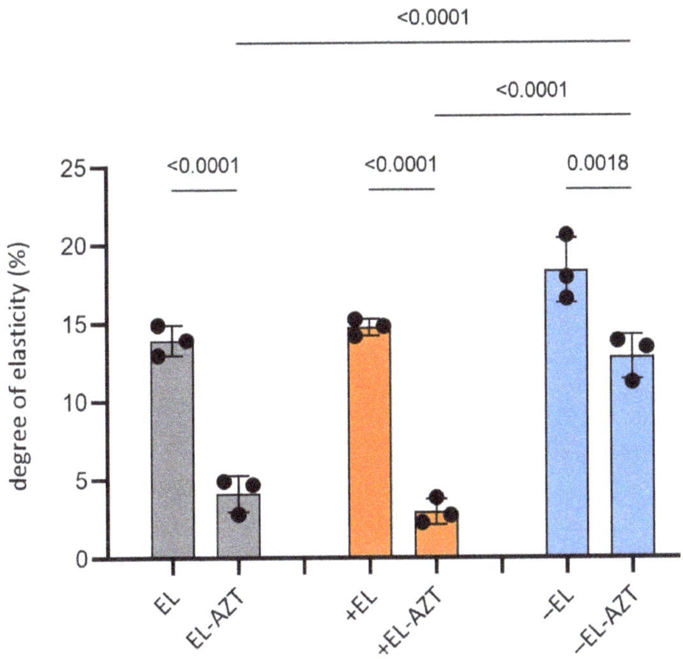

Figure 3. Degree of bilayer elasticity (E) of the different surface-charged liposomes. Data are mean ± S.D and individual values. ($n = 3$). Statistically significant differences between the corresponding empty liposomes and AZT-containing liposomes and between different AZT containing liposomes are shown. Statistical analysis was performed by one-way ANOVA with post-hoc Tukey test. EL, empty neutral liposomes; +EL, empty cationic liposomes; −EL, empty anionic liposomes; EL-AZT, neutral liposomes with azithromycin; +EL-AZT, cationic liposomes with azithromycin; −EL-AZT, anionic liposomes with azithromycin.

Separation of the non-encapsulated AZT was performed by ultracentrifugation method, which was previously reported to be similarly effective for separation AZT-liposomes as minicolumn centrifugation method [18]. Following the separation of the non-encapsulated AZT from the liposomally-encapsulated AZT, 25 and 37% of the drug was detected in cationic and anionic AZT-liposomes, respectively, while about 32% AZT was encapsulated in neutral liposomes (Figure 4). A tendency of higher AZT encapsulation in the anionic liposomes relative to the neutral and cationic AZT-liposomes is most likely a consequence of the interaction between the drug and the negatively charged EPG [33]. In comparison to the conventional liposomes, characterized by the rigid bilayers [25,27], the increased membrane deformability of the elastic liposomes could result in the leakage of the drug during storage. Therefore, to avoid these negative consequences, antichlamydial studies were performed with AZT-liposomes in which the non-encapsulated AZT was not removed from the liposomally-encapsulated drug. It is worth to note that the presence of the non-encapsulated AZT in AZT-liposomal dispersions contributes to the long-term stability of the nanoformulations by preventing leakage of the encapsulated drug from the vesicles.

Drug leakage is considered a limitation; hence many commercial liposomal formulations are available as lyophilized products, which are converted to liposomal dispersions before use [11,34].

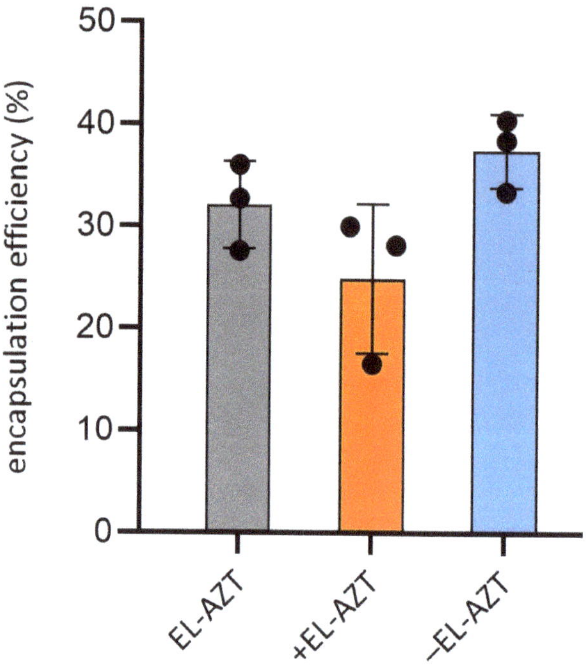

Figure 4. Encapsulation efficiency of AZT in the neutral (EL-AZT), cationic (+EL-AZT) and anionic (−EL-AZT) elastic AZT-liposomes. Data are mean ± S.D. and individual values (n = 3). Statistics analysis was performed by one-way ANOVA with post-hoc Tukey test. EL-AZT, neutral liposomes with azithromycin; +EL-AZT, cationic liposomes with azithromycin; −EL-AZT, anionic liposomes with azithromycin.

3.2. *Cytotoxicity Measurement of the of AZT-Liposomes on HeLa 229 Cells*

Chlamydiae are obligate intracellular bacteria; hence the viability of the host cells may have an impact on the chlamydial growth. We performed an MTT assay to test whether the liposomes had a non-specific antichlamydial effect due to host cell cytotoxicity. MTT assay was performed with the AZT-liposomes at AZT concentrations ranging from 1 to 0.00156 μg/mL (Figure 5), the corresponding free liposomes and free AZT. Cell viability was assessed after 48 h of incubation. Cationic liposomes (+EL-AZT) had the highest impact on cell viability; the highest concentration without significant impact on cell viability was 0.125 μg/mL. It should be noted that this concentration was at least 32- up to 128-fold higher than the corresponding MIC and MBC values for cationic AZT-liposomes (Figures 6 and 7), thus confirming their biocompatibility. The anionic (−EL-AZT) liposomes had a moderate impact on the cell viability with the highest non-toxic concentration of AZT 0.5 μg/mL, while the neutral (EL-AZT) liposomes had no significant impact on the cell viability at the tested concentrations. Empty liposomes and AZT alone did not decrease the cell viability in any of the tested concentrations.

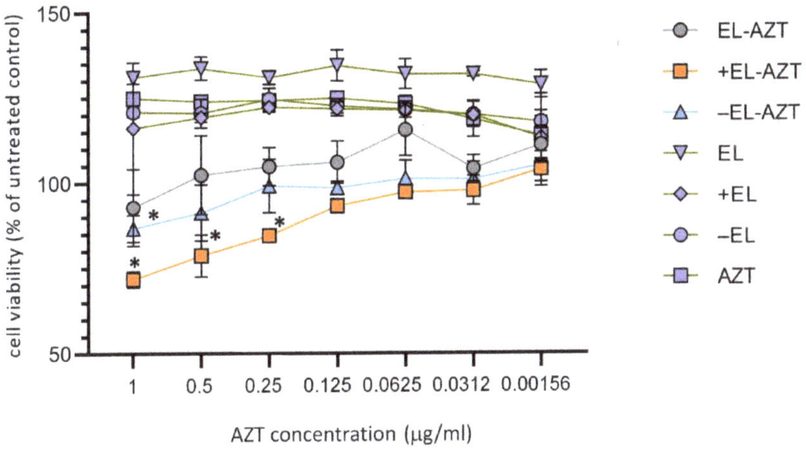

Figure 5. MTT assay of the HeLa 229 cells treated with the neutral, anionic and cationic AZT-liposomes. Viability of the liposomally-treated cells were compared to the untreated controls. Data are mean ± S.D. ($n = 3$). * Statistically significant MTT values of AZT-liposome treated cells compared to the untreated control cells (t-test, $p < 0.05$). EL-AZT, neutral liposomes with azithromycin; +EL-AZT, cationic liposomes with azithromycin; −EL-AZT, anionic liposomes with azithromycin; EL, neutral liposomes without azithromycin; +EL, cationic liposomes without azithromycin; −EL, anionic liposomes with azithromycin; AZT, free azithromycin.

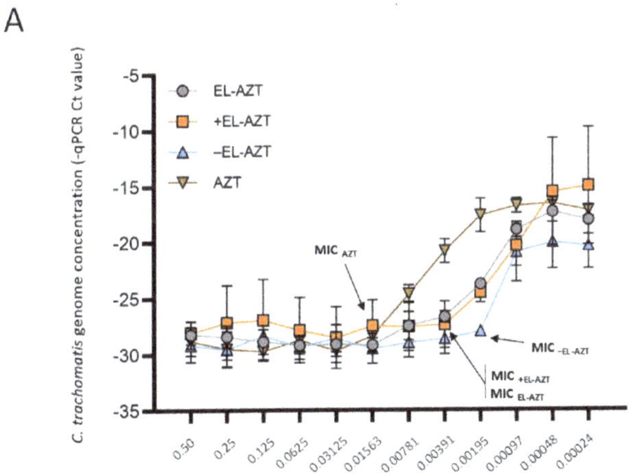

Figure 6. *Cont.*

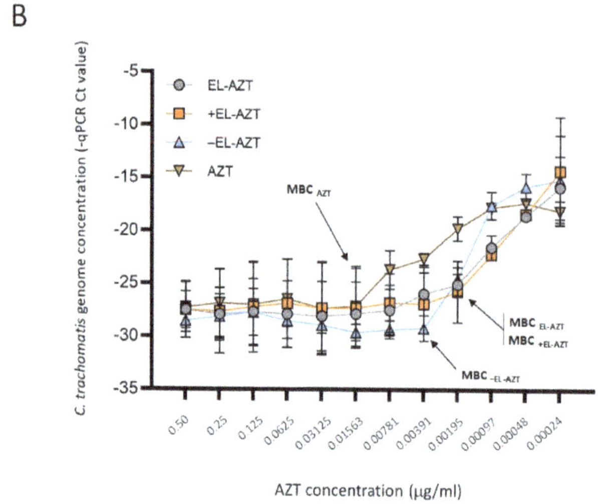

Figure 6. Impact of different types of AZT-liposomes on the growth of *C. trachomatis* serovar D. (**A**) Direct *C. trachomatis* growth in the presence of free AZT and liposomally-encapsulated AZT. (**B**) Recoverable *C. trachomatis* growth. Data are mean −qPCR Ct levels ± S.D. (n = 3). EL-AZT, neutral liposomes with azithromycin; +EL-AZT, cationic liposomes with azithromycin; −EL-AZT, anionic liposomes with azithromycin.

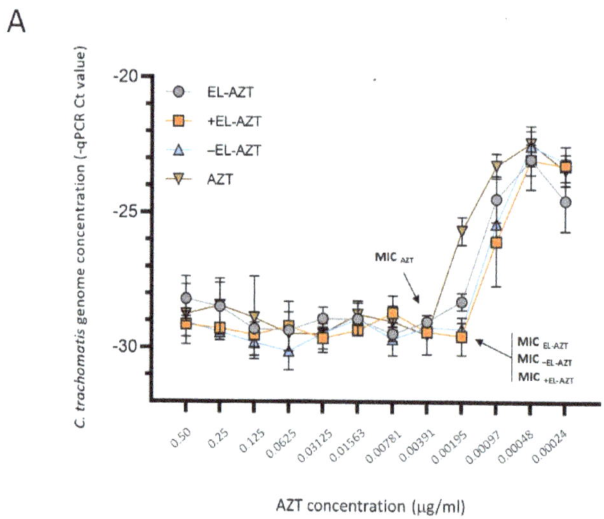

Figure 7. *Cont.*

B

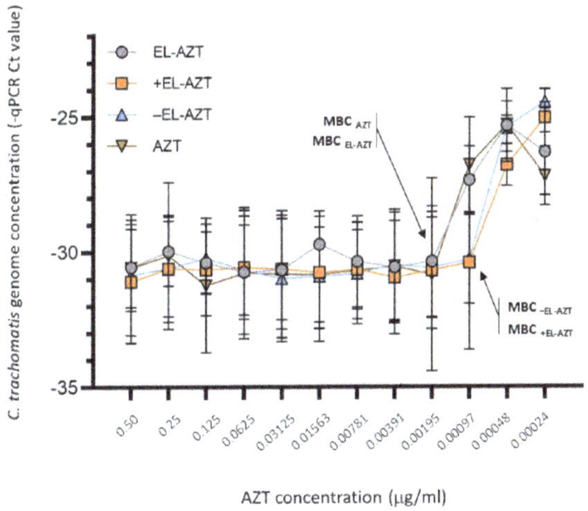

Figure 7. Impact of the AZT-liposomes on the growth of *C. trachomatis* serovar F clinical isolate. (**A**) Direct *C. trachomatis* growth in the presence of free AZT and encapsulated AZT. (**B**) Recoverable *C. trachomatis* growth. Data are mean −qPCR Ct levels ± S.D. (n = 3). EL-AZT, neutral liposomes with azithromycin; +EL-AZT, cationic liposomes with azithromycin; −EL-AZT, anionic liposomes with azithromycin.

3.3. Impact of AZT-Liposomes on C. trachomatis Growth

To evaluate the antimicrobial action of AZT-liposomes against *C. trachomatis* serovar, D/UW-3/CX laboratory strain and serovar F clinical isolate, infected HeLa 229 cells were treated with a 1:2 dilution series of AZT-liposomes (0.5–0.00024 μg/mL). Chlamydial genome content was measured 48 h post infection by direct qPCR and the MIC values were determined as described previously [26]. As shown in Figure 6A,B, all AZT-liposomes efficiently inhibited the growth of *C. trachomatis* serovar D. MIC values for anionic, cationic and neutral AZT-liposomes were lower compared to the 0.01563 μg/mL MIC for free AZT (Figure 6A). Among the tested liposomes, the most effective were the anionic liposomes with an 8-fold lower MIC than the MIC determined for free AZT. Neutral and cationic liposomes were equally effective in inhibiting the growth of *C. trachomatis* serovar D with a MIC value 4-fold lower in comparison to that obtained with free AZT (Figure 6A). The recoverable *Chlamydia* was also measured by centrifugation of the infected HeLa 229 cell lysates onto McCoy cells followed by the measurement of the chlamydial genome concentration 48 h post infection. The MBC value was calculated based on the recoverable chlamydial growth in McCoy cells in a similar manner to how the MIC was calculated. All the AZT-liposomes were effective in killing *C. trachomatis*; the MBCs for AZT-liposomes were lower compared to the 0.01563 μg/mL MBC of free AZT (Figure 6B). Neutral (EL-AZT) and cationic (+EL-AZT) liposomes exhibited the strongest activity against *C. trachomatis* serovar D with an MBC 8-fold lower than the MBC of the free AZT, while the anionic (−EL-AZT) liposomes had a 4-fold lower MBC relative to the free AZT (Figure 6B).

AZT-liposomes were also evaluated to determine their potential for inhibiting the growth and killing of the *C. trachomatis* serovar F clinical isolate (Figure 7A,B). Overall, the clinical isolate was more susceptible to AZT than the laboratory strain. Free AZT inhibited growth of serovar F at 0.00391 μg/mL (Figure 7A), in comparison to 0.01563 μg/mL determined for *C. trachomatis* serovar D. There were no differences in the MIC values between the different types of AZT-liposomes against serovar F clinical isolates. All the AZT-liposomes were equally effective with a MIC of 0.00195 μg/mL, which was 2-fold lower than the MIC determined for free AZT (Figure 7A). When AZT-liposomes were assessed for MBCs, those embedding only neutral phospholipid (EL-AZT) demonstrated the same 0.00195 μg/mL MBC value as free AZT, while cationic and anionic liposomes exhibited 2-fold lower MBC levels than the free AZT (Figure 7B). In vitro antichlamydial studies were performed up to 3 months after liposomal preparations. No microbiological contamination of the samples was observed in that period, nor did the storage affect the results. There were no significant differences between the results for the same type of the AZT-liposomes examined.

4. Discussion

In this study, we have tested several novel AZT-liposomal preparations to improve the efficacy of AZT against *C. trachomatis*. Our laboratory previously developed a direct qPCR method to quantitatively measure *C. trachomatis* growth as an alternative of the commonly used immunofluorescence detection and manual counting of chlamydial inclusions [26]. This method was used to measure the chlamydial genome content 48 h post infection in HeLa cells and the genome content of the recoverable *C. trachomatis* in McCoy cells. Hence, instead of direct infectious unit (IFU) and recoverable IFU, the two terms, direct growth and recoverable growth, were used. The measured MIC values refer to the inhibition of direct growth, while MBC values are characteristic of the inability of recovery—the inhibition of recoverable growth. Generally, our results showed that liposomal encapsulation enhanced the AZT activity by decreasing the MIC and MBC levels. MTT assays showed that the MIC (and MBC) concentrations were ~2 orders of magnitudes lower than the cytotoxic concentrations of the (charged) liposomes; therefore, the decrease of MIC levels were not due to a specific effect of the liposomes on the host cells. First, we measured the free AZT MIC and MBC for the laboratory strain *C. trachomatis* serovar D. The MIC value was 0.01563 μg/mL, which was similar to, or lower than the previously measured 0.03–0.5 μg/mL AZT values [35,36]. The MBC value was identical to the MIC for the laboratory strain in accordance with the previously measured similar AZT MIC and MBC values for *C. trachomatis* serovar D and other serovars [36,37]. Liposomal encapsulation of AZT decreased its MIC and MBC values even 4–8-fold. Interestingly, for neutral and cationic liposomes the MBC value was 2-fold lower than the MIC value. This was probably because during the MIC measurement the first sub-MIC concentration allowed a persistent state with ongoing chlamydial DNA synthesis, but these bacteria are not infectious and therefore the recoverable growth (represented by the MBC) at this concentration could not be detected. Indeed, Belland et al. have previously showed that the persistent, non-infectious form of *C. trachomatis* has intact DNA synthesis, but the bacterium cannot go through the normal division and re-differentiation to the infectious elementary body form [38]. As for the clinical serovar F strain, the free AZT MIC value was 2-fold lower than for the laboratory strain, while the MBC value was 2-fold lower than the MIC value, most likely due to the persistence inducing effect described above. The liposomal encapsulation had a moderate, but observable effect on the growth and recoverable growth of the clinical isolate, with a 2-fold lower MIC and MBC values. A notable exception was the neutral AZT-liposomes, which decreased the MIC value 2-fold, but could not change the MBC value compared to free AZT. Overall, these data show that there are differences between serovars and/or laboratory and clinical strains in the antibiotic susceptibility and in the efficacy of the liposomally-encapsulated antibiotic. Phenotypic differences between serotypes and clinical vs. laboratory strains have been shown before. A study by Thomas et al. on six different cell cultures demonstrated the propensity of the reference *C. trachomatis* type strain D/UW-3/CX to produce significantly higher amounts of infectious progeny in comparison to five clinical isolates from cervical swabs. Moreover, these clinical isolate used has also responded to AZT, penicillin and iron deprivation in a more pronounced manner than the reference strain revealing the potential for developing a persistence state more habitually in vivo [39].

Altogether, the surface charged liposomes were the most effective at improving the efficacy of AZT against both serovars. Although we expected cationic AZT-liposomes as the most effective due to opposite surface charge to the cell membrane, permitting extended contact on the cell surface, they were equally efficient as anionic AZT-liposomes. This result could be due to the lower content of DODAB (cationic lipid) in bilayers of cationic liposomes that do not provide strong interaction with the cell membrane, as well as lower encapsulation efficiency (25%) compared to the highest encapsulation achieved with anionic liposomes (37%). We assume that in this non-specific targeting approach, AZT-liposomes accumulate near the cells, serving as a depot and allowing higher local drug concentration. The released AZT passively diffuse across the cell membrane following concentration gradient and as a weak base accumulate inside the infected cells. A certain portion of the liposomes is likely to be internalized via endocytosis. The results of this study were different from our previous study, where the examined liposomes could not increase antichlamydial efficacy of AZT [19]. The difference in antichlamydial activity of the currently tested elastic AZT-liposomes compared to the deformable propylene glycol AZT-liposomes from our preceding study [19] is probably due to the presence of Tween 80, providing liposomal bilayer elasticity as well as increased permeation and internalization of AZT to the infected HeLa 229 cells. It has already been confirmed that deformable liposomes improved the delivery of the encapsulated drug deeper in the skin [27] and vaginal tissue [19]. The ratio of the edge activator inside the liposomal membrane as well as the type of the edge activator used influenced the membrane elasticity and permeation ability of elastic liposomes. By increasing the edge activator ratio, the membrane elasticity increases, and the release rate of the encapsulated drug can also be enhanced [40,41]. Therefore, by balancing the ratio of the edge activator and bilayer-building phospholipids, suitable liposomal nanoformulations could be obtained. Compared to the previous research [19], the edge activator in this study was present in a higher ratio (25%), contributing to enhanced elasticity. Another reason for the improved antichlamydial activity could be associated with the penetration enhancing effect of non-ionic surfactants, including Tween 80 [42,43], increasing the cellular delivery of AZT. However, further studies should be done to confirm these assumptions.

Our in vitro studies show the possibility that the elastic AZT-liposomes can be more effective for the treatment of *C. trachomatis* infections than the free AZT. They are biodegradable and biocompatible with the cervical cells, and significantly increase the antichlamydial activity of AZT. Additional studies such as drug release and storage stability measurements are needed to confirm the optimal liposomal preparation for the treatment of chlamydial infections.

Author Contributions: Conceptualization, Ž.V. and D.P.V.; methodology, Ž.V., D.P.V., L.J. and A.B.; investigation, Ž.V., D.P.V., L.J., A.B., J.V., T.M., S.L.-S. and Z.C.; writing—original draft preparation, Ž.V., D.P.V., L.J., A.B., K.B. and V.E.; writing—review and editing, Ž.V., D.P.V., L.J., T.M., K.B., A.B. and V.E.; supervision, Ž.V. and D.P.V.; and funding acquisition, Ž.V., D.P.V. and K.B. All authors have read and agreed to the published version of the manuscript.

Funding: A.B. and D.P.V. were supported by the EFOP-3.6.1-16-2016-00008 European Union—Hungary grant. D.P.V. was supported by the Albert Szent-Györgyi Research Grant of University of Szeged.

Institutional Review Board Statement: Not applicable.

Informed Consent Statement: Not applicable.

Data Availability Statement: The article contains all the data.

Acknowledgments: Ž.V. is thankful to Phospholipid Research Center for financing research with AZT-liposomes, and to Lipoid (Ludwigshafen, Germany) and PLIVA (Zagreb, Croatia) for donation of phospholipids and AZT, respectively.

Conflicts of Interest: The authors declare no conflict of interest.

References

1. Trachoma Key Facts (World Health Organization). Available online: https://www.who.int/News-Room/Fact-Sheets/Detail/Trachoma (accessed on 20 December 2021).
2. Smolarczyk, K.; Mlynarczyk-Bonikowska, B.; Rudnicka, E.; Szukiewicz, D.; Meczekalski, B.; Smolarczyk, R.; Pieta, W. The Impact of Selected Bacterial Sexually Transmitted Diseases on Pregnancy and Female Fertility. *Int. J. Mol. Sci.* **2021**, *22*, 2170. [CrossRef]
3. Gladue, R.P.; Snider, M.E. Intracellular Accumulation of Azithromycin by Cultured Human Fibroblasts. *Antimicrob. Agents Chemother.* **1990**, *34*, 1056–1060. [CrossRef]
4. Challenges in Treating Chlamydia Trachomatis, Including Rectal Infections: Is It Time to Go Back to Doxycycline?—S. Lena Kang-Birken. 2021. Available online: https://journals.sagepub.com/doi/10.1177/10600280211029945 (accessed on 5 July 2021).
5. Kong, F.Y.S.; Tabrizi, S.N.; Fairley, C.K.; Phillips, S.; Fehler, G.; Law, M.; Vodstrcil, L.A.; Chen, M.; Bradshaw, C.S.; Hocking, J.S. Higher Organism Load Associated with Failure of Azithromycin to Treat Rectal Chlamydia. *Epidemiol. Infect.* **2016**, *144*, 2587–2596. [CrossRef]
6. Shao, L.; You, C.; Cao, J.; Jiang, Y.; Liu, Y.; Liu, Q. High Treatment Failure Rate Is Better Explained by Resistance Gene Detection than by Minimum Inhibitory Concentration in Patients with Urogenital Chlamydia Trachomatis Infection. *Int. J. Infect. Dis. IJID Off. Publ. Int. Soc. Infect. Dis.* **2020**, *96*, 121–127. [CrossRef] [PubMed]
7. Sandoz, K.M.; Rockey, D.D. Antibiotic Resistance in Chlamydiae. *Future Microbiol.* **2010**, *5*, 1427–1442. [CrossRef]
8. Vanić, Ž.; Jøraholmen, M.W.; Škalko-Basnet, N. Nanomedicines for the Topical Treatment of Vulvovaginal Infections: Addressing the Challenges of Antimicrobial Resistance. *Adv. Drug Deliv. Rev.* **2021**, *178*, 113855. [CrossRef] [PubMed]
9. Drulis-Kawa, Z.; Dorotkiewicz-Jach, A. Liposomes as Delivery Systems for Antibiotics. *Int. J. Pharm.* **2010**, *387*, 187–198. [CrossRef]
10. Rukavina, Z.; Vanić, Ž. Current Trends in Development of Liposomes for Targeting Bacterial Biofilms. *Pharmaceutics* **2016**, *8*, E18. [CrossRef]
11. Allen, T.M.; Cullis, P.R. Liposomal Drug Delivery Systems: From Concept to Clinical Applications. *Adv. Drug Deliv. Rev.* **2013**, *65*, 36–48. [CrossRef]
12. Sachetelli, S.; Khalil, H.; Chen, T.; Beaulac, C.; Sénéchal, S.; Lagacé, J. Demonstration of a Fusion Mechanism between a Fluid Bactericidal Liposomal Formulation and Bacterial Cells. *Biochim. Biophys. Acta* **2000**, *1463*, 254–266. [CrossRef]
13. Ferreira, M.; Ogren, M.; Dias, J.N.R.; Silva, M.; Gil, S.; Tavares, L.; Aires-da-Silva, F.; Gaspar, M.M.; Aguiar, S.I. Liposomes as Antibiotic Delivery Systems: A Promising Nanotechnological Strategy against Antimicrobial Resistance. *Mol. Basel Switz.* **2021**, *26*, 2047. [CrossRef] [PubMed]
14. Wang, Z.; Ma, Y.; Khalil, H.; Wang, R.; Lu, T.; Zhao, W.; Zhang, Y.; Chen, J.; Chen, T. Fusion between Fluid Liposomes and Intact Bacteria: Study of Driving Parameters and in Vitro Bactericidal Efficacy. *Int. J. Nanomed.* **2016**, *11*, 4025–4036. [CrossRef]
15. Vanić, Z.; Holæter, A.-M.; Skalko-Basnet, N. (Phospho)Lipid-Based Nanosystems for Skin Administration. *Curr. Pharm. Des.* **2015**, *21*, 4174–4192. [CrossRef] [PubMed]
16. Hussain, A.; Singh, S.; Sharma, D.; Webster, T.J.; Shafaat, K.; Faruk, A. Elastic Liposomes as Novel Carriers: Recent Advances in Drug Delivery. *Int. J. Nanomed.* **2017**, *12*, 5087–5108. [CrossRef] [PubMed]
17. Li, W.-Z.; Hao, X.-L.; Zhao, N.; Han, W.-X.; Zhai, X.-F.; Zhao, Q.; Wang, Y.-E.; Zhou, Y.-Q.; Cheng, Y.-C.; Yue, Y.-H.; et al. Propylene Glycol-Embodying Deformable Liposomes as a Novel Drug Delivery Carrier for Vaginal Fibrauretine Delivery Applications. *J. Control. Release Off. J. Control. Release Soc.* **2016**, *226*, 107–114. [CrossRef]
18. Rukavina, Z.; Šegvić Klarić, M.; Filipović-Grčić, J.; Lovrić, J.; Vanić, Ž. Azithromycin-Loaded Liposomes for Enhanced Topical Treatment of Methicillin-Resistant Staphyloccocus Aureus (MRSA) Infections. *Int. J. Pharm.* **2018**, *553*, 109–119. [CrossRef]
19. Vanić, Ž.; Rukavina, Z.; Manner, S.; Fallarero, A.; Uzelac, L.; Kralj, M.; Amidžić Klarić, D.; Bogdanov, A.; Raffai, T.; Virok, D.P.; et al. Azithromycin-Liposomes as a Novel Approach for Localized Therapy of Cervicovaginal Bacterial Infections. *Int. J. Nanomed.* **2019**, *14*, 5957–5976. [CrossRef]
20. Bonicelli, M.G.; Giansanti, L.; Ierino, M.; Mancini, G. Interaction of Cationic Liposomes with Cell Membrane Models. *J. Colloid Interface Sci.* **2011**, *355*, 1–8. [CrossRef]
21. Ternullo, S.; Gagnat, E.; Julin, K.; Johannessen, M.; Basnet, P.; Vanić, Ž.; Škalko-Basnet, N. Liposomes Augment Biological Benefits of Curcumin for Multitargeted Skin Therapy. *Eur. J. Pharm. Biopharm. Off. J. Arb. Pharm. Verfahr. EV* **2019**, *144*, 154–164. [CrossRef]
22. Jurstrand, M.; Falk, L.; Fredlund, H.; Lindberg, M.; Olcén, P.; Andersson, S.; Persson, K.; Albert, J.; Bäckman, A. Characterization of Chlamydia Trachomatis Omp1 Genotypes among Sexually Transmitted Disease Patients in Sweden. *J. Clin. Microbiol.* **2001**, *39*, 3915–3919. [CrossRef]
23. Sabet, S.F.; Simmons, J.; Caldwell, H.D. Enhancement of Chlamydia Trachomatis Infectious Progeny by Cultivation of HeLa 229 Cells Treated with DEAE-Dextran and Cycloheximide. *J. Clin. Microbiol.* **1984**, *20*, 217–222. [CrossRef] [PubMed]
24. Pavelić, Z.; Skalko-Basnet, N.; Jalsenjak, I. Characterisation and in Vitro Evaluation of Bioadhesive Liposome Gels for Local Therapy of Vaginitis. *Int. J. Pharm.* **2005**, *301*, 140–148. [CrossRef]
25. Vanić, Ž.; Hurler, J.; Ferderber, K.; Golja Gašparović, P.; Škalko-Basnet, N.; Filipović-Grčić, J. Novel Vaginal Drug Delivery System: Deformable Propylene Glycol Liposomes-in-Hydrogel. *J. Liposome Res.* **2014**, *24*, 27–36. [CrossRef] [PubMed]
26. Eszik, I.; Lantos, I.; Önder, K.; Somogyvári, F.; Burián, K.; Endrész, V.; Virok, D.P. High Dynamic Range Detection of Chlamydia Trachomatis Growth by Direct Quantitative PCR of the Infected Cells. *J. Microbiol. Methods* **2016**, *120*, 15–22. [CrossRef]

27. Palac, Z.; Engesland, A.; Flaten, G.E.; Škalko-Basnet, N.; Filipović-Grčić, J.; Vanić, Ž. Liposomes for (Trans)Dermal Drug Delivery: The Skin-PVPA as a Novel in Vitro Stratum Corneum Model in Formulation Development. *J. Liposome Res.* **2014**, *24*, 313–322. [CrossRef] [PubMed]
28. Perrett, S.; Golding, M.; Williams, W.P. A Simple Method for the Preparation of Liposomes for Pharmaceutical Applications: Characterization of the Liposomes. *J. Pharm. Pharmacol.* **1991**, *43*, 154–161. [CrossRef]
29. Pavelić, Z.; Skalko-Basnet, N.; Schubert, R.; Jalsenjak, I. Liposomal Gels for Vaginal Drug Delivery. *Methods Enzymol.* **2004**, *387*, 287–299. [CrossRef]
30. Pavelić, Z.; Skalko-Basnet, N.; Filipović-Grcić, J.; Martinac, A.; Jalsenjak, I. Development and in Vitro Evaluation of a Liposomal Vaginal Delivery System for Acyclovir. *J. Control. Release Off. J. Control. Release Soc.* **2005**, *106*, 34–43. [CrossRef]
31. Lee, E.H.; Kim, A.; Oh, Y.-K.; Kim, C.-K. Effect of Edge Activators on the Formation and Transfection Efficiency of Ultradeformable Liposomes. *Biomaterials* **2005**, *26*, 205–210. [CrossRef]
32. Vanić, Ž.; Hafner, A.; Bego, M.; Škalko-Basnet, N. Characterization of Various Deformable Liposomes with Metronidazole. *Drug Dev. Ind. Pharm.* **2013**, *39*, 481–488. [CrossRef]
33. Van Bambeke, F.; Montenez, J.P.; Piret, J.; Tulkens, P.M.; Courtoy, P.J.; Mingeot-Leclercq, M.P. Interaction of the Macrolide Azithromycin with Phospholipids. I. Inhibition of Lysosomal Phospholipase A1 Activity. *Eur. J. Pharmacol.* **1996**, *314*, 203–214. [CrossRef]
34. Chang, H.-I.; Yeh, M.-K. Clinical Development of Liposome-Based Drugs: Formulation, Characterization, and Therapeutic Efficacy. *Int. J. Nanomed.* **2012**, *7*, 49–60. [CrossRef]
35. Agacfidan, A.; Moncada, J.; Schachter, J. In Vitro Activity of Azithromycin (CP-62,993) against Chlamydia Trachomatis and Chlamydia Pneumoniae. *Antimicrob. Agents Chemother.* **1993**, *37*, 1746–1748. [CrossRef] [PubMed]
36. Donati, M.; Rodrìguez Fermepin, M.; Olmo, A.; D'Apote, L.; Cevenini, R. Comparative In-Vitro Activity of Moxifloxacin, Minocycline and Azithromycin against *Chlamydia* spp. *J. Antimicrob. Chemother.* **1999**, *43*, 825–827. [CrossRef] [PubMed]
37. Welsh, L.E.; Gaydos, C.A.; Quinn, T.C. In Vitro Evaluation of Activities of Azithromycin, Erythromycin, and Tetracycline against Chlamydia Trachomatis and Chlamydia Pneumoniae. *Antimicrob. Agents Chemother.* **1992**, *36*, 291–294. [CrossRef]
38. Belland, R.J.; Nelson, D.E.; Virok, D.; Crane, D.D.; Hogan, D.; Sturdevant, D.; Beatty, W.L.; Caldwell, H.D. Transcriptome Analysis of Chlamydial Growth during IFN-Gamma-Mediated Persistence and Reactivation. *Proc. Natl. Acad. Sci. USA* **2003**, *100*, 15971–15976. [CrossRef]
39. Thomas, M.; Lawrence, A.; Kroon, S.; Vodstrcil, L.A.; Phillips, S.; Hocking, J.S.; Timms, P.; Huston, W.M. Chlamydial Clinical Isolates Show Subtle Differences in Persistence Phenotypes and Growth in Vitro. *Access Microbiol.* **2021**, *3*, 000204. [CrossRef]
40. Aboud, H.M.; Ali, A.A.; El-Menshawe, S.F.; Elbary, A.A. Nanotransfersomes of Carvedilol for Intranasal Delivery: Formulation, Characterization and in Vivo Evaluation. *Drug Deliv.* **2016**, *23*, 2471–2481. [CrossRef]
41. Elsayed, M.M.A.; Abdallah, O.Y.; Naggar, V.F.; Khalafallah, N.M. Lipid Vesicles for Skin Delivery of Drugs: Reviewing Three Decades of Research. *Int. J. Pharm.* **2007**, *332*, 1–16. [CrossRef]
42. Cserháti, T. Alkyl Ethoxylated and Alkylphenol Ethoxylated Nonionic Surfactants: Interaction with Bioactive Compounds and Biological Effects. *Environ. Health Perspect.* **1995**, *103*, 358–364. [CrossRef]
43. Nokhodchi, A.; Shokri, J.; Dashbolaghi, A.; Hassan-Zadeh, D.; Ghafourian, T.; Barzegar-Jalali, M. The Enhancement Effect of Surfactants on the Penetration of Lorazepam through Rat Skin. *Int. J. Pharm.* **2003**, *250*, 359–369. [CrossRef]

pharmaceutics

MDPI

Article

The Biological Effects of Novel Nutraceuticals with Curcuminoids and Other Plant-Derived Immunomodulators and Pre-Probiotics

Adina Elena Răducanu [1], Bianca-Maria Tihăuan [1,2,*], Ioana Cristina Marinaş [2,3], Oana Teodora Ciupercă [4,5], Carmen Elena Ţebrencu [5,6], Elena Ionescu [5,6] and Tatiana Onisei [1]

1 Research and Development Department, The National Institute for Research and Development in Food Bioresources, Str. Dinu Vintilă, No.6, 021102 Bucharest, Romania; adina.raducanu@bioresurse.ro (A.E.R.); tatiana.onisei@bioresurse.ro (T.O.)

2 Research Institute of the University of Bucharest—ICUB, 91–95 Spl. Independentei, 50567 Bucharest, Romania; ioana.cristina.marinas@gmail.com

3 Physico-Chemistry Laboratory, National Institute of Research & Development for Food Bioresources—IBA Bucharest, 5 Băneasa Ancuța, 020323 Bucharest, Romania

4 Faculty of Chemical Engineering and Environmental Protection "Cristofor Simionescu", "Gheorghe Asachi" Technical University of Iași, 73, Prof. Dr. Docent D. Mangeron Street, 700050 Iași, Romania; oana.ciuperca@plantavorel.ro

5 Research and Processing Centre for Medicinal Plants "PLANTAVOREL" S.A., 46, Cuza Voda Street, 610019 Piatra Neamţ, Romania; carmen@plantavorel.ro (C.E.Ţ.); elenaionescu@plantavorel.ro (E.I.)

6 Academy of Romanian Scientists, 54, Splaiul Independentei, 050094 Bucharest, Romania

* Correspondence: bianca.tihauan@bioresurse.ro; Tel.: +40-733229834

Abstract: An effective and well-balanced immune system is pivotal for maintaining health. Diet and nutrition can affect the functioning of numerous immune parameters, with direct repercussions on homeostasis. Since our immune functions are indispensable in defending the body against pathogens and thus play a vital role in maintaining health, modulating immune response may well serve as the basis for the development of plant-based functional foods and novel nutraceuticals. This concept is currently utilized in attempts to prevent or mitigate inflammatory reactions via the development of targeted food products or active ingredients since an extended number of phytoconstituents (such as curcuminoids) are associated with beneficial effects on immunity. Immunomodulatory plant-based dietary supplements are considered effective in improving immune functions and reducing the incidence of immunological disorders or imbalances. Therefore, the main focus of this study was to evidence the beneficial biological effects such as antioxidant and antimicrobial, as well as nutritional status, biocompatibility and cell proliferation capacity and immunomodulation of two novel nutraceuticals. The first nutraceutic was based on curcuminoids and other actives from *Trigonella foenum- graecum* (seeds), *Chelidonium majus* L. (aerial parts), *Taraxacum officinale* L. (roots), vitamins (C, D3, A, E) and minerals (zinc) whereas the second one was made of probiotics such as *Lactobacillus acidophilus* and *Bifidobacterium animalis subsp. Lactis* combined with actives from *Helianthus tuberosus* (tubers) and *Psyllium/Plantago ovata* (husk) as herbal prebiotics.

Keywords: novel nutraceuticals; plant-based immunomodulators; pre-probiotics; antimicrobial; cytotoxic

1. Introduction

Our immune system acts as a protective surveillance grid being able to interpret changes in the world around it and respond appropriately. At its finest, it is capable of distinguishing self from non-self and distinguishing harmless non-self from dangerous non-self. During recent years, the implication of nutrition and dietary supplementation on boosting the immune system capacities has gained interest and might become even more important in the prevention of disease when the interplay between nutritional processes and immune system is better understood. Therefore, nutritional immunology is a rapidly developing field [1–4].

Studies over the past decade bear out a direct correlation between diet and nutrition on the functioning capacity of numerous immune parameters. This concept is currently utilized in attempts to prevent or mitigate inflammatory reactions via the development of targeted food products or ingredients. Therefore, food and nutrition in general play a critical role in supporting and maintaining immune homeostasis for various groups of individuals [5–8].

Nevertheless, the research into the possible role of functional foods and nutraceuticals in mitigating immune function is still in its infancy. Therefore, controversies around health claims will remain the rule rather than the exception, especially for nutraceutical compounds [9]. Even though the accurate definition of nutraceuticals is not yet quite well outlined, it can be said that nutraceutical compounds are health-enhancing products that improve mental and physical activities of the body. Their potential role in minimizing risk factors for various diseases are key factors of their success in commercialization [10]. Nutraceutical products are simply a hybrid between drug and food [11].

On the other hand, this terminology is a broader term that includes minerals, vitamins, amino-acids, botanicals or herbs [12], therefore, both dietary supplements and fortified foods can be classified as nutraceuticals [13]. The terminology of nutraceutical was defined by the foundation for innovation in medicine in (New York, USA) in 1989 [11]. Defelice's definition in 1995 was: "A food or parts of food that provide medical or health benefits, including the prevention and/or treatment of disease" [14]. This term originated from two terminologies: "nutrition" and "pharmaceutical" [11].

As we know by now, modulation of the immune system refers to a certain alteration in the immune response in the form of stimulation, amplification, expression or inactivation of some stage of the immune response. Therefore, an immunomodulator is a material used to induce an effect on the immune system. As a general rule, a product (active ingredient, dietary supplement, etc.) which exhibits a modifying immune system response to a threat, is an immunomodulator. They modulate and arm the immune system by keeping them in a highly prepared state against any threat [15,16].

Modulating immune responses may serve as the basis for the development of functional foods, although constant research on better understanding physiological, immunological and biochemical mechanisms associated with nutrients is necessary [17]. This approach can be a potential route; nevertheless, it is essential to keep in mind that functional foods and nutraceuticals should be regarded as food products with specific health benefits, above those of average consumption levels and aiming at the specific needs of individuals [8,17–19].

Over the last few years, a plethora of dietary supplements with natural actives have been widely investigated for their possible use for immunomodulation via various food components, including probiotics, prebiotics, polysaccharides, vitamins, minerals, fatty acids, peptides, amino acids or polyphenols [20]. Moreover, the scientific literature supports the correlation between use of immunomodulatory and probiotic ingredients with prevention of various diseases [1,21–23].

Therefore, the main focus of this study was to assess the immunomodulatory properties of two novel nutraceutical formulations, one based on curcuminoids and other actives from *Trigonella foenum- graecum* (seeds), *Chelidonium majus* L. (aerial parts), *Taraxacum officinale* L. (roots), vitamins (C, D3, A, E) and minerals (zinc) and one with probiotics such as *Lactobacillus acidophilus* and *Bifidobacterium animalis subsp. Lactis* combined with actives from *Helianthus tuberosus* (tubers) and *Plantago ovate* (husk) as herbal prebiotics.

2. Materials and Methods

2.1. Phytochemical Screening of Selected Plant Actives for Novel Nutraceuticals

In order to obtain the novel nutraceuticals, a phytochemical screening of actives was performed on the following species: *Trigonella foenum- graecum* (seeds), *Chelidonium majus* L. (aerial parts), *Taraxacum officinale* L. (roots), *Helianthus tuberosus* (tubers) and *Psyllium/Plantago ovata* (husk).

For the species *Trigonella foenum- graecum* (seeds), *Chelidonium majus* L. (aerial parts), *Taraxacum officinale* L. (roots), first were assessed the humidity (method from Romanian Pharmacopoeia 10th Edition—IX.C.15), granulometry, density and flow, and secondly, qualitative, semiquantitative and quantitative physico-chemical assays were performed in order to highlight the presence of active principles with immunomodulatory potential.

Qualitative determination of saponins, aminoacids, phenolic acids and flavones was performed by HPTLC (high performance thin layer chromatography). For the semiquantitative determination of diosgenin from *Trigonella foenum—graecum* (seeds) a HPTLC-UV densitometric method was used. Quantitative assessment of total chelidonine alkaloids from *Chelidonium majus* L.(aerial parts) was performed by UV-VIS spectrophotometry [24].

The species *Helianthus tuberosus* (tubers) and *Psyllium/Plantago ovata* (husk) were analysed in order to identify and quantify the phytocompounds with prebiotic action. Therefore, the qualitative evaluation of saccharides in these two herbal prebiotics was performed using specific tests (the reaction with thymol) and HPTLC identification. The presence with high and medium intensity of saccharides was determined. The HPTLC evaluation of *Helianthus tuberosus* (tubers) evidenced the presence of mono and oligosaccharides such as D-fructose, D-lactose, D-galactose, D-sucrose, D-glucose and inulin. The quantitative evaluation of *Helianthus tuberosus* tubers using UV-VIS spectrophotometric determination revealed a content of total sugars expressed in fructose of 15.62% g/g. These results were relevant for the obtaining of innovative formulas between *Helianthus tuberosus* tubers and *Psyllium* husk [25]. The action of these herbal prebiotics in association with probiotics can ensure the maintaining of a normal intestinal function and health.

2.2. Formulas of the Novel Nutraceuticals with Immunomodulatory Properties

The two novel nutraceuticals, one based on curcuminoids and other actives from *Trigonella foenum—graecum* (seeds), *Chelidonium majus* L. (aerial parts), *Taraxacum officinale* L. (roots), vitamins (C, D3, A, E) and minerals (zinc) and one with probiotics such as *Lactobacillus acidophilus* and *Bifidobacterium animalis* subsp. *Lactis* combined with actives from *Helianthus tuberosus* (tubers) and Psyllium/*Plantago ovate* (husk) as herbal prebiotics, were obtained in the laboratory of Research and Processing Centre for Medicinal Plants PLANTAVOREL S.A., Piatra-Neamţ, Romania. The formulas of these nutraceuticals resulted from the performed formulation tests. These tests had as a result the selection of an optimal formula capable to ensure a content of phytocompounds which can generate an immune response.

The optimum formula for the tested Immunomodulatory product consisted of 80% actives from which 18% represents curcumin extract with curcuminoids 95% and 20% inactive ingredients, whilst the Pre-Probiotic product formula had 95% actives and 5% inactive ingredients.

To start with a successful formulation for a novel nutraceutical, there are factors to consider such as: an appropriate dosage of the ingredients in order to guarantee the efficiency, validation by scientific studies, and the novelty element of the proposed formula. Thus, the selection of the species *Trigonella foenum- graecum* (seeds), *Chelidonium majus* L. (aerial parts), *Taraxacum officinale* L. (roots) as herbal active ingredients in a immunomodulatory formula is the right choice for obtaining a novel nutraceutical [24,26–30]. In order to provide enough nutrients to meet immune systems requirements and to modulate various physiological functions, curcuminoids, vitamins and minerals can be used. Hence, the combination of *Trigonella foenum- graecum* (seeds), *Chelidonium majus* L. (aerial parts), *Taraxacum officinale* L. (roots) with curcumin extract, C, D_3, A, E vitamins and minerals such as zinc represents the novel formula for a dietary supplement with immunomodulatory effect. The presence of micronutrients in the novel formula improves the bioavailability of the product. For the proper conditioning of this formula, inactive ingredients (excipients and adjuvants) were used in proportions to ensure an optimal flow of the mixture.

The gut immune system is influenced by many factors, including dietary components and commensal bacteria. Strategies that try to restore the normal gut microbiota have been extensively studied and new therapeutic approaches have appeared and are represented by probiotics (live microorganisms) associated with herbal prebiotics. *Helianthus tuberosus* (tubers) and *Psyllium/Plantago ovata* (husk) are an consistent source of fibers and important herbal prebiotics [25,31–34] that mixed with probiotics from genus *Lactobacillus acidophylus* and *Bifidobacterium animalis* confer the health of the intestine. This pre-probiotic formula for a plant-based dietary supplement ensures health benefits on the intestinal microenvironment.

2.3. Health Claims and Nutritional Status

According to Regulation (EC) No 1924/2006 (nutrition and health claims made on foods), (EU) No 432/2012 (vitamins and minerals), and EFSA regulations regarding health claims of botanicals, there are several health claims associated with ingredients used in our two nutraceuticals. Their importance is underlined by legislation support, as it facilitates rapid association of active and more efficient development of formulas.

The health claims [35,36] of our main ingredients support our formula's rationale, and for the Immunomodulatory nutraceutical, they are:

Chelidonium majus: contributes to the maintenance of normal liver function and additionally supports the digestion and the body's purification—ID 2238.

Curcuma longa: curcumin and its derivatives are currently used as oral supplements in the management of various dermatological and medical conditions [37]; moreover, it helps maintain the efficacy of the immune system and to maintain resistance to allergies. It also has significant antioxidant properties—ID 4009; According to ID 2751, curcuma prevents the accumulation of fats and facilitates their clearance by the liver, maintaining the health of the liver—ID 3098; is registered to provide cell protection—ID 3748; also helps maintain the health of joints and bones—ID 4012, keeps the skin healthy—ID 4008 and supports heart function and blood circulation—ID 4010;

Trigonella foenum-graecum: main benefits are in providing support and balance to the metabolism of fats, contributing to normal glucose and insulin metabolism—ID 4186; it helps with the physiological pH balance of the stomach—ID 3623, through balanced diets helps the control of lipidic metabolism (cholesterol and triglycerides)—ID 3832; Antioxidants can protect from free radicals and helps in case of foods intake deficiency or increased amount of nutrients. Protection against the free radicals action due to stress, alcoholics, UV exposure or polluted ambiance conditions—ID 3935; is also is a physical and mental tonic—ID 4494.

Taraxacum officinale: confers protection against free radicals action due to stress, alcoholics, UV exposure or polluted ambiance conditions—ID 3828, helps maintain urinary tract function by maintaining normal urinary flow—ID 3827; It has prebiotic activities, contributes to the gastrointestinal well-being by its prebiotic effects—ID 4315; also helps the physiological pH balance of the stomach—ID 3609.

According to (EU) No 432/2012, vitamins C, E, A, D and zinc all contribute to the normal function of the immune system and to the protection of cells from oxidative stress [38].

As for the Pre-Probiotic formula:

Topinambur/*Helianthus tuberosus*: is actively implicated into the growth, development and maintenance of body function—ID 4412; A generally restorative product for strengthening of the immune system, activates body's antioxidant protection system, increases body's resistance against the harmful effects of the environment, infections and other unfavorable factors—ID 2449.

Psyllium/*Plantago ovata*: its main effect are correlated to lowering cholesterol and blood lipids level—ID 4461; It helps to control blood levels of cholesterol—ID 2106, contributes to intestinal transit and intestinal function and helps to maintain a healthy bowel and facilitate intestinal transit—ID 3932.

2.4. Evaluation of Key Physico-Chemical Parameters and Antioxidative Status

(a) Sample preparation

The nutraceuticals were solubilized in 70% ethanol to obtain a final concentration of 10 mg dry product/mL. The extracts were obtained by an ultrasound-assisted method for 20 min. The method involves the use of ultrasounds, with frequencies ranging from 20 kHz to 2000 kHz, allowing cell lysis. The extract thus obtained was centrifuged for 15 min at 10,000 rpm to remove insoluble material.

(b) Total phenol content

The total phenol content was measured spectrophotometrically by the method described by Singleton and Rossi in 1965 and modified by other authors [39,40]. Over 0.1 mL of sample or standard solution (gallic acid) was added 0.9 mL of H_2O d.d. After adding 0.1 mL of Folin Ciocâlteu reagent, stir for 5 min, then add 1 mL of 7% Na2CO3 and make up to a final volume of 2.5 mL with H_2O d.d. After incubation at room temperature in dark place for 60 min, the absorbance values at 765 nm were recorded. The calibration curve was performed for gallic acid concentrations between 0–300 μg/mL (R^2 = 0.9969).

(c) Total flavonoid content

The total flavonoid content was determined by the aluminum chloride method [41] which consisted of treating 0.1 mL of sample/standard solution (quercetin) with 0.1 mL of 10% sodium acetate. A total of 0.12 mL of 2.5% $AlCl_3$ solution was added to this mixture and made up to a final volume of 1 mL with 70% ethanol. After stirring, this solution was allowed to stand for 45 min in the dark and the absorbance of the mixture was read at 430 nm. The calibration curve was performed for quercetin concentrations in the range of 0–200 μg/mL (R^2 = 0.996).

(d) DPPH

The measurement of DPPH radical uptake activity was performed by an adapted method after Madhu, 2013 [42]. The reaction mixture consisted of adding 100 μL of sample/standard and 100 μL of 0.3 mM DPPH radical solution in 50% ethanol. The absorbance reading was performed at λ = 517 nm after 30 min of reaction, using a UV-VIS spectrophotometer. Ethanol was used as a control. The concentrations used for the Trolox calibration curve were in the range of 5–80μM (R^2 = 0.9968).

(e) CUPRAC

The CUPRAC method is based on the reduction in the cupric complex, neocuproine (Cu (II) -Nc) by antioxidants in cuprous form (Cu (I) -Nc). Copper ion reduction was performed according to a method adapted from Meng et al., 2011 [43], as follows: 20 μL of sample/standard solutions were mixed with 60 μL CuSO4 (5 mM), neocuproine 60 μL (3.75 mM) and distilled water (560 μL), reaching a final volume of 700 μL. After 30 min, the absorbance was measured at 450 nm. The standard Trolox solutions required for the calibration curve were between: 1–0 mM (R^2 = 0.9962).

(f) FRAP

The FRAP method is based on the ability of antioxidants to reduce the tripyridyltriazine-Fe^{3+} (Fe (III) -TPTZ) complex to the blue-colored tripyridyltriazine-Fe^{2+} (Fe (II) -TPTZ) complex. The determination of the antioxidant power of iron reduction was performed by the method described by Benzie and Strain, 1999 [44] with some modifications [45]. The following solutions were prepared: 300 mM acetate buffer, pH 3.6, 10 mM TPTZ stock solution in 40 mM HCl and 20 mM $FeCl_3$ solution in distilled water. Prepare the FRAP reagent by mixing 300 mM acetic acid-sodium acetate buffer, pH 3.6 with 10 mM TPTZ solution and 20 mM $FeCl_3$ solution (10:1:1). The FRAP reagent is kept on a water bath at 37 °C until the time of analysis. At 10 μL sample/standard solution, 190 μL FRAP reagent and incubate for 30 min at 37 °C. After incubation, the absorbance at λ = 593 nm is read. The calibration curve was performed for the concentration range 50–250 μM Trolox/mL (R^2 = 0.9938).

2.5. Assessment of Antimicrobial Activity

Evaluation of antimicrobial activity of samples was performed by determining the logarithmic and percentage reduction in reference microorganisms.

Three reference strains from the American Type Culture Collection (ATCC, Manassas, VA, USA) were used for testing (*Staphylococcus aureus* ATCC 6538, *Escherichia coli* ATCC 8739, and *Candida albicans* ATCC 10231). The microbial suspensions of 1.5 × 10^8 CFU/mL, obtained from fresh 15 to 18 h cultures, developed on solid medium, were adjusted according to McFarland standard 0.5, respectively, McFarland 1 for microfungi, and then serially diluted to 10^5. Samples were weighed and the inoculum volume was adjusted according to the mass of the samples. The samples were placed in contact with the microbial inoculum for 30 min, thoroughly spun on a vortex, and afterwards 5 decimal serial dilutions were carried out in order to determine the logarithmic and percentage reduction in the microbial populations. A total of 10 µL in triplicate were inoculated in spot on Muller Hinton solid medium, respectively, Sabouraud for microfungi. After 18–24 h of incubation at 36 ± 2 °C the plates were read by counting the colonies.

The logarithmic reduction was calculated using the formula:

Logarithmic reduction = lg A/B

A = no. of viable organisms before treatment; B = no. of viable organisms after treatment

The percentage reduction in microbial populations was calculated using the formula:

$P = (1 - 10^{-L}) \times 100$

where P is the percentage reduction and L the logarithmic reduction

2.6. Determination of Intracellular Toxicity

The L929 murine fibroblasts (ECACC—European Collection of Authenticated Cell Cultures) were selected as a model for intracellular toxicity assessment of novel products. L929 cells were cultivated in DMEM (Dulbecco's Modified Eagle Medium, Sigma-Aldrich) media supplemented with 10% FBS (Fetal Bovine Serum, Sigma–Aldrich) and 1% Pen/Strep (penicillin/streptomycin solution, 50 µg/mL, Sigma–Aldrich) for 24 h at 37 °C, 95% humidity with 5% CO_2. After 24 h, the cells were washed with PBS (Phosphate Buffered Solution, Sigma–Aldrich), harvested using trypsin (Sigma–Aldrich) and counted using Trypan Blue (Sigma–Aldrich, St. Louis, MO, USA) and a hemocytometer. The seeding density for the MTT and LDH assays was optimized at 5 × 10^4.

Samples were prepared by a conventional heating technique in PBS. This mixture was mechanically stirred until it was well homogenized, then heated to extraction temperature (i.e., 35–40 °C) and rigorously stirred throughout the extraction time (i.e., 5–150 min). The extracts were then carefully filtered using a vacuum pump filter.

The MTT assay evaluates the cellular metabolic activity, therefore is a good indicator of cell viability, proliferation and cytotoxicity. This colorimetric assay is based on the reduction in a yellow tetrazolium salt (3-(4,5-dimethylthiazol-2-yl)-2,5-diphenyltetrazolium bromide or MTT) to purple formazan crystals by metabolically active cells [46,47].

Cells seeded at 5 × 10^4 density in a clear 96 well cell culture plate were treated with the novel products in the following concentrations: 200 mg/mL, 100 mg/mL 25 mg/mL, 3.12 mg/mL, 0.2 mg/mL and 0.01 mg/mL. Afterwards, cells were incubated for 24 h at 37 °C, 95% humidity with 5% CO_2. After 24 h of exposure to novel products, cells were incubated for 4 h with MTT reagent (Roche) at 37 °C, 95% humidity with 5% CO_2. After incubation, cells were treated with MTT solvent (Roche) for 15 min at room temperature. Absorbance was measured using FlexStation 3 (Molecular Devices Company, Sunnyvale, CA, USA) at OD = 570 nm.

The LDH assay involves the assessment of cell death by quantification of plasma membrane damage. This increase in the amount of enzyme activity in the supernatant directly correlates to the amount of formazan formed. Therefore, the amount of color formed in assay is proportional to the number of lysed cells [47]. With the LDH Cytotoxicity Detection Kit (Roche, Basel, Switzerland), LDH activity was measured in culture supernatants using FlexStation 3 (Molecular Devices Company, Sunnyvale, CA, USA) at

492 nm with a 600 nm wavelength reference. Cells were prepared and treated alike as for the MTT assay.

2.7. Evaluation of Immunomodulatory Activity

To investigate the immune-modulatory activity of the novel compounds, RAW murine macrophages were cultivated in Dulbecco's Modified Eagle's medium (DMEM) supplemented with 10% FBS. Compounds were diluted 1:5 in the cell culture medium and the stimulation of macrophages was done for 24 h. Supernatants were collected and used for cytokine quantification. Murine IL-10 and IL-6 were quantified using ELISA (Thermo Scientific, Waltham, MA, USA) following the manufacturer's instructions.

2.8. Statistical Assesment of Data

All tests were performed in triplicate and the results were expressed as ±SD (standard deviation). The statistical analysis was performed by Unpaired t-test (n = 2) using the GraphPadPrism Software (San Diego, CA, USA).

3. Results

3.1. Evaluation of Key Physico-Chemical Parameters and Antioxidative Status

The flavonoid content represents approximately 14.21% of the total polyphenols for the immunomodulatory product and 2.85% for the pre- and probiotic product (Table 1). In fact, the polyphenol content determined by the Folin Ciocalteu method probably comes from the following ingredients: Chelidonium majus (rich in quercetin, caffeic acid and neochlorogenic acid) [48], Trigonella foenum-graecum (apigenin, luteolin and kaempferol as aglycones and hydroxycinnamic acids mostly dominated by caffeic acid derivatives) [49], Curcuma longa (chlorogenic acid, curcuminoids, isorhamnetin, digalloyl-hexoside) [50] extract and vitamin C for the first product while for the second product the continuum of polyphenols and flavonoids, respectively, is given by Helianthus tuberosus (chlorogenic and dicaffeoylquinic acids) [51] and Plantago ovata (gallic acid and rutin) [52].

Table 1. Total content of polyphenols and flavonoids expressed as mg GAE/g sample and mg QE/g sample, respectively.

Sample	Total Polyphenols (mg GAE/g)	Total Flavonoids (mg QE/g)	% Flavonoids from Total Polyphenols
Immunomodulator	144.07 ± 4.21	20.472 ± 2.005	14.21
Pre-Probiotic	1.13 ± 0.26	0.032 ± 0.009	2.85

Flavonoids are secondary metabolites of plants, with numerous pharmacological functions: antioxidant, antimutagenic, antibacterial, anti-inflammatory, antiallergenic, enzymatic modulators and antitumor. These phytocompounds are in the form of free aglycones or in the form of glycosides. Flavonoids are classified into flavones, flavanols, isoflavones, flavonols, flavanones, flavanonols and chalcones. The various structures of flavonoids have led to numerous modulating properties on the immune system response. These compounds can act both synergistically and antagonistically [53].

The Pre-/Probiotic product has a mixture of plant materials poor in phenolic compounds, so it correlates with reduced antioxidant capacity.

The higher values obtained by the DPPH method highlight the double mechanism involving both proton transfer (HAT) and electron transfer (SET) in the case of the immunomodulatory product [54]. Meanwhile, the pre- and probiotic product involves a predominantly SET antioxidant mechanism, but considerably lower than the immunomodulatory one (Table 2).

Table 2. Antioxidant activity of the Immunomodulatory and Pre-Probiotic products.

Sample	DPPH (μM Trolox/mg)	CUPRAC (μM Trolox/mg)	FRAP (μM Trolox/mg)
Immunomodulator	2956.63 ± 292.12	1001.13 ± 30.82	918.91 ± 2.97
Pre-Probiotic	7.47 ± 0.04	33.59 ± 0.76	18.25 ± 0.55

The evaluation of the antioxidant activity of phenolic compounds is performed mainly by chemical methods. Antioxidant testing methods, including Iron Reducing Antioxidant Power (FRAP), Copper Reducing Antioxidant Capacity (CUPRAC), 2,2-diphenyl-1-picrylhydrazyl (DPPH) and photochemiluminescence have been developed as rapid assessment and screening tools. According to the literature data [48,55–57], it has been observed that the antioxidant activity of the immunomodulatory product is higher than the extracts from individual plant materials, thus, the components act synergistically. None of these in vitro methods are relevant to the physiology itself in vivo. Phenolic compounds generally have low bioavailability. Cellular determinations can provide physiologically relevant information. However, the rapid metabolism and low solubility of phenolic compounds often lead to low bioavailability in vivo. The plasma concentration of phenolic compounds is generally low, generating insufficient antioxidant activities for a significant effect [58].

3.2. Antimicrobial Activity

The methodology for evaluating antimicrobial activity aims to assess the safety and effectiveness of products. Testing of antimicrobial properties involves testing the ability of microorganisms to survive under the effect of a given antimicrobial agent (or with antimicrobial potential), at a certain concentration and for a certain period of time.

Although our novel products do not aim to obtain an antimicrobial effect after administration, the existence of this quality has a number of advantages, both for the consumer and for the quality of the product.

Samples were evaluated in the presence of *Staphylococcus aureus* ATCC 6538, *Escherichia coli* ATCC 8739, and *Candida albicans* ATCC 10,231 strains. The results on logarithmic reduction are shown in Figure 1a. Logarithmic factor 1, which represents a 99% efficiency in bacterial reduction, was considered as a control. For the Immunomodulator product, in the presence of *S.aureus* ATCC 6538 a reduction of 0.47 log was obtained, and for the Pre-Probiotic product a reduction of 0.55 log, this having a more pronounced antimicrobial effect.

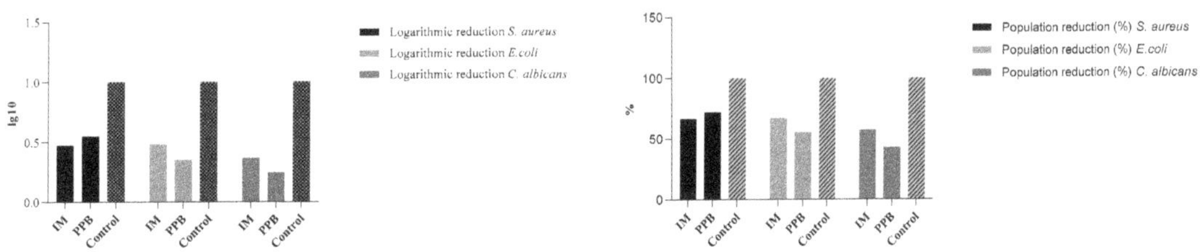

(a) (b)

Figure 1. (**a**) Logarithmic reduction and (**b**) Population reduction in *Staphylococcus aureus* ATCC 6538, *Escherichia coli* ATCC 8739, and *Candida albicans* ATCC 10,231 strains by tested samples represented by novel nutraceuticals.

In the case of Gram-negative strain *E. coli* ATCC 8739, results show a logarithmic reduction of 0.48 log for the Immunomodulatory product and of 0.35 log for the Pre-Probiotic product. Both products also significantly reduced *E. coli* populations (55–66%) (Figure 1b).

The reduced efficiency of the two products on the Gram-negative *E. coli* strain may be due to the structural differences between Gram-positive (are more susceptible to antimicrobial treatments) and Gram-negative microorganisms (presence of outer membrane, lipopolysaccharides, murein, etc.).

The logarithmic reduction (Figure 1a) and the percentage reduction (Figure 1b) in the case of the fungal strain *C. albicans* ATCC 10,321 is modest, the Immunomodulatory product being more efficient (0.36 log and 57.1% reduction).

3.3. Determination of Intracellular Toxicity

The analysis of the biological action on the in vitro cell model of the two novel products was performed by evaluating cellular viability, proliferation and membrane integrity by MTT and LDH assays.

In the case of the Immunomodulatory product, concentrations of 200 mg/mL, 100 mg/mL and 25 mg/mL significantly reduced the percentages of cell viability (Figure 2a,b). Lower concentrations favor cell viability and proliferation. In the case of the Pre-Probiotic product, the percentages of cell viability are high, all the concentrations tested do not negatively influence the viability and cell proliferation.

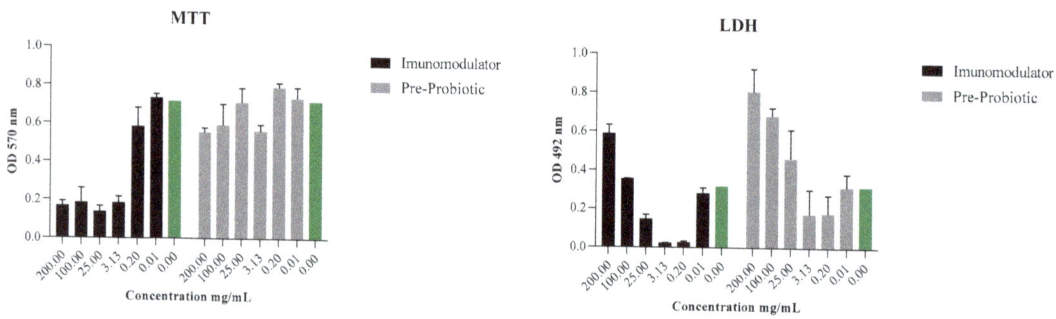

(a) (b)

Figure 2. (**a**) MTT and (**b**) LDH assessment of novel nutraceuticals on L929 murine fibroblasts.

The MTT test (Figure 2a) performed on the L-929 fibroblasts highlights, in the case of the Immunomodulator product, the cytotoxicity of the concentrations 200 mg/mL, 100 mg/mL and 25 mg/mL; the decrease in the concentrations of this product is directly proportional to the increased viability rate. The Pre-Probiotic product favors cell proliferation, having at lower concentrations the effect of a nutritious substrate, the viability of the cells not being affected.

Evaluation of cytotoxicity by LDH test (Figure 2b) confirms the results obtained by MTT test; in the case of the Immunomodulatory product, the administration of concentrations higher than 25 mg/mL increases the degree of toxicity at the cellular level.

3.4. Evaluation of Immunomodulatory Activity

The immune system is a remarkably developed defense system which guards against attacking factors. It is capable of producing diverse molecules and cells able to distinguish and reduce unlimited changes from external and unwanted agents [17].

IL-10, an important anti-inflammatory cytokine, plays a critical role in the control of immune responses, it opposes the switch to the metabolic program induced by inflammatory stimuli in macrophages [59]. It is notable that IL-10 effectively down-regulates proinflammatory cytokines, such as IL-1, IL-6, and TNF-α [60]. As for IL-6 (pro- and anti-inflammatory cytokine), is a very interesting, pleiotropic cytokine with complex roles in inflammation and metabolic disease [61]. Stenvinkel, P et.al., mention that bio-incompatibility and oxidative stress are factors that may be associated with increased circulating IL-6.

Our study showed that, after treatment with novel formulated products, the IL-10 levels were significantly increased (Figure 3a), especially by the *Pre-Probiotic* supplement. Conversely, the Immunomodulatory supplement yielded IL-10 levels similar to the control group. IL-6 levels were slightly increased (Figure 3b), compared with the negative control, especially for the *Pre-Probiotic* product.

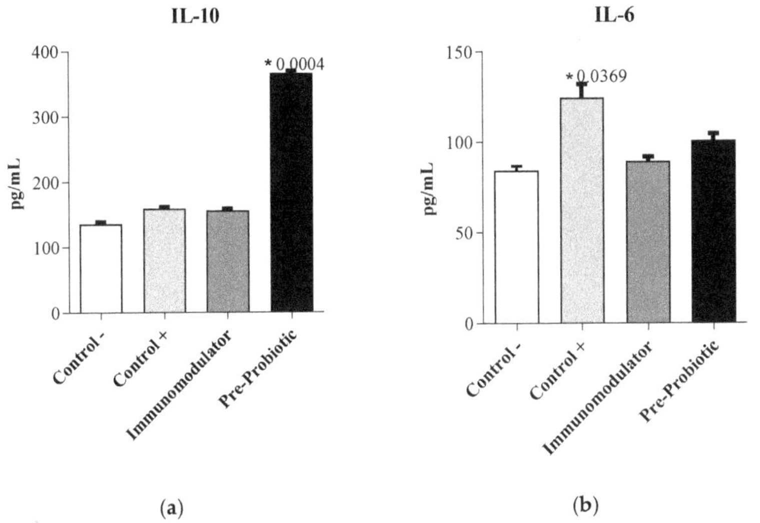

Figure 3. Levels of anti-inflammatory (**a**) IL-10 cytokine and pro-inflammatory (**b**) IL-6 cytokine of RAW murine macrophages treated with novel formulated products. Control − are untreated cells and Control + is positive control of cells treated with a suspension of *E. coli* 10^6. * $p < 0.05$.

As mentioned by other authors, coupling aggressive inflammatory factors with the delayed secretion of IL-10 ensures that any inflammatory response will securely be down-regulated after some time. The increased levels of IL-10 and slightly reduced levels of IL-6 showcase good indicators for initial phase immunostimulation.

Assessing the ratio of IL-6 to IL-10 can be constructive in evaluating cellular response after exposure to stress or trauma-causing factors [62]. For the Pre-Probiotic product (Figure 4a), we can remark the increased levels of IL-10 and reduced levels of IL-6, indicating good immunostimulation; the potent immunosuppressive cytokine, IL-10 was activated properly and inhibits the production of the proinflammatory cytokine IL-6. As for the Immunomodulator product, keeping in mind that the yielded IL-10 levels were similar to the control group, and the IL-6 levels were slightly higher than the negative control, we can attest a case of delayed secretion of IL-10 and a small-scale immunostimulation.

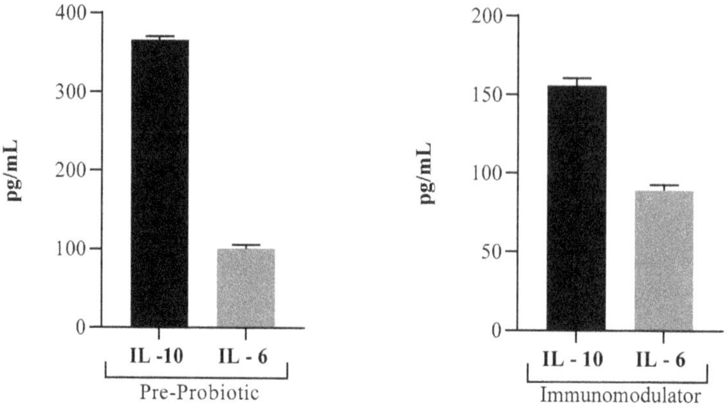

(a) (b)

Figure 4. The ratio of IL-6 to IL-10 for the novel formulated products; (**a**) Pre-Probiotic; (**b**) Immunomodulator.

4. Discussion

Immunomodulation via nutrition has become an interesting approach in the research world. Such products modulate and arm the immune system by keeping them in a highly prepared state against any threat.

As stated above and mentioned by other authors as well, there is a controversy over a specific definition and set of regulations to define the nutraceutical compounds, however, nutraceutical compounds are health-enhancing products that improve mental and physical activities of the body [10,14].

This study evaluated two novel nutraceuticals with Immunomodulatory and Pre-Probiotic properties derived from curcuminoids, vitamins, minerals and plant actives from *Trigonella foenum- graecum*(seeds), *Chelidonium majus* L. (aerial parts), *Taraxacum officinale* L. (roots), *Helianthus tuberosus* (tubers) and Psyllium/*Plantago ovata*(husk).

The Immunomodulatory product has a higher content of polyphenols and flavonoids than the Pre- and probiotic one, correlating with the specific ingredients of each product. Thus, the product with a higher content of vegetable ingredients proved to be rich in polyphenols, flavonoids and better antioxidant activity.

Antioxidant activity has been demonstrated by three chemical methods involving different mechanisms of action, but in vitro determinations on eukaryotic cells are needed to highlight reduced intracellular oxygen reactive species concentrations, improved metabolism and cell proliferation.

According to the results obtained from the microbiological analysis, it can be seen that the antimicrobial activity of Immunomodulatory and Pre-Probiotic products is more pronounced against Gram-positive strain *S. aureus* ATCC 6538 and more modest (bacteriostatic/fungistatic) against Gram-negative strain *E. coli* ATCC 8739 and the fungal strain *C. albicans* ATCC 10321. The samples showed moderate efficiency (bacteriostatic/fungistatic) from an antimicrobial point of view. However, further investigations are needed using alternative methods to confirm the effects obtained.

Evaluation of the biological activity on the in vitro cellular model of the Immunomodulatory and Pre-Probiotic products, with the help of MTT and LDH tests, showed percentages of increased viability, the composition and formulation of the products proved to be well balanced and suitable for consumption at lower concentrations. As the results indicate, it should be borne in mind that in the case of the Immunomodulatory product, the cytotoxicity of concentrations of 200 mg/mL, 100 mg/mL and 25 mg/mL is quite high (70% reduction in cell viability); the decrease in the concentrations of this product is directly correlated to the viability rate. The Pre-Probiotic product favors cell proliferation, having at lower concentrations the effect of a nutritive substrate, the viability of the cells not being affected.

As for the immunomodulatory assessment, we know that cytokines are small secreted proteins released by different types of cells with specific effects on cellular signaling and communication via binding to their receptors on the cell surface [15].

The capacity of the Pre-Probiotic product to upregulate IL-10 was demonstrated. As other authors report, pre-probiotics have a beneficial impact on immunomodulation [63], emphasizing the fact that IL-10 (among other cytokines) inhibit pro-inflammatory cytokines, chemokines, and chemokine receptors, responsible for intestinal inflammation [63]. As for the expression of IL-6, studies have revealed that it favors the clonal expansion of IgA B lymphocytes and stimulates the production of antibodies such as IgM, IgG, and reduced secretion of IgE [64].

Recently, Karamese et al. [65] administrated a mixture of *Lactobacillus* and *Bifidobacterium* species to rats for evaluation of the immunomodulatory effects of probiotics. Their study demonstrated that modulation or regulation of immune responses is evident through the upregulation of IL-10 (an anti-inflammatory cytokine) and the downregulation of TNF-α and IL-6 (proinflammatory cytokines). Moreover, it has been reported that the application of probiotics leads to significant increases in IgA and IgG concentrations in rats.

Evaluating the ratio of IL-6 to IL-10 for both products, we observed that the effect of the Pre-Probiotic product was noticed rapidly (at 24 h), whereas it is possible that the Immunomodulator product requires a longer contact time for higher stimulation.

5. Conclusions

In conclusion, the novel formulated nutraceuticals present great potential for their designated purposes, also being good candidates for further in vivo studies. The evaluation of effectiveness has proven to be in correlation with some of their associated health claims approved by Regulation (EC) No 1924/2006 on nutrition and health claims made on foods, (EU) No 432/2012 on vitamins and minerals, and EFSA regulations regarding health claims of botanicals. For the *Probiotic* formula, as a cautionary mention, it should be kept in mind that the European Commission has indicated that the term *probiotic* implies a health claim and therefore cannot be used in conjunction with an approved health claim. According to the Guide for the implementation of Regulation 1924/2006 (Chapter III. 1) it is specified that if the name of a substance implies an indication of a functionality or an effect on health, it is considered a health mention. EFSA has so far not approved any health claims for pre and probiotics, so product labels cannot include messages regarding the health
benefits of pre/probiotics.

As for future perspectives, further in vivo studies are envisioned. We seek to demonstrate and confirm on suitable animal models (mice) the beneficial effects of obtained products in order to safely place them onto current market. Certainly, preliminary human clinical studies, conducted on a specific group of patients (i.e., patients with Inflammatory Bowel Diseases, or with immuno-sensitive systems (i.e., allergies)) would increase the consumer compliance, sustain associated health claims and diminish the gap between pharma grade products and functional foods. Moreover, all this could lead to assembling dossiers for Novel Food application according to EFSA.

In our vision, as a regulatory organism for dietary supplements placement on the Romanian market, manufacturers need to push forward and contribute more to the research community with additional testing of products. This could benefit both parties, helping improve current legislation and mitigate for higher placement on the therapeutic market of dietary supplements.

Author Contributions: Conceptualization, A.E.R., B.-M.T., I.C.M. and T.O.; methodology, I.C.M., B.-M.T. and C.E.Ţ.; data curation, I.C.M., B.-M.T., T.O., C.E.Ţ. and O.T.C.; writing—original draft preparation, A.E.R., B.-M.T., I.C.M., T.O., O.T.C. and C.E.Ţ.; writing—review and editing, T.O., C.E.Ţ. and E.I.; supervision, T.O., C.E.Ţ. and E.I.; project administration, A.E.R., T.O. and E.I. All authors have read and agreed to the published version of the manuscript.

Funding: This work was supported by a European funding programme POC-A1-A1.2.3-G-2015, ID_40_404, MySMIS code 105509, "Exploiting expertise in agro-food research by transferring knowledge to the private environment in order to obtain safe and nutritionally optimized food products" acronym EXPERTAL, Financing Agreement no. 57/05.09.2017, Subsidiary Contract Type D no. 13/21.12.2018 IBA—CCPPM PLANTAVOREL S.A.

Institutional Review Board Statement: Not applicable.

Informed Consent Statement: Not applicable.

Data Availability Statement: Not applicable.

Conflicts of Interest: The authors declare no conflict of interest. The funders had no role in the design of the study; in the collection, analyses, or interpretation of data; in the writing of the manuscript, or in the decision to publish the results.

References

1. Sun, J.; de Vos, P. Immunomodulatory Functions of Nutritional Ingredients in Health and Disease. *Front. Immunol.* **2019**, *10*, 50. [CrossRef]
2. High, K.P. Nutritional Strategies to Boost Immunity and Prevent Infection in Elderly Individuals. *Clin. Infect. Dis.* **2001**, *33*, 1892–1900. [CrossRef]
3. Saeed, F.; Nadeem, M.; Ahmed, R.S.; Tahir Nadeem, M.; Arshad, M.S.; Ullah, A. Studying the Impact of Nutritional Immunology Underlying the Modulation of Immune Responses by Nutritional Compounds — A Review. *Food Agric. Immunol.* **2016**, *27*, 205–229. [CrossRef]
4. Marcos, A.; Nova, E.; Montero, A. Changes in the Immune System Are Conditioned by Nutrition. *Eur. J. Clin. Nutr.* **2003**, *57*, S66–S69. [CrossRef]
5. Pang, G.; Xie, J.; Chen, Q.; Hu, Z. How Functional Foods Play Critical Roles in Human Health. *Food Sci. Hum. Wellness* **2012**, *1*, 26–60. [CrossRef]
6. Wu, D.; Lewis, E.D.; Pae, M.; Meydani, S.N. Nutritional Modulation of Immune Function: Analysis of Evidence, Mechanisms, and Clinical Relevance. *Front. Immunol.* **2019**, *9*, 3160. [CrossRef] [PubMed]
7. Faria, A.M.C.; Gomes-Santos, A.C.; Gonçalves, J.L.; Moreira, T.G.; Medeiros, S.R.; Dourado, L.P.A.; Cara, D.C. Food Components and the Immune System: From Tonic Agents to Allergens. *Frontiers in Immunology* **2013**, *4*, 102. [CrossRef] [PubMed]
8. De Rosa, V.; Galgani, M.; Santopaolo, M.; Colamatteo, A.; Laccetti, R.; Matarese, G. Nutritional Control of Immunity: Balancing the Metabolic Requirements with an Appropriate Immune Function. *Seminars in Immunology* **2015**, *27*, 300–309. [CrossRef]
9. Nasri, H.; Baradaran, A.; Shirzad, H.; Kopaei, M.R. New Concepts in Nutraceuticals as Alternative for Pharmaceuticals. *Int. J. Prev. Med.* **2014**, *5*, 1487–1499. [PubMed]
10. Gulati, O.P.; Berry Ottaway, P. Legislation Relating to Nutraceuticals in the European Union with a Particular Focus on Botanical-Sourced Products. *Toxicology* **2006**, *221*, 75–87. [CrossRef]
11. Andlauer, W.; Fürst, P. Nutraceuticals: A Piece of History, Present Status and Outlook. *Food Res. Int.* **2002**, *35*, 171–176. [CrossRef]
12. Dickinson, A. History and Overview of DSHEA. *Fitoterapia* **2011**, *82*, 5–10. [CrossRef] [PubMed]
13. Bass, I.S. *Dietary Supplement Regulation: A Comprehensive Guide*; FDLI: Washington, DC, USA, 2011.
14. DeFelice, S.L. The Nutraceutical Revolution: Its Impact on Food Industry R&D. *Trends in Food Science and Technology* **1995**, *6*, 59–61.
15. Abood, W.N. Immunomodulatory and Natural Immunomodulators. *J. Allergy Inflamm.* **2017**, *2*.
16. Childs, C.E.; Calder, P.C.; Miles, E.A. Diet and Immune Function. *Nutrients* **2019**, *11*, 1933. [CrossRef]
17. Wichers, H. Immunomodulation by Food: Promising Concept for Mitigating Allergic Disease? *Anal. Bioanal. Chem.* **2009**, *395*, 37–45. [CrossRef]
18. Kaminogawa, S.; Nanno, M. Modulation of Immune Functions by Foods. *Complement. Altern. Med.* **2004**, *1*, 3. [CrossRef]
19. Ló Pez-Varela, S.; González-Gross, M.; Marcos, A. Functional Foods and the Immune System: A Review. *Eur. J. Clin. Nutr.* **2002**, *56*, 29–33. [CrossRef] [PubMed]
20. Hachimura, S.; Totsuka, M.; Hosono, A. Immunomodulation by Food: Impact on Gut Immunity and Immune Cell Function. *Biosci. Biotechnol. Biochem.* **2018**, *82*, 584–599. [CrossRef]
21. Polak, E.; Stępień, A.E.; Gol, O.; Tabarkiewicz, J. Potential Immunomodulatory Effects from Consumption of Nutrients in Whole Foods and Supplements on the Frequency and Course of Infection: Preliminary Results. *Nutrients* **2021**, *13*, 1157. [CrossRef] [PubMed]
22. Toumi, R.; Samer, A.; Soufli, I.; Rafa, H.; Touil-Boukoffa, C. Role of Probiotics and Their Metabolites in Inflammatory Bowel Diseases (IBDs). *Gastroenterol. Insights* **2021**, *12*, 56–66.

23. Toumi, R.; Soufli, I.; Rafa, H.; Belkhelfa, M.; Biad, A.; Touil-Boukoffa, C. Probiotic Bacteria Lactobacillus and Bifidobacterium Attenuate Inflammation in Dextran Sulfate Sodium-Induced Experimental Colitis in Mice. *Int. J. Immunopathol. Pharmacol.* **2014**, *27*, 615–627. [CrossRef]
24. Ciupercă Oana, T.; Ţebrencu Carmen, E.; Onisei Tatiana, R.A. Innovative Association of Medicinal Species with Potential for Use in Food Supplements with Immunomodulatory Effects. In Proceedings of the International Symposium ISB-INMA Agricultural and Mechanical Engineering, Bucharest, Romania, 31 October–1 November 2019; pp. 684–694.
25. Ciupercă Oana, T.; Ţebrencu Carmen, E.; Răducanu, A.; Pop Anca Lucia, O.T. Phytochemical Evaluation of Some Herbal Prebiotics and Their Ptential Use in Food Supplements. In Proceedings of the Inaugural Edition of the Virtual Congresswith International Participation Nutrition, Diet Therapy &Food Safety in the Context of the Covid-19 NutriTerra, Bucharest, Romania, 28–29 May 2020; pp. 305–312.
26. Kim, D.H.; Lee, J.H.; Park, S.; Oh, S.s.; Kim, S.; Kim, D.W.; Park, K.H.; Kim, K.D. 6-Acetonyl-5,6-Dihydrosanguinarine (ADS) from Chelidonium Majus L. Triggers Proinflammatory Cytokine Production via ROS-JNK/ERK-NFκB Signaling Pathway. *Food Chem. Toxicol.* **2013**, *58*, 273–279. [CrossRef]
27. Chung, H.-S.; An, H.-J.; Jeong, H.-J.; Won, J.-H.; Hong, S.-H.; Kim, H.-M. Water Extract Isolated from Chelidonium Majus Enhances Nitric Oxide and Tumour Necrosis Factor-α Production via Nuclear Factor-KB Activation in Mouse Peritoneal Macrophages. *J. Pharm. Pharmacol.* **2004**, *56*, 129–134. [CrossRef] [PubMed]
28. Song, J.Y.; Yang, H.O.; Pyo, S.N.; Jung, I.S.; Yi, S.Y.; Yun, Y.S. Immunomodulatory Activity of Protein-Bound Polysaccharide Extracted from Chelidonium Majus. *Arch. Pharm. Res.* **2002**, *25*, 158–164. [CrossRef] [PubMed]
29. Kour, K.; Bani, S.; Sangwan, P.L.; Singh, A. Upregulation of Th1 Polarization by Taraxacum Officinale in Normal and Immune Suppressed Mice. *Curr. Sci.* **2016**, *111*, 671–685. [CrossRef]
30. Ramendra, P.; Pandey; Chanchal, A.; Vohra, R.; Elesela, S.; Bhushan, L.; Kumar, S.; Kumar, S.; Ahmad, S.; Ramendra, C.; et al. Gelatin Biopolymer: A Journey from Micro to Nano. *J. Pharm. Res.* **2014**, *8*, 1387–1397.
31. Szewczyk, A.; Zagaja, M.; Bryda, J.; Kosikowska, U.; Stępień-Pyśniak, D.; Winiarczyk, S.; Andres-Mach, M. Topinambur—New Possibilities for Use in a Supplementation Diet. *Ann. Agric. Environ. Med.* **2019**, *26*, 24–28. [CrossRef] [PubMed]
32. Jalanka, J.; Major, G.; Murray, K.; Singh, G.; Nowak, A.; Kurtz, C.; Silos-Santiago, I.; Johnston, J.M.; de Vos, W.M.; Spiller, R. The Effect of Psyllium Husk on Intestinal Microbiota in Constipated Patients and Healthy Controls. *Int. J. Mol. Sci.* **2019**, *20*, 433. [CrossRef]
33. EMA/HMPC/199775/2012. *Assessment Report on Plantago Ovata Forssk., Seminis Tegumentum*; European Medicine Agency: Amsterdam, The Netherlands, 2013.
34. European Medicines Agency. *European Medicines Agency Post-Authorisation Evaluation of Medicines for Human Use*; EMA Website: Amsterdam, The Netherlands, 2013.
35. EUR-Lex—32006R1924—RO—EUR-Lex. Available online: https://eur-lex.europa.eu/legal-content/EN/ALL/?uri=celex%3A32006R1924 (accessed on 3 April 2021).
36. Botanicals | European Food Safety Authority. Available online: https://www.efsa.europa.eu/en/topics/topic/botanicals (accessed on 3 April 2021).
37. Nistico, S.; Tamburi, F.; Bennardo, L.; Dastoli, S.; Schipani, G.; Caro, G.; Fortuna, M.C.; Rossi, A. Treatment of Telogen Effluvium Using a Dietary Supplement Containing *Boswellia Serrata, Curcuma Longa*, and *Vitis Vinifera*: Results of an Observational Study. *Dermatol. Ther.* **2019**, *32*, e12842. [CrossRef]
38. EUR-Lex—32012R0432—RO—EUR-Lex. Available online: https://eur-lex.europa.eu/legal-content/EN/ALL/?uri=CELEX%3A32012R0432 (accessed on 3 April 2021).
39. Kim, D.O.; Jeong, S.W.; Lee, C.Y. Antioxidant Capacity of Phenolic Phytochemicals from Various Cultivars of Plums. *Food Chem.* **2003**, *81*, 321–326. [CrossRef]
40. Singleton, V.L.; Rossi, J.A. Colorimetry of Total Phenolics with Phosphomolybdic-Phosphotungstic Acid Reagents. *Am. J. Enol. Vitic.* **1965**, *16*, 144–158.
41. Formagio, A.S.N.; Kassuya, C.A.L.; Neto, F.F.; Volobuff, C.R.F.; Iriguchi, E.K.K.; Vieira, M.d.C.; Foglio, M.A. The Flavonoid Content and Antiproliferative, Hypoglycaemic, Anti-Inflammatory and Free Radical Scavenging Activities of Annona Dioica St. Hill. *BMC Complement. Altern. Med.* **2013**, *13*, 14. [CrossRef]
42. Madhu, G.; Bose, V.C.; Aiswaryaraj, A.S.; Maniammal, K.; Biju, V. Defect Dependent Antioxidant Activity of Nanostructured Nickel Oxide Synthesized through a Novel Chemical Method. *Colloids Surfaces A Physicochem. Eng. Asp.* **2013**, *429*, 44–50. [CrossRef]
43. Meng, J.; Fang, Y.; Zhang, A.; Chen, S.; Xu, T.; Ren, Z.; Han, G.; Liu, J.; Li, H.; Zhang, Z.; et al. Phenolic Content and Antioxidant Capacity of Chinese Raisins Produced in Xinjiang Province. *Food Res. Int.* **2011**, *44*, 2830–2836. [CrossRef]
44. Benzie, I.F.F.; Strain, J.J. Ferric Reducing/Antioxidant Power Assay: Direct Measure of Total Antioxidant Activity of Biological Fluids and Modified Version for Simultaneous Measurement of Total Antioxidant Power and Ascorbic Acid Concentration. *Methods Enzymol.* **1999**, *299*, 15–27.
45. Thaipong, K.; Boonprakob, U.; Crosby, K.; Cisneros-Zevallos, L.; Hawkins Byrne, D. Comparison of ABTS, DPPH, FRAP, and ORAC Assays for Estimating Antioxidant Activity from Guava Fruit Extracts. *J. Food Compos. Anal.* **2006**, *19*, 669–675. [CrossRef]
46. Tihăuan, B.-M.; Berca, L.M.; Adascalului, M.; Sanmartin, A.M.; Nica, S.; Cimponeriu, D.; Duță, D. Experimental in Vitro Cytotoxicity Evaluation of Plant Bioactive Compounds and Phytoagents: A Review. *Rom Biotechnol Lett.* **2020**, *25*, 1832–1842.
47. Bopp, S.K.; Lettieri, T. Comparison of Four Different Colorimetric and Fluorometric Cytotoxicity Assays in a Zebrafish Liver Cell Line. *BMC Pharmacol.* **2008**, *8*, 8. [CrossRef] [PubMed]

48. Wojdyło, A.; Oszmiański, J.; Czemerys, R. Antioxidant Activity and Phenolic Compounds in 32 Selected Herbs. *Food Chem.* **2007**, *105*, 940–949. [CrossRef]
49. Es-Safi, N.E.; Gómez-Cordovés, C. Characterization of Flavonoid Glycosides from Fenugreek (Trigonella Foenum-Graecum) Crude Seeds by HPLC-DAD-ESI/MS Analysis. *Int. J. Mol. Sci.* **2014**, *15*, 20668–20685.
50. Sabir, S.M.; Zeb, A.; Mahmood, M.; Abbas, S.R.; Ahmad, Z.; Iqbal, N. Phytochemical Analysis and Biological Activities of Ethanolic Extract of Curcuma Longa Rhizome. *Brazilian J. Biol.* **2021**, *81*, 737–740. [CrossRef] [PubMed]
51. Showkat, M.M.; Falck-Ytter, A.B.; Strætkvern, K.O. Phenolic Acids in Jerusalem Artichoke (*Helianthus Tuberosus* L.): Plant Organ Dependent Antioxidant Activity and Optimized Extraction from Leaves. *Molecules* **2019**, *24*, 3296. [CrossRef]
52. Talukder, P.; Talapatra, S.; Ghoshal, N.; Sen Raychaudhuri, S. Antioxidant Activity and High-Performance Liquid Chromatographic Analysis of Phenolic Compounds during in Vitro Callus Culture of Plantago Ovata Forsk. and Effect of Exogenous Additives on Accumulation of Phenolic Compounds. *J. Sci. Food Agric.* **2016**, *96*, 232–244. [CrossRef] [PubMed]
53. Hosseinzade, A.; Sadeghi, O.; Biregani, A.N.; Soukhtehzari, S.; Brandt, G.S.; Esmaillzadeh, A. Immunomodulatory Effects of Flavonoids: Possible Induction of T CD4+ Regulatory Cells through Suppression of MTOR Pathway Signaling Activity. *Front. Immunol.* **2019**, *10*, 51. [CrossRef] [PubMed]
54. Francenia Santos-Sánchez, N.; Salas-Coronado, R.; Villanueva-Cañongo, C.; Hernández-Carlos, B. Antioxidant Compounds and Their Antioxidant Mechanism. In *Antioxidants*; IntechOpen: London, UK, 2019.
55. Koley, T.K.; Maurya, A.; Tripathi, A.; Singh, B.K.; Singh, M.; Bhutia, T.L.; Tripathi, P.C.; Singh, B. Antioxidant Potential of Commonly Consumed Underutilized Leguminous Vegetables. *Int. J. Veg. Sci.* **2019**, *25*, 362–372. [CrossRef]
56. Nelson, M.L. Modulation of Antibiotic Efflux in Bacteria. *Curr. Med. Chem. Anti Infective Agents* **2002**, *1*, 35–54. [CrossRef]
57. Simion, I.M.; Moț, A.C.; Găceanu, R.D.; Pop, H.F.; Sârbu, C. Characterization and Classification of Medicinal Plant Extracts According to Their Antioxidant Activity Using High-Performance Liquid Chromatography and Multivariate Analysis. *Stud. Univ. Babeș-Bolyai Chem.* **2020**, *65*, 71–82. [CrossRef]
58. Zhang, H.; Tsao, R. Dietary Polyphenols, Oxidative Stress and Antioxidant and Anti-Inflammatory Effects. *Curr. Opin. Food Sci.* **2016**, *8*, 33–42. [CrossRef]
59. Ip, W.K.E.; Hoshi, N.; Shouval, D.S.; Snapper, S.; Medzhitov, R. Anti-Inflammatory Effect of IL-10 Mediated by Metabolic Reprogramming of Macrophages. *Science* **2017**, *356*, 513–519. [CrossRef]
60. Stenvinkel, P.; Ketteler, M.; Johnson, R.J.; Lindholm, B.; Pecoits-Filho, R.; Riella, M.; Heimbürger, O.; Cederholm, T.; Girndt, M. IL-10, IL-6, and TNF-α: Central Factors in the Altered Cytokine Network of Uremia—The Good, the Bad, and the Ugly. *Kidney Int.* **2005**, *67*, 1216–1233. [CrossRef] [PubMed]
61. Fuster, J.J.; Walsh, K. The Good, the Bad, and the Ugly of Interleukin-6 Signaling. *EMBO J.* **2014**, *33*, 1425–1427. [CrossRef] [PubMed]
62. Taniguchi, T.; Koido, Y.; Aiboshi, J.; Yamashita, T.; Suzaki, S.; Kurokawa, A. The Ratio of Interleukin-6 to Interleukin-10 Correlates with Severity in Patients with Chest and Abdominal Trauma. *Am. J. Emerg. Med.* **1999**, *17*, 548–551. [CrossRef]
63. Azad, M.A.K.; Sarker, M.; Wan, D. Immunomodulatory Effects of Probiotics on Cytokine Profiles. *BioMed. Res. Int.* **2018**, *2018*. [CrossRef]
64. Maldonado Galdeano, C.; De Moreno De Leblanc, A.; Vinderola, G.; Bibas Bonet, M.E.; Perdigón, G. Proposed Model: Mechanisms of Immunomodulation Induced by Probiotic Bacteria. *Clin. Vaccine Immunol.* **2007**, *14*, 485–492. [CrossRef]
65. Li, C.Y.; Lin, H.C.; Lai, C.H.; Lu, J.J.Y.; Wu, S.F.; Fang, S.H. Immunomodulatory Effects of Lactobacillus and Bifidobacterium on Both Murine and Human Mitogen-Activated T Cells. *Int. Arch. Allergy Immunol.* **2011**, *156*, 128–136. [CrossRef]

pharmaceutics

MDPI

Article

Stability Evaluation of Extemporaneously Compounded Vancomycin Ophthalmic Drops: Effect of Solvents and Storage Conditions

Christopher Ross [1], Basir Syed [1], Joanna Pak [1], Vishal Jhanji [2], Jason Yamaki [1] and Ajay Sharma [1,*]

[1] Chapman University School of Pharmacy, Chapman University Irvine, Room 267, 9401 Jeronimo Road, Irvine, CA 92618, USA; chross@chapman.edu (C.R.); bsyed@chapman.edu (B.S.); jpak@chapman.edu (J.P.); yamaki@chapman.edu (J.Y.)

[2] Department of Ophthalmology School of Medicine, University of Pittsburgh School of Medicine, Pittsburgh, PA 15213, USA; jhanjiv@upmc.edu

* Correspondence: sharma@chapman.edu; Tel.: +1-714-516-5498

Abstract: Vancomycin is the drug of choice for methicillin-resistant *Staphylococcus aureus* keratitis and other ocular infections. Vancomycin ophthalmic drops are not commercially available and require compounding. The present study was designed to investigate the stability of vancomycin ophthalmic drops in normal saline, phosphate-buffered saline (PBS), and balanced salt solution (BSS) while stored at room temperature or under refrigeration. Vancomycin ophthalmic drops (50 mg/mL) were aseptically prepared from commercially available intravenous powder using PBS, BSS, and saline. Solutions were stored at room temperature and in a refrigerator for 28 days. The vancomycin stability was tested by a microbiology assay and high-performance liquid chromatography HPLC analysis immediately after formulation and at days 7, 14, and 28 after storage at room temperature or under refrigeration. The pH, turbidity was also tested. Vancomycin formulations in PBS, BSS and normal saline had initial pH of 5; 5.5; 3 respectively. The formulation in PBS developed turbidity and a slight decrease in pH upon storage. Microbiological assay did not show any change in zone of inhibition with any of the formulation upon storage either at room temperature or under refrigeration. HPLC analysis did not detect any decrease in vancomycin concentration or the accumulation of degraded products in any of the formulations upon storage either at room temperature or under refrigeration. Vancomycin ophthalmic drops prepared using PBS, BSS, and normal saline were stable up to the tested time point of 28 days, irrespective of their storage temperature.

Keywords: MRSA; keratitis; vancomycin; compounding; stability

1. Introduction

Keratitis is an infection of the cornea caused by bacteria, fungi, herpes virus, or acanthamoeba. It is a sight-threatening medical emergency requiring prompt and effective intervention. *Staphylococcus aureus* is the second most common pathogen causing bacterial keratitis, although the exact incidence of infection varies by geographic location [1,2]. Methicillin-resistant S. aureus (MRSA) are strains of S. aureus resistant to the all antibiotics within the β-lactam class, except ceftaroline [3–5]. The acquisition of the mecA or mecC genes, which lead to alternative penicillin-binding proteins, account for the most common mechanism of resistance that results in the methicillin-resistant phenotype [6–9]. Although historically considered a hospital-acquired infection, MRSA clones are increasingly being encountered among community infections [10–12]. MRSA is known to cause a variety of ocular infections, including keratitis [13–15]. The incidence of MRSA ocular infections in the United States has been rising in recent years, with multiple studies reporting that MRSA accounts for roughly 30–40% of all S. aureus cases of keratitis [16–18].

The American Academy of Ophthalmology recommends 1.2%, 2.5%, or 5% vancomycin ophthalmic drops for keratitis caused by MRSA [19]. Alternatively, fluoroquinolone ophthalmic drops, which are commercially available, have been anecdotally used for MRSA keratitis, but ocular MRSA infections respond poorly, and the incidence of fluoroquinolone resistance remains a serious concern [20,21]. On the contrary, MRSA-associated keratitis isolates are largely responsive to vancomycin, which is the gold standard MRSA treatment for a wide range of different infection types, and resistance is rare [21,22]. A recent study has reported that 5% vancomycin ophthalmic drops have a higher efficacy in treating MRSA keratitis compared to lower concentrations [22]. However, vancomycin ophthalmic drops are not available commercially and require preparation by a licensed compounding pharmacy. The compounding pharmacy is required to be a member of the Pharmacy Compounding Accreditation Board 152 and designated by the FDA as a 503A and/or 503B facility. The American Academy of Ophthalmology's preferred practices pattern guidelines recommend the use of sterile normal saline for compounding vancomycin and its subsequent storage under refrigeration [19]. However, case reports and circumstantial evidence show that, besides normal saline, phosphate-buffered saline (PBS) and balanced salt solution (BSS) have also been used for compounding vancomycin. The requirement of compounding vancomycin ophthalmic drops leads to an increased financial cost for the patient, and the drops are typically recommended to be discarded after seven days despite the possibility of a longer shelf-life. Additionally, some patients with keratoprosthetics will require the use of ophthalmic vancomycin for longer durations [23,24]. These clinical needs necessitate the evaluation of the stability of compounded vancomycin ophthalmic drops. There are no comprehensive studies that evaluate the stability of compounded vancomycin ophthalmic drops in normal saline, PBS, and BSS with storage conditions either under refrigeration or at room temperature. Therefore, the present study was designed to investigate the stability of compounded vancomycin ophthalmic drops in normal saline, PBS, and BSS while stored at room temperature or under refrigeration, using the United States Pharmacopeia-recommended microbiological assay [25] and a high-performance liquid chromatography (HPLC) analysis.

2. Materials and Methods

2.1. Preparation of Vancomycin Ophthalmic Drops

Vancomycin eye drops were prepared aseptically from commercially available lyophilized powder of vancomycin hydrochloride (Mylan Pharmaceuticals, Morgantown, WV, USA) intended for intravenous use. A 50-mg/mL solution of vancomycin hydrochloride was prepared using three different vehicles: 1× phosphate-buffered saline (PBS), balanced salt solution (BSS), and normal saline. Each solution was then aliquoted into two equal parts with storage at room temperature and at 4 °C in a refrigerator.

The formulations were tested for pH and turbidity immediately after formulation and then at days 7, 14, and 28 after storage.

2.2. Stability Testing of Vancomycin Ophthalmic Drops

The stability of vancomycin ophthalmic drops prepared in three different vehicles and stored under two different conditions were tested using a HPLC chemical analysis and microbiological assay. The assays were performed using freshly prepared solutions immediately after formulation and then at days 7, 14, and 28 after the preparation and storage.

(a) Chemical Analysis of vancomycin using HPLC method

The HPLC method was used to quantify the vancomycin concentrations in the aliquots of the three ophthalmic formulations stored under two different conditions immediately after formulation and then at days 7, 14, and 28 after preparation and storage. The HPLC chromatograms were also run to screen for any additional peaks indicative of degradation products. A freshly prepared solution of vancomycin was used as a standard. The vancomycin ophthalmic formulation samples were diluted to be analyzed at

1-mg/mL concentrations. The sample concentration was calculated against five-point (2, 1, 0.5, 0.25, and 0.125 mg/mL) vancomycin standard curve concentrations. The HPLC analysis was done on a Hitachi LaChrome Ultra,(Hitachi High-Tech, Dallas, TX, USA) equipped with an autosampler and UV/Vis detector, using Vydac 218tp54, 250 × 4.6 mm, 5-um particle size, and C18 column. Mobile phase A, 0.1% trifluoroacetic acid (TFA) in water, was prepared by mixing 1 mL of TFA to 1 L of water and sonicating for 10 min. Mobile phase B, 0.1% TFA in acetonitrile, was prepared by mixing 1 mL of TFA to 1 L of acetonitrile and sonicating for 10 min. The flow rate was 1.0 mL/min with the following gradient method: hold at 10% mobile phase B+ 90% mobile phase A for 4 min, 10–35% mobile phase B + 65% mobile phase A for 7 min, and hold at 10% mobile phase B + 90% mobile phase A for 4 min. A 20-uL sample injection volume of the samples was used, and detection was performed at the 280-nm wavelength.

(b) Microbiological Assay

The United States Pharmacopeia method was used for the microbiological assay of vancomycin [25]. The media containing peptone (6 g/L), yeast extract (3 g/L), beef extract (1.5 g/L), agar (15 g/L), and distilled water (1 L) was used to prepare agar plates. Eight-millimeter holes were punched in the inoculated agar plates. Aliquots were obtained from each of the three ophthalmic formulations stored under two different conditions immediately after formulation and then at days 7, 14, and 28 after preparation and storage. These aliquots were diluted to an obtained a 1-µg/50-µL concentration of vancomycin, and 50 µL of each of these ophthalmic formulation samples was added to the punched hole in the agar plate. A freshly prepared solution of 1-µg/50-µL vancomycin was used as a standard. The plates were then incubated for 24 h at 37 °C, and the zone of inhibition in mm was noted down. The formulations were prepared twice, and each of the samples was analyzed in duplicate. The presented data is a replicate of $n = 4$.

2.3. Statistical Analysis

The data are presented as mean ± standard error of mean. Statistical analysis was performed using GraphPad Prism software (GraphPad Prism, Version 8, San Diego, CA, USA). The data were analyzed using one-way ANOVA, followed by Dunnett's post-hoc test. A p-value of <0.05 was considered statistically significant.

3. Results

3.1. Appearance and PH of Vancomycin Ophthalmic Formulations

Vancomycin formulated in PBS and stored at room temperature or under refrigeration remained clear for 14 days after preparation, but subsequently, it developed visible turbidity under both the storage conditions. The BSS and normal saline formulations stored at room temperature or under refrigeration remained visibly clear up to the tested time point of 28 days.

Table 1 shows the pH of the vancomycin formulations immediately after preparation and then at various days after storage at room temperature or under refrigeration.

Table 1. The pH of vancomycin formulations in phosphate-buffered saline (PBS), balanced salt solution (BSS), and normal saline immediately after formulation and then at days 7, 14, and 28 after storage at room temperature or under refrigeration. RT: room temperate.

Solvent	Day 0	Day 7	Day 14	Day 28
PBS—RT	5.0 ± 0	4.5 ± 0	4.5 ± 0	4.5 ± 0
PBS—4 °C	5.0 ± 0	4.5 ± 0	4.5 ± 0	4.5 ± 0
BSS—RT	5.5 ± 0	5.5 ± 0	5.5 ± 0	5.5 ± 0
BSS—4 °C	5.5 ± 0	5.5 ± 0	5.5 ± 0	5.5 ± 0
Saline—RT	3.0 ± 0	3.0 ± 0	3.0 ± 0	3.0 ± 0
Saline—4 °C	3.0 ± 0	3.0 ± 0	3.0 ± 0	3.0 ± 0

Vancomycin formulated in PBS, BSS, and normal saline had an initial pH of 5, 5.5, and 3, respectively. The pH of the PBS formulation decreased to 4.5 after seven days in both the samples stored at room temperature and under refrigeration. After this initial decrease, no further decrease in pH was noted in this PBS formulation up to the tested time point of 28 days. On the other hand, the BSS and normal saline formulations did not show any change in pH with storage either at room temperature or under refrigeration.

3.2. HPLC Quantification of Vancomycin Ophthalmic Formulations

Figure 1 shows the HPLC quantification of the PBS, BSS, and normal saline formulations of vancomycin immediately after preparation and then at various days after storage at room temperature or under refrigeration. The HPLC analysis did not detect any significant decrease in the vancomycin concentration after storage in any of these formulations. Furthermore, the stability of vancomycin was not affected by the storage temperature, since no significant difference in the vancomycin concentrations was noted whether the formulations were stored at room temperature or under refrigeration.

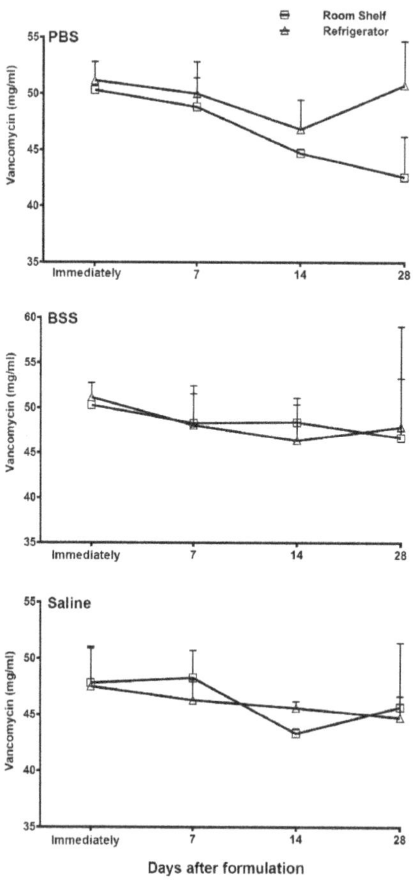

Figure 1. Quantification of vancomycin (mg/mL) in the phosphate-buffered saline (PBS), balances salt solution (BSS), and normal saline formulations analyzed by high-performance liquid chromatography (HPLC) immediately after formulation and then at days 7, 14, and 28 after storage at room temperature or under refrigeration. The data is presented as mean ± standard error of mean (SEM).

The HPLC chromatograms showed a vancomycin peak with an average elution time of 9.5 min in all these formulations when analyzed immediately after preparation.

Subsequently, an identical vancomycin peak with a similar elution time was also detected in these formulation samples upon analysis. There were no additional peaks in the HPLC chromatogram of the vancomycin formulations upon storage either at room temperature or under refrigeration compared to the freshly prepared vancomycin standard, suggesting that vancomycin did not have any degradation product formations upon storage in any of the three formulations stored either at room temperature or under refrigeration.

3.3. Microbiology Assay of Vancomycin Ophthalmic Formulations

A change in the bactericidal activity of antibiotics in microbiology assays is a reliable and activity-based indicator of a loss of potency. Therefore, we next tested the potency of vancomycin in the ophthalmic formulations using the microbiological assay method recommended by the United States Pharmacopeia [25]. Figure 2 shows the zone of inhibition of vancomycin in the formulations prepared using PBS, BSS, and normal saline. The zone of inhibition assay was performed in these formulations immediately after preparation, and then at 7 days, 14 days, and 28 days after compounding. No significant decrease in the zone of inhibition was observed in any of the formulations tested at the various time points upon storage when compared to the zone of inhibition analyzed in these formulations immediately upon preparation. Therefore, the microbiological assay data suggest that the vancomycin formulations did not lose potency upon storage either at room temperature or under refrigeration.

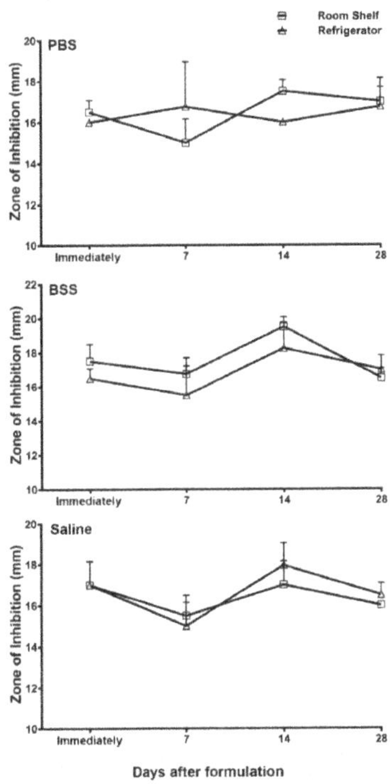

Figure 2. Zone of inhibition (in mm) of the PBS, BSS, and normal saline vancomycin formulations analyzed by a microbiology assay immediately after formulation and then at days 7, 14, and 28 after storage at room temperature or under refrigeration. The data is presented as mean ± standard error of mean (SEM).

4. Discussion

The topical application of vancomycin is the sight-saving treatment for keratitis caused by vancomycin-susceptible bacteria resistant to other antibiotic classes [20–22,26–28]. Topical vancomycin has also been used for preventing devastating ocular infections in patients who receive keratoprosthetics [23,24,29]. The treatment duration for keratitis typically requires approximately one to two weeks of application, and keratoprosthetics prophylaxis can last for a much longer duration [20–24,26–29]. There are currently no commercially available vancomycin ophthalmic formulations, thus necessitating compounding by a licensed pharmacy. Vancomycin ophthalmic drops need to be compounded under sterile conditions. These requirements make vancomycin ophthalmic drops expensive. The repeated compounding of vancomycin drops for the duration of a treatment further adds to the cost and the inconvenience of making additional trips to the pharmacy for picking up the drops, which can adversely affect patient adherence. The results of the present study demonstrate that compounded vancomycin eye drops are stable up to the tested time points of 28 days. Based on our data, it is feasible to compound larger quantities of vancomycin ophthalmic drops and then aliquot them into multiple smaller sterile containers for dispensing to the patient. This approach will reduce the cost while still maintaining the sterility if the patient is adequately counseled on opening and using one container at a time.

The American Academy of Ophthalmology's preferred practice pattern guidelines recommend the use of sterile normal saline for compounding vancomycin [19]. However, a variety of other solvents have also been used. The differences in ionic composition, pH, and buffering capacity of these solvents can affect the physicochemical stability of vancomycin. In the present study, we used a microbiological assay, as well as a HPLC-mediated chemical analysis, to compare the stability of vancomycin in three different formulations. Microbiological testing is a functional assay providing direct evidence of bactericidal activity. If an antibiotic has conformational changes, it may cause a decrease in its potency without causing any detectable changes in the HPLC analysis. On the other hand, a slight chemical degradation may not affect the antimicrobial activity of an antibiotic in the microbiology assay, but the accumulation of degradation products could be a concern for potential toxicity. The HPLC analysis is a sensitive assay to detect any additional peaks arising from the degrading products. Using these two assays, the results of the present study demonstrate that the stability of vancomycin is not affected by the type of solvent, since the formulations in the three different solvents did not show any significant decrease or dissimilarity in stability over time. Additionally, no extra peaks of impurities were detected in any of the formulations by the HPLC analysis, suggesting that vancomycin does not undergo degradation in these solvents.

Further, we tested whether the storage temperature can affect the stability of vancomycin for any of these ophthalmic formulations. The results of the present study demonstrated no significant difference in vancomycin stability in any of the three formulations based on their storage condition at room temperature or under refrigeration. Formulations or drugs that require refrigeration impose many practical challenges, such as their transportation from the pharmacy to a patient's home and the storage while the patient is at work or is traveling. Based on the results of the present study, it may be recommended that patients can carry and store the compounded ophthalmic drops at room temperature, avoiding excessive heat or direct sunlight as is generally recommended. Our results suggest that there may not be a need for refrigeration over the concerns of losing the potency while storing vancomycin ophthalmic drops at room temperature. Vancomycin is commonly used in the antimicrobial therapy to treat hospital-acquired MRSA, and the common route of administration is a slow intravenous infusion in 5% dextrose or normal saline. The stability of vancomycin at room temperature during a slow intravenous infusion is a concern. Therefore, several previous studies have tested the stability of vancomycin for a shorter duration, ranging from 1 h, 48 h, and 17 days when diluted in 5% dextrose or saline at ambient temperature [30–35]. In agreement with the results of the present study, these studies also demonstrated that vancomycin is stable when stored at room temperature. The results of the present study extend that stability for a longer duration of 28 days. Further, the data from these previous studies also showed that plastic tubing or iv bags do not cause vancomycin adsorption [30–35].

Although our study did not detect any differences in the stability of vancomycin over 28 days of storage either at room temperature or refrigeration, we did observe a difference in the pH and turbidity. It is worthwhile to note that vancomycin formulated in PBS developed a visible turbidity, with a slight decrease in pH in storage both at room temperature and under refrigeration. Vancomycin is available as a hydrochloride salt, and its dissolution results in an acidic solution. As anticipated, the formulation in normal saline offered no buffering, and this formulation had the most acidic pH. On the other hand, both BSS and PBS offered some buffering capacity, and the pH of the vancomycin formulations in these solvents were significantly less acidic. Acidic ophthalmic formulations may cause a stinging sensation upon instillation, and our results suggest that vancomycin formulated in normal saline may be less comfortable to the patient compared to the formulations in PBS and BSS.

5. Conclusions

In conclusion, the results of the present study demonstrate that vancomycin ophthalmic drops prepared using PBS, BSS, and normal saline are stable up to the tested time point of 28 days irrespective of their storage at room temperature or under refrigeration.

Author Contributions: Conceptualization, A.S., V.J. and J.Y.; methodology, J.P., C.R., B.S.; validation, C.R., B.S.; formal analysis, A.S., C.R., B.S.; resources, A.S.; writing—original draft preparation, A.S.; writing—review and editing, A.S., V.J.; supervision, A.S.; project administration, A.S.; funding acquisition, A.S. All authors have read and agreed to the published version of the manuscript.

Funding: This research received no external funding.

Conflicts of Interest: The authors declare no conflict of interest.

References

1. Mah, F.S.; Davidson, R.; Holland, E.J.; Hovanesian, J.; John, T.; Kanellopoulos, J.; Shamie, N.; Starr, C.; Vroman, D.; Kim, T. Current knowledge about and recommendations for ocular methicillin-resistant *Staphylococcus aureus*. *J. Cataract Refract. Surg.* **2014**, *40*, 1894–1908. [CrossRef]
2. Otri, A.M.; Fares, U.; Al-Aqaba, M.A.; Miri, A.; Faraj, L.A.; Said, D.G.; Maharajan, S.; Dua, H.S. Profile of sight-threatening infectious keratitis: A prospective study. *Acta Ophthalmol.* **2013**, *91*, 643–651. [CrossRef]
3. Saravolatz, L.D.; Stein, G.E.; Johnson, L.B. Ceftaroline: A novel cephalosporin with activity against methicillin-resistant *Staphylococcus aureus*. *Clin. Infect. Dis.* **2011**, *52*, 1156–1163. [CrossRef] [PubMed]
4. White, B.P.; Barber, K.E.; Stover, K.R. Ceftaroline for the treatment of methicillin-resistant *Staphylococcus aureus* bacteremia. *Am. J. Health Syst. Pharm.* **2017**, *74*, 201–208. [CrossRef]
5. Hassoun, A.; Linden, P.K.; Friedman, B. Incidence, prevalence, and management of MRSA bacteremia across patient populations—A review of recent developments in MRSA management and treatment. *Crit. Care* **2017**, *21*, 211. [CrossRef] [PubMed]
6. Ballhausen, B.; Kriegeskorte, A.; Schleimer, N.; Peters, G.; Becker, K. The mecA homolog mecC confers resistance against β-lactams in *Staphylococcus aureus* irrespective of the genetic strain background. *Antimicrob. Agents Chemother.* **2014**, *58*, 3791–3798. [CrossRef]
7. Fuda, C.; Suvorov, M.; Vakulenko, S.B.; Mobashery, S. The basis for resistance to beta-lactam antibiotics by penicillin-binding protein 2a of methicillin-resistant *Staphylococcus aureus*. *J. Biol. Chem.* **2004**, *279*, 40802–40806. [CrossRef]
8. Peacock, S.J.; Paterson, G.K. Mechanisms of Methicillin Resistance in *Staphylococcus aureus*. *Annu. Rev. Biochem.* **2015**, *84*, 577–601. [CrossRef] [PubMed]
9. Llarrull, L.I.; Fisher, J.F.; Mobashery, S. Molecular basis and phenotype of methicillin resistance in *Staphylococcus aureus* and insights into new beta-lactams that meet the challenge. *Antimicrob. Agents Chemother.* **2009**, *53*, 4051–4063. [CrossRef]
10. Klevens, R.M.; Morrison, M.A.; Nadle, J.; Petit, S.; Gershman, K.; Ray, S.; Harrison, L.H.; Lynfield, R.; Dumyati, G.; Townes, J.M. Invasive methicillin-resistant *Staphylococcus aureus* infections in the United States. *JAMA* **2007**, *298*, 1763–1771. [CrossRef] [PubMed]
11. Peterson, J.C.; Durkee, H.; Miller, D.; Maestre-Mesa, J.; Arboleda, A.; Aguilar, M.C.; Relhan, N.; Flynn, H.W., Jr.; Amescua, G.; Parel, J.M. Molecular epidemiology and resistance profiles among healthcare- and community-associated *Staphylococcus aureus keratitis* isolates. *Infect. Drug Resist.* **2019**, *12*, 831–843. [CrossRef]

12. Ong, S.J.; Huang, Y.C.; Tan, H.Y.; Ma, D.H.; Lin, H.C.; Yeh, L.K.; Chen, P.Y.; Chen, H.C.; Chuang, C.C.; Chang, C.J. *Staphylococcus aureus* keratitis: A review of hospital cases. *PLoS ONE* **2013**, *8*, e80119. [CrossRef] [PubMed]
13. Blomquist, P.H. Methicillin-resistant *Staphylococcus aureus* infections of the eye and orbit. *Trans. Am. Ophthalmol. Soc.* **2006**, *104*, 322–345. [PubMed]
14. Freidlin, J.; Acharya, N.; Lietman, T.M.; Cevallos, V.; Whitcher, J.P.; Margolis, T.P. Spectrum of eye disease caused by methicillin-resistant *Staphylococcus aureus*. *Am. J. Ophthalmol.* **2007**, *144*, 313–315. [CrossRef]
15. Solomon, R.; Donnenfeld, E.D.; Perry, H.D.; Rubinfeld, R.S.; Ehrenhaus, M.; Wittpenn, J.R., Jr.; Solomon, K.D.; Manche, E.E.; Moshirfar, M.; Matzkin, D.C. Methicillin-resistant *Staphylococcus aureus* infectious keratitis following refractive surgery. *Am. J. Ophthalmol.* **2007**, *143*, 629–634. [CrossRef] [PubMed]
16. Asbell, P.A.; Colby, K.A.; Deng, S.; McDonnell, P.; Meisler, D.M.; Raizman, M.B.; Sheppard, J.D., Jr.; Sahm, D.F. Ocular TRUST: Nationwide antimicrobial susceptibility patterns in ocular isolates. *Am. J. Ophthalmol.* **2008**, *145*, 951–958. [CrossRef] [PubMed]
17. Asbell, P.A.; Sahm, D.F.; Shaw, M.; Draghi, D.C.; Brown, N.P. Increasing prevalence of methicillin resistance in serious ocular infections caused by *Staphylococcus aureus* in the United States: 2000 to 2005. *J. Cataract Refract. Surg.* **2008**, *34*, 814–818. [CrossRef]
18. Vola, M.E.; Moriyama, A.S.; Lisboa, R.; Vola, M.M.; Hirai, F.E.; Bispo, P.J.; Höfling-Lima, A.L. Prevalence and antibiotic susceptibility of methicillin-resistant *Staphylococcus aureus* in ocular infections. *Arq. Bras. Oftalmol.* **2013**, *76*, 350–353. [CrossRef]
19. Lin, A.; Rhee, M.K.; Akpek, E.K.; Amescua, G.; Farid, M.; Garcia-Ferrer, F.J.; Varu, D.M.; Musch, D.C.; Dunn, S.P.; Mah, F.S. Bacterial Keratitis Preferred Practice Pattern. *Ophthalmology* **2019**, *126*, 1–55. [CrossRef]
20. Chang, V.S.; Dhaliwal, D.K.; Raju, L.; Kowalski, R.P. Antibiotic Resistance in the Treatment of *Staphylococcus aureus* Keratitis: A 20-Year Review. *Cornea* **2015**, *34*, 698–703. [CrossRef] [PubMed]
21. Durrani, A.F.; Atta, S.; Bhat, A.K.; Mammen, A.; Dhaliwal, D.; Kowalski, R.P.; Jhanji, V. Methicillin-resistant *Staphylococcus aureus* keratitis: Initial treatment, risk factors, clinical features, and treatment outcomes. *Am. J. Ophthalmol.* **2020**, *214*, 119–126. [CrossRef] [PubMed]
22. Romanowski, E.G.; Romanowski, J.E.; Shanks, R.M.Q.; Yates, K.A.; Mammen, A.; Dhaliwal, D.K.; Jhanji, V.; Kowalski, R.P. Topical vancomycin 5% is more efficacious than 2.5% and 1.25% for reducing viable methicillin-resistant *Staphylococcus aureus* in infectious keratitis. *Cornea* **2020**, *39*, 250–353. [CrossRef]
23. Durand, M.L.; Dohlman, C.H. Successful prevention of bacterial endophthalmitis in eyes with the Boston keratoprosthesis. *Cornea* **2009**, *28*, 896–901. [CrossRef] [PubMed]
24. Behlau, I.; Martin, K.V.; Martin, J.N.; Naumova, E.N.; Cadorette, J.J.; Sforza, J.T.; Pineda, R., 2nd; Dohlman, C.H. Infectious endophthalmitis in Boston keratoprosthesis: Incidence and prevention. *Acta Ophthalmol.* **2014**, *92*, e546–e555. [CrossRef]
25. Biological tests and assays: Antibiotics, microbial assay. *U. S. Pharmacop.* **2018**, *41*, 134–151.
26. Austin, A.; Schallhorn, J.; Geske, M.; Mannis, M.; Lietman, T.; Rose-Nussbaumer, J. Empirical treatment of bacterial keratitis: An international survey of corneal specialists. *BMJ Open Ophthalmol.* **2017**, *2*, e000047. [CrossRef]
27. Tena, D.; Rodríguez, N.; Toribio, L.; González-Praetorius, A. Infectious Keratitis: Microbiological Review of 297 Cases. *Jpn. J. Infect. Dis.* **2019**, *72*, 121–123. [CrossRef]
28. Akova Budak, B.; Baykara, M.; Kvanç, S.A.; Yilmaz, H.; Cicek, S. Comparing the ocular surface effects of topical vancomycin and linezolid for treating bacterial keratitis. *Cutan. Ocul. Toxicol.* **2016**, *35*, 126–130. [CrossRef] [PubMed]
29. Davies, E.; Chodosh, J. Infections after keratoprosthesis. *Curr. Opin. Ophthalmol.* **2016**, *27*, 373–377. [CrossRef]
30. Das Gupta, V.; Stewart, K.R.; Nohria, S. Stability of vancomycin hydrochloride in 5% dextrose and 0.9% sodium chloride injections. *Am. J. Hosp. Pharm.* **1986**, *43*, 1729–1731. [CrossRef]
31. Raverdy, V.; Ampe, E.; Hecq, J.D.; Tulkens, P.M. Stability and compatibility of vancomycin for administration by continuous infusion. *J. Antimicrob. Chemother.* **2013**, *68*, 1179–1182. [CrossRef]
32. Peterlini, M.A.; Barbosa, A.; Pedreira, M. Stability of vancomycin hydrochloride solutions in high concentration and extended time of infusion. *Intensive Care Med. Exp.* **2015**, *3* (Suppl. S1), A717. [CrossRef]
33. Godet, M.; Simar, J.; Closset, M.; Hecq, J.; Braibant, M.; Soumoy, L.; Gillet, P.; Jamart, J.; Bihin, B.; Galanti, L. Stability of Concentrated Solution of Vancomycin Hydrochloride in Syringes for Intensive Care Units. *Pharm. Technol. Hosp. Pharm.* **2018**, *3*, 23–30. [CrossRef]
34. d'Huart, É.; Vigneron, J.; Charmillon, A.; Clarot, I.; Demoré, B. Physicochemical Stability of Vancomycin at High Concentrations in Polypropylene Syringes. *Can. J. Hosp. Pharm.* **2019**, *72*, 360–368. [CrossRef] [PubMed]
35. Masse, M.; Genay, S.; Martin Mena, A.; Carta, N.; Lannoy, D.; Barthélémy, C.; Décaudin, B.; Odou, P. Evaluation of the stability of vancomycin solutions at concentrations used in clinical services. *Eur. J. Hosp. Pharm.* **2020**, *27*, e87–e92. [CrossRef] [PubMed]

pharmaceutics MDPI

Article

Preparation of Ligand Brush Nanocapsules for Robust Self-Controlled Antimicrobial Activity with Low Cytotoxicity at Target pH and Humidity

Eunjoo Koh and Yong Taek Lee *

Department of Chemical Engineering, Faculty of Chemical Engineering, Kyung Hee University, Deogyeong-daero, Giheung-gu, Yongin-si 17104, Korea; ejkoh@khu.ac.kr
* Correspondence: yongtlee@khu.ac.kr

Abstract: This study prepared nanocapsules (NCs) with excellent self-controlled antimicrobial activity at pH 6–7 and humidity 45–100%, conditions in which most bacterial and fungal strains thrive. The nanocapsule substrate (NC@SiO_2) was 676 nm in diameter, and the ligand-grafted capsule (NC@SiO_2-g-MAA) was 888 nm. The large surface area and outer ligand brush of the NCs induced a rapid, self-controlled antibacterial response in the pH and humidity conditions needed for industrial and medical applications. Ligand-brush NCs containing an anionic antimicrobial drug had a rapid release effect because of the repellent electrostatic force and swelling properties of the ligand brushes. Controlled release of the drug was achieved at pH 6 and humidity of 45% and 100%. As many carboxylic acid groups are deprotonated into carboxylic acids at pH 5, the NC@SiO_2-g-MAA had a high negative charge density. Carboxylic acid groups are anionized (–COO^-) at pH 6 and above and push each other out of the capsule, expanding the outer shell as in a polymer brush to create the release behavior. The surface potential of the NC intermediate (NC@SiO_2-MPS) was −23.45 [mV], and the potential of the capsule surface decreased to −36.4 [mV] when the MAA ligand brushes were grafted onto the surface of the capsule intermediate. In an antimicrobial experiment using *Escherichia coli*, a clear zone of 13–20 mm formed at pH 6, and the *E. coli* was eradicated completely at pH 6 and pH 7 when the humidity was 100%.

Keywords: controlled drug release; self-antibacterial activity; ligand brush nanocapsule; pH and humidity response; cytotoxicity

1. Introduction

Polymeric, pH-responsive nanocapsules (NCs) [1–5] have been applied in a variety of biotech and environmental processes, such as pharmaceuticals [6–10], antibacterial packing materials, enzymes [11] and cosmetics. Temperature-sensitive polymers with a low critical solution temperature undergo phase separation, changing from hydrophilic to hydrophobic at certain temperatures. For example, hydrophilicity is retained at the normal temperature of the human body, 37 °C, but it changes to hydrophobicity at higher temperatures (about 42 °C), causing the nanoparticles to break through interactions among them and leading to instantaneous drug release. That method cannot be used with non-biodegradable material that does not decompose in the body. On the other hand, pH-sensitive polymers can release drug agents in response to pH changes without external help [12,13]. Change in the pH of blood and cells does occur, but this alteration from 7.4 is small and does not exceed ±0.04. Cells use endocytosis to absorb certain particles and form endosomes whose pH (6.0) is lower than that of the cytosol (7.15). In addition, the pH around cancer tumors is lower than that of the surrounding tissue because of the organic acid produced by cancer cell metabolism. In polymer nanoparticles, pH differences can cause phase shifts that prompt aggregation or expansion of the nanoparticle surface to induce drug release [14–18]. Therefore, effective self-antibacterial activity in the presence of rapid pH changes can be obtained by using ligand- and pH-sensitive functional groups on NCs [12,19,20]. Triclosan is a typical disinfectant material commonly used as a germicide/microbicide/fungicide. As drug carrier systems have developed to reduce the side effects in non-target sites through use of ligands, functional groups, and pH signals, triclosan has gained therapeutic efficiency. As a typical pH-responsive material, methalic acid (MAA) is widely used as a drug carrier that can be swollen/collapsed at particular pH values via ionization of its –COOH groups [21–26]. A smart pH-responsive polymer can be used to fabricate a multifunctional nanocomposite that acts as a drug container and carrier in vivo and in vitro [27,28].

The objectives of this study are fast and immediate antibacterial effect and excellent biocompatibility. We have sought out antibacterial activity and biocompatibility on the NCs. *E. coli* was picked on the antibacterial effect, and also biocompatibility was verified by the cytotoxicity test of the epithelial cells. Furthermore, we prepared pH-sensitive NCs via Pickering encapsulation technology [29–32]. These NCs release their cargo above pH 5 when the humidity is high, conditions in which most bacteria, including *Escherichia coli* and fungi, are active. In other words, we designed humidity- and pH-sensitive ligand brush–grafted [33–37] NCs [5,38–40] for use as a drug carrier. The grafted ligands serve as an outer shell to carry and control the release of an antibacterial drug (triclosan). The NC substrate [41–44] was modified with humidity- and pH-sensitive ligand brushes by applying 3-(trimethoxysilyl)propyl methacrylate (MPS) [45,46] to produce NC@SiO_2-MPS intermediates that were further modified into ligand-brush NC nanospheres by grafting on the carboxylic group of MAA [47,48]. Anionic triclosan was used as a model drug, and controlled release was achieved at pH 4, pH 6 and pH 7. Moreover, we defined the mechanism for the controlled release of a negatively charged drug from the anionic outer-shell NCs [49]. The size of the NCs was controlled to 1.0 μm to ensure the proper surface area. Further antibacterial testing was performed to investigate the pH sensitivity of the anionic drug delivery system and determine its self-antifungal and -antimicrobial activities and its cytotoxicity toward *E. coli* at different pH and humidity levels. The developed self-healing antibacterial NCs can be used in various fields such as medical disposal antibacterial patches, infectious disease protective clothing, hemostatic patches, water filters, export packaging materials (high humidity areas), food and industrial packaging materials.

2. Experimental

2.1. Materials

Tetraethyl orthosilicate (TEOS), ammonium hydroxide (28~30%), polyvinylpyrrolidone (PVP, MW = 40,000), 3-(trimethoxysilyl)propylmethacrylate (MPS), sodium dodecylbenzenseulfonate, poly(methalic acid, sodium salt) solution (MAA), *N,N′*-methylenebisacrylamide (BIS), pentanol, sodium citrate, ammonium persulfate (APS), ammonium hydroxide, absolute ethanol, triclosane (CAS No. 638-67-5), and buffer solutions (pH 4, 6) were purchased from Sigma-Aldrich (Yongin-si, Korea). The cell resource is human lung epithelial cells (A549 cells) that were cultured in the Specially Materials Lab. of Kyung Hee University.

2.2. Preparation of Self-Healing Antibacterial NCs

2.2.1. Synthesis of Self-Assembly NCs

PVP (2.5 *g*) was dispersed in 25 mL of pentanol and stirred at 1 g force for 30 min at 60 °C. Immediately after stirring, the sample was transferred to a prepared bath (60 °C, 1 L) and sonicated for 1 h. After sonication, the PVP-pentanol mixture was placed in a centrifuge with water (2.5 g), sodium citrate solution (0.016 g of sodium citrate in 243.16 μL of water), ethanol (1.8936 g), and ammonium hydroxide (250 μL) and vortex-stirred for 2 min at 408× *g* force. Next, 1 mL of TEOS was added and continuously vortexed for 2 min at 408× *g* force and then overnight [23]. The supernatant was collected, rinsed with ethanol, centrifuged in the same manner, and dried overnight. The dried NCs were then ground to powder, as shown in Figure 1a [50–55].

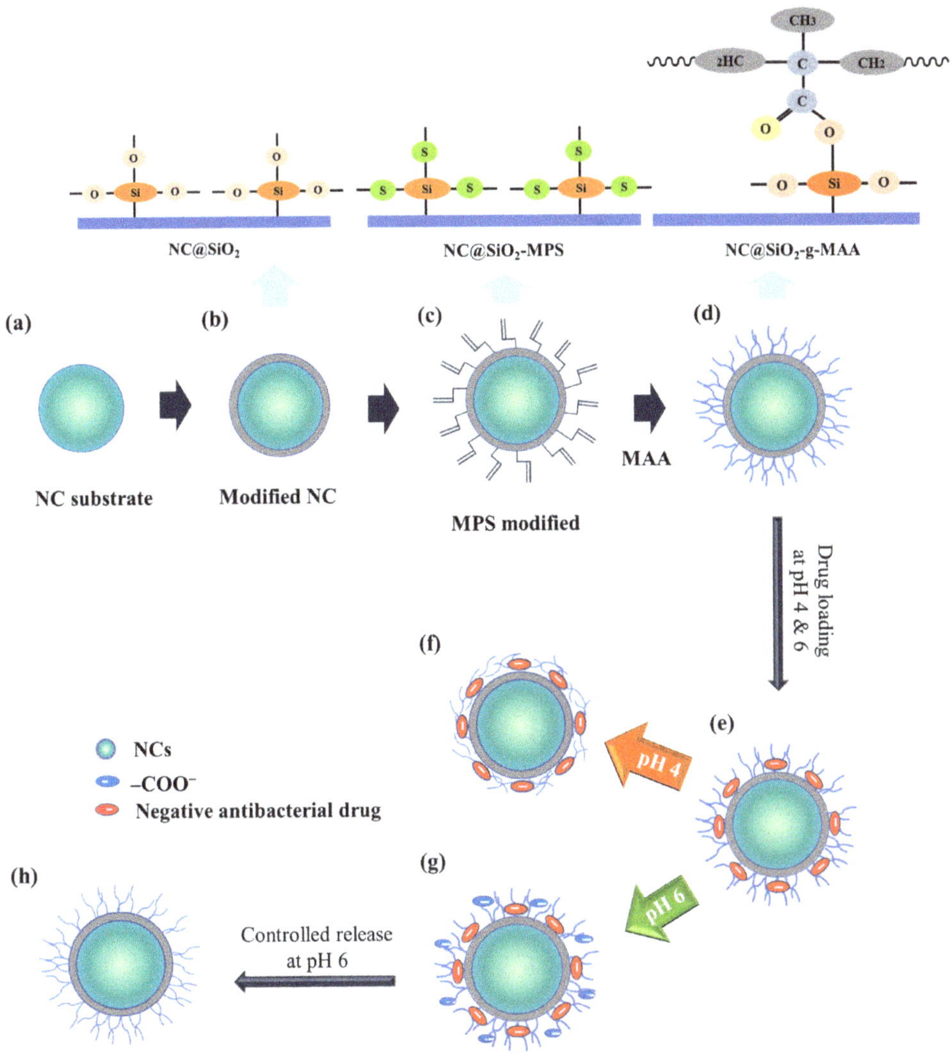

Figure 1. Chemical structure as it was modified to form pH- and humidity-sensitive self-antimicrobial NCs; (**a**) formation of NC substrate, (**b**) TEOS modification, (**c**) surface modification with MPS, (**d**) MAA ligand brushes grafted on the NC surface, (**e**) drug-loaded NC, (**f**) controlled drug release at pH 4, (**g**) controlled drug release at pH 6, and (**h**) pH triggered drug release at pH 6.

2.2.2. Surface Modification

Ethanol (100 mL) and MPS (0.5 mL) were added and stirred at 6× g force for 24 h at 25 °C to coat the NC surfaces (100 mg). After centrifugation at 6523× *g* force for 10 min, the sample was washed with 20 wt.% ethanol. The NCs were dispersed in solution (50 mL of water, 5 mg of sodium dodecyl benzene sulfonate) and stirred for 1 h at 6 *g* force, after which MAA solution (200 mg) and BIS (15 mg) were added. Then, the mixture was bubbled with N_2 gas for 30 min, 10 mg of APS was quickly added [47,56], and the liquid was stirred at 6× *g* force and 75 °C for 4 h. After centrifugation at 6523× *g* force for 10 min, the sample was washed with distilled water and dried at 40 °C for 24 h, as shown in Figure 1b,c.

2.2.3. Modification of pH-Sensitive Ligand-Brush NCs

As described in Figure 1, two 50 mg samples of nanospheres were dispersed in 10 mL of pH 4 and 6 buffer, for a total of four samples. To those samples, 10 mg of triclosan was added and stirred at 1 g force for 24 h at 25 °C in a water bath. After centrifugation at 408× *g* force for 10 min and drying at 40 °C for 24 h, the prepared NCs were dispersed in buffer (pH 4 and pH 6) at 50 mg/10 mL. The mixture was stirred continuously at 1× *g* force and 37 °C for 24 h. After centrifugation for 10 min at 408× *g* force, the supernatant was collected, and the pellet was dried at 40 °C for 24 h and stored (Figure 1c,d). The humidity- and pH-sensitivity of the self-antimicrobial NCs occurred through the following steps. The MPS modification induced a negative charge on the NC surface due to the equilibrium of the Si–OH group. The ionized silanol group (Si–OH) was crosslinked by the oxygen bridge of the MAA, which induced agglomeration on the grafted ligand brushes on the NC surface. The ligand-brush NCs had a suitable surface area and a release mechanism easily controlled by the swelling of the ligand brushes. All the grafted, hybrid, self-assembled NCs exhibited an excellent surface area that enhanced the loading capacity for the drug and affected the NCs' antimicrobial activity. The NC@SiO_2 was modified with humidity- and pH-sensitive brushes by applying MPS to prepare NC intermediates (NC@SiO_2-MPS) that were modified into ligand-brush NC@SiO_2-g-MAA nanospheres by grafting them with the carboxylic group of MAA, as shown in Figure 1c,d. The C=O bonds and Si–O–Si bonds of the polymer formed rapidly at high temperature and dissolved in the solvent before the polymer solidified, as shown in Figure 1d. However, if the crosslinked polymer is slowly added dropwise at a mild temperature (50 °C), capsules of uniform size with a smooth surface can be obtained, and agglomeration of the polymer and excessive tangling of the Si–O molecular chains can be prevented.

2.3. *Antimicrobial Evaluation with E. coli*

The antimicrobial activity of the NCs was determined by the disk diffusion test with *E. coli* [57–60]. The antimicrobial activity values were measured at pH 4 and pH 6 on a filter paper disk (control sample (C.S.)), NCs with no drug, and NCs with triclosan. All of the NC plates and the C.S. had a diameter of 10 mm. The bacterial strains were cultivated at 37 °C and 45% humidity for 12 h. After incubation, the bacterial solution was transferred evenly across the surface of an agar plate with a diameter of 90 mm in a sterile environment. For the 45% humidity condition, the concentration of growth medium (LB) in the agar plate was 0.08 g/L. The prepared 10-mm plates containing 55 mg of NCs were placed on top of the agar plates, and the C.S. and NC plates were placed in an incubator at 37 °C for 24 h. Then, the clear zone around each NC plate was measured to determine the diameter of inhibition. For the 100% humidity condition, the optical density of the cell suspension was determined by using a Synergy HTX multi-mode reader (BioTek, Winooski, VT, USA) to measure the concentration of *E. coli*. This test was repeated three times.

2.4. *Cytotoxicity Test with Lung Epithelial Cells*

The sample was diluted, and A549 human lung epithelial cells were dispensed (3.12×10^5 cells in 10 mL of medium). The dispensed cells were cultured for 24 h, and then the sample was exposed to the test material/solution for 5 min under sonication. The samples were cultured for an additional 24 h, and then cell proliferation was observed with an optical microscope. The three test liquids were buffer solution (pH 6) for the control sample, and pristine NCs and drug-loaded NCs in buffer solution (pH 6). For the cytotoxicity test, when the number of surviving cells survives more than 85% of the initial number of cells, it can be analyzed that there is no cytotoxicity and good biocompatibility. The concentration of drug extracted from the capsule was 100 µg/mL, and cell adhesion activity was compared with the control sample.

2.5. Characterization

The morphological structures of the prepared nanofiber membranes were observed with a field emission scanning electron microscope (SEM, LEO SUPRA 55, Carl Zeiss, Oberkochen, Germany) and transmission electron microscopy (TEM, JEM-2100F (Tokyo, Japan, JEOL)), and the surface elemental compositions of the NCs were analyzed by X-ray photoelectron spectroscopy (XPS, K-Alpha (Thermo Electron, Billerica, MA, USA)). The particle size was measured using a particle size analyzer (Otsuka (ELSZ-2000ZS), Chiyoda, Japan). Particle sizes and zeta potentials were measured on the NC surfaces at 25 °C in deionized water. We used the conversion equation for zeta potential (mV) proposed by Smoluchowski and calculated it using Contin to indicate the surface electrons on the nanoparticles. The surface area was calculated using the Brunauer, Emmett, and Teller (BET, BELSORP-Max, BEL, Osaka, Japan) method. Surface chemical components were characterized using a Fourier transform infrared spectrometer (FT-IR; Spectrum One System, PerkinElmer, Waltham, MA, USA) over a wavelength range of 500–4000 cm^{-1}, according to the shift in chemical absorption peaks. The controlled release of the drug agent was confirmed using UV-Vis spectrometry (CARY 300 Bio, Santa Clara, CA, USA). The A549 cells were tested using a cell viability assay kit (EZ-Cytox, Dogen, Seoul, Korea), and live cells were captured using an optical microscope (Olympus, CKX41, Hamburg, Germany).

3. Results and Discussion

3.1. Surface Morphology

Ligand-brush NCs have a suitable surface area and a release mechanism easily controlled by the swelling of the ligand brushes. All our grafted, hybrid, self-assembled NCs exhibited an excellent surface area that maximized the loading capacity of the drug and affected their antimicrobial activity. Figure 2a shows the surface formation and surface composition at a reaction temperature of 40 °C, which induced mild chemical reactions between the Si–O–Si and C=O and produced round, uniformly sized NCs. If the crosslinked polymer is slowly added dropwise at a mild temperature (40–50 °C), capsules of uniform size with a smooth surface can be obtained (Figure 2a,b). At 65 °C and 80 °C, NC encapsulation was prevented by agglomeration of the polymer and excessive tangling of the Si–O molecular chains (Figure 2c,d). The pH-sensitive NCs capable of storing and transporting drugs were prepared at different temperatures to confirm their uniformity, size, and surface shape by SEM (Figure 2). NC formation was performed between 40 °C and 80 °C to confirm the shapes at different temperatures. Capsules prepared at 50 °C showed a smooth surface and a relatively uniform size (Figure 2a,b) because the C=O and Si–O–Si bonds of the polymer form rapidly at 40–50 °C and are dissolved in the solvent before the polymer solidifies.

The surface morphology, particle size, and surface potential of the NCs were evaluated at each modification step and differed with the formation conditions. Figure 2a shows entangled ligand brushes caused by inconsistent target humidity and pH conditions. Figure 2(a1) shows the surface morphology and shape of the NCs used as supports. The NC@SiO_2 formed by the Pickering method produced a weak bumpy pattern on the surface due to the adsorption and reaction of the polymer molecules. The ligand brushes were grafted onto the NC@SiO_2-g-MPS by reacting them with the carboxylic group of MAA. Therefore, the NC@SiO_2-MPS shows a raspberry-like surface because the hydrogen bonds have the silanol group (Si–OH) and (C=O)–OH structure (Figure 2(a2)). This NC@SiO_2-MPS intermediate sphere can be combined with MAA. At that step, the (C=O)–OH bond of the NC@SiO_2-MPS reacts with the carboxyl group (–COOH) of the MAA, leading to self-assembly of the NC@SiO_2-g-MAA. The raspberry shape was also formed by the ligand-brush grafting and resulted in the $-COO^-$ groups of MAA shown in Figure 2(a3). The preparation yield of the nanocapsules was 45%, as calculated by Equation (1) [61]:

$$\text{Preparation yield } (\%) = \frac{\text{Nanocapsule weight}}{\text{Theoretical weight of all components}} \times 100 \quad (1)$$

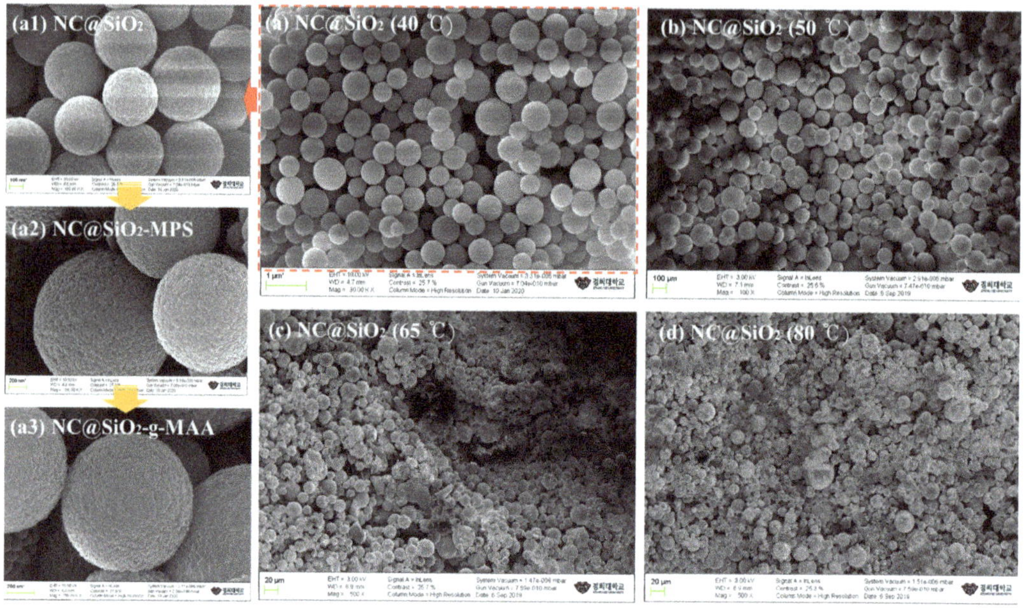

Figure 2. Aspects and shapes of NCs prepared at temperatures from 40 °C to 80 °C; (**a**) NC encapsulation at 40 °C, (**a1**) NC substrate, (**a2**) surface composition with MPS, (**a3**) ligand brush NC ($NC@SiO_2$-g-MAA), (**b**) NC encapsulation at 50 °C, (**c**) NC encapsulation at 65 °C, and (**d**) NC encapsulation at 80 °C.

3.2. Analysis of Surface Components and Surface Area

As the capsules prepared under each condition used silica and polymer, spherical capsules were clearly distinguished in the TEM analysis. As shown in Figure 3, the surface composition, shape and size, and area of the NCs prepared at 40 °C were confirmed by SEM, EDS, and BET measurements, respectively, revealing well-formed NCs. A relatively uniform spherical capsule was observed by TEM: Figure 3(a1) shows the smooth shape of the surface of the $NC@SiO_2$ substrate via Si–O–Si and C=O bonding. The EDS microanalysis was performed to provide qualitative determinations of the elemental composition. Figure 3(a4) shows the SEM image of the shape and size of the capsule surface, and the EDS mapping analysis of $NC@SiO_2$ confirmed the presence of Si and O, as shown in Figure 3(a2,a3). The Si–O–Si bonds were the major constituents of the capsule surface. The shell elements of the NCs contained Si, C, and O, as shown in Figure 3(b2–b4). The surface area of the $NC@SiO_2$ was 347.03 cm^2/g, an increase of 3470 times compared with the surface area of the 1.0 g spherical capsule (Figure 3(a5)).

Figure 4b show the XPS and XRD patterns of the NCs and nanocomposites. All the intense peaks correspond with the surface elemental composition. The constituents of the NC surface were confirmed by the XPS analysis, as were the chemical bonds of the Si group and the specific polymer peaks. As the XPS analysis results show, the atomic percentages (At. %) of Si, N, O and C were 21.25, 4.07, 42.27 and 31.71, respectively, as shown in Figure 4a. The X-ray powder diffraction pattern of the prepared NCs shows an amorphous peak at $2\theta = 21.8°$ caused by the equivalent Bragg angle marks. Martinez et al. (2006) prepared amorphous SiO_2 by the sol-gel procedure. In their work, a Si–O–Si peak was centered at $2\theta = 23°$. The peak can be shifted to lower 2θ values depending on the molar ratio (R) of water to APS (R = 5 or 11.66). The XRD analysis indicated a crystal structure, which Figure 4b shows as one broadened XRD peak for amorphous silica centered at a 2θ value close to our measurement. The pure Si peak appearing at 28.5° is caused by chemical interactions between the silanol group (Si–O) and the carboxyl group

(–COOH), and the peak of the Si was shifted by polymerization according to the oxygen (O) element. The Si–O–Si bond was found at 23°.

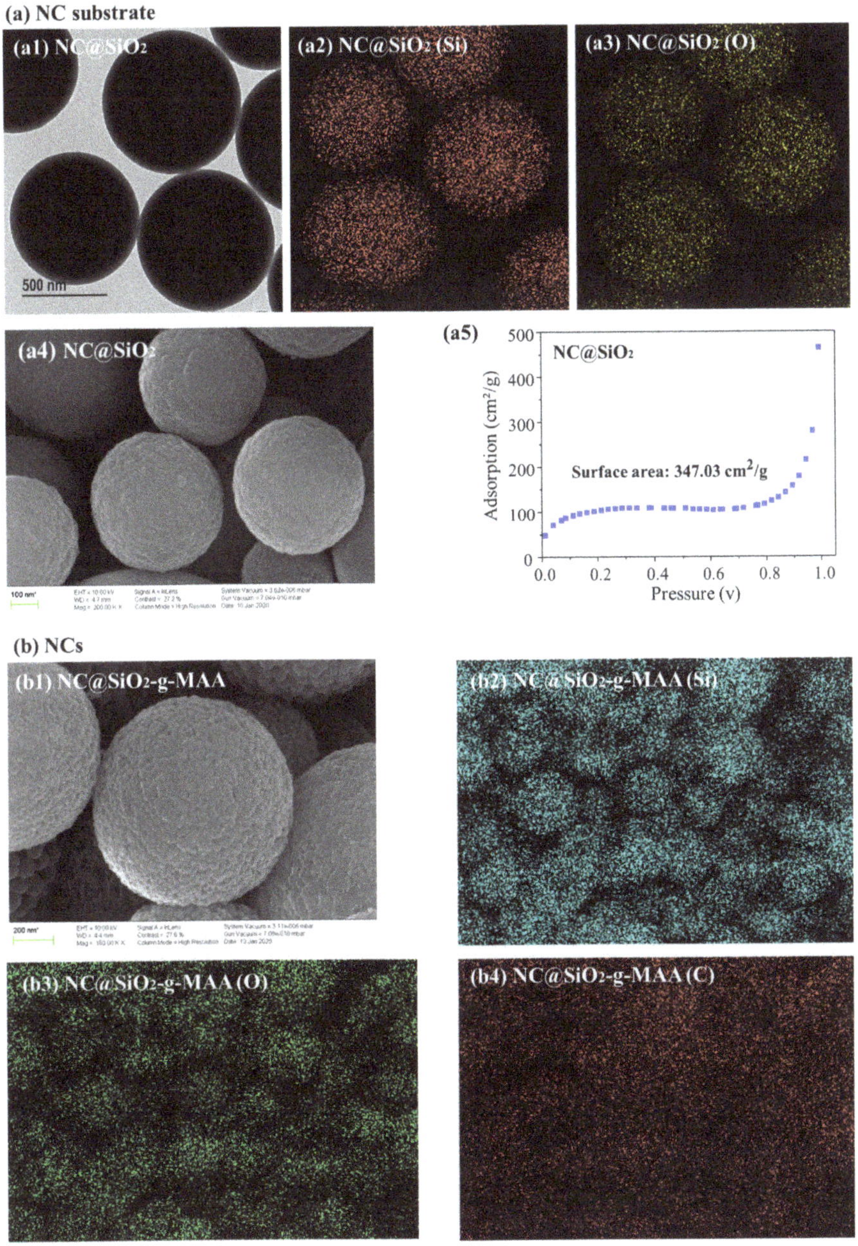

Figure 3. Analysis of morphology, surface shape, composition, and surface area of the (**a**) NC substrate and (**b**) self-antimicrobial NCs; (**a1**) shape analysis of the NC substrate (NC@SiO_2) by TEM, (**a2**) Si group determination, (**a3**) O group determination, (**a4**) surface morphology by SEM, (**a5**) surface area by BET analysis, (**b1**) surface morphology by SEM on the NCs, (**b2**) Si group determination, (**b3**) O group determination, and (**b4**) C group determination.

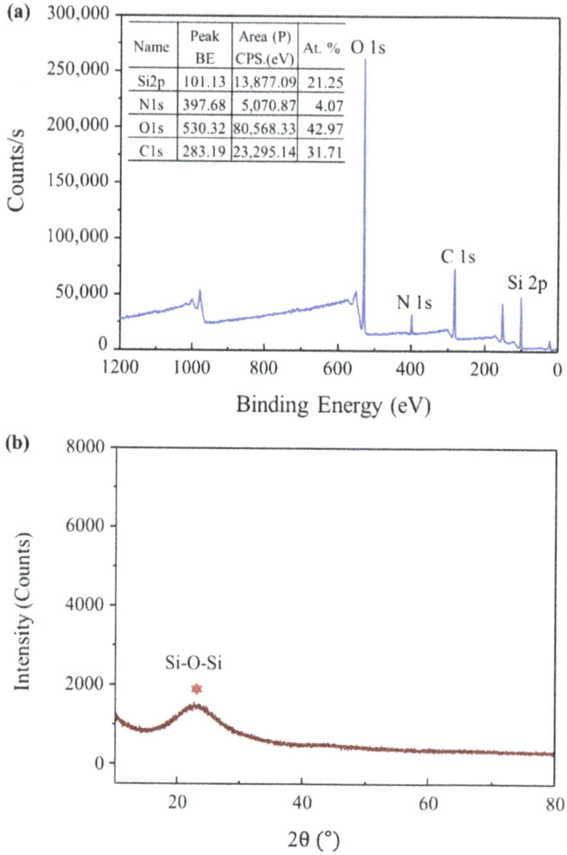

Figure 4. (**a**) XPS and (**b**) XRD analyses confirm the chemical component formation rate in the NCs.

3.3. Analysis of Zeta Potentials and NC Sizes

Controlled drug release can be initiated when the ligand brushes on the NC surface produce a repellent electrostatic force. We investigated the electric potential of NCs at each modification step and observed the zeta potential and mobility, as shown in Table 1. The anionic zeta potential value of the NCs was further increased by the carboxyl groups of the MAA polymer. The zeta potential was negative, indicating an increase in negative charges on the particle surface due to equilibrium of the zeta potential of Si–O–Si, C=O, and $-COO^-$ on the NC surfaces. The surface electron value of the NC@SiO_2 was −37.43 because of the Si-O groups in the TEOS, and the zeta value increased after MPS modification (−23.45 mV). The MAA surface modification decreased the surface zeta potential to −36.41 as a result of the $-COO^-$ groups of MAA [62,63]. That negative charge produced higher surface energy. The ionized silanol groups (Si–O) could be crosslinked by the oxygen bridge of the MAA and induce agglomeration of the ligand-brush chemical structure on the NC surfaces. As seen in Table 1, the mean particle diameter of the NC series was calculated with the standard deviation (S.D.) as 676.3 (±119.6) nm for NC@SiO_2 and 761.2 (±100.1) nm for the NC@SiO_2-MPS modified NCs, and the mean diameter of the surface-grafted NCs was 887.8 (±141.2) nm. The mean diameter of the NC increased under the influence of hydrogen bonding and brushes grafted to the NC surface. The uniformity of the NC diameter was excellent, producing a uniform thickness when used as a coating or a spray, and the amount of drug loading was relatively constant, allowing consistent antibacterial and bactericidal activity at any area of the sample.

Table 1. Zeta potential values and mean diameter of NCs at pH 7.

Sample	Zeta Potential [mV]	Mobility [cm^2/Vs]	Diameter [nm]	Std. Dev. [nm]
NC@SiO_2	−37.43	-2.92×10^{-4}	676.3	119.6
NC@SiO_2-MPS	−23.45	-1.86×10^{-4}	761.2	100.1
NC@SiO_2-g-MAA (NCs)	−36.41	-2.84×10^{-4}	887.8	141.2

3.4. Analysis of Drug Loading and Controlled Release of NCs

The FT-IR spectra in Figure 5 show surface modification, drug loading, and drug-controlled release at different pH conditions (pH 4 and 6). In this study, loading tests of the anionic triclosan were carried out at pH 6. The peaks corresponding to triclosan at pH 4 indicate that triclosan was not released in that condition, whereas the triclosan-loaded NCs demonstrated controlled released in pH 6 buffer solution due to the repellent electrostatic force shown in Figure 5a. In the triclosan spectra, the hydroxyl group (–OH) peak occurred at 3312 cm^{-1}, C–Cl absorption bands were observed in a frequency range of 722–570 cm^{-1}, and two strong CH wagging bands were observed for the CH_2Cl group in the 1300–1150 cm^{-1} region. The spectra for the pristine NCs and ligand-brush NCs without triclosan loading are shown in Figure 5b. The $-COO^-$ component of the MAA was observed on the ligand brushes grafted onto the NCs, and a peak corresponding to Si–O–Si was not imaged after the grafting modification. On the contrary, the drug-loaded NCs had absorption bands at 3450 cm^{-1} (OH) and 1630 cm^{-1} (H_2O). On the grafted NC surfaces, SiO_2 was present at 1080 cm^{-1} (Si–O–Si), 800 cm^{-1} (Si–O–Si) and 965 cm^{-1} (Si–OH), as shown in Figure 5a. The intensity of the Si–O–Si peak decrease is due to the ligand brushes on the surface of the NC@SiO_2-MPS. The MAA polymer might induce a negatively charged electron density on the capsule surface. After modification with MPS, a new absorption band at 1705 cm^{-1} was observed and assigned to the vibration of C=O from MPS. Thus, the NC@SiO_2 was successfully functionalized for drug loading. When the anionic antibacterial drug was loaded onto the NCs at pH 6, specific peaks were identified for each modification, allowing us to observe all the changing components.

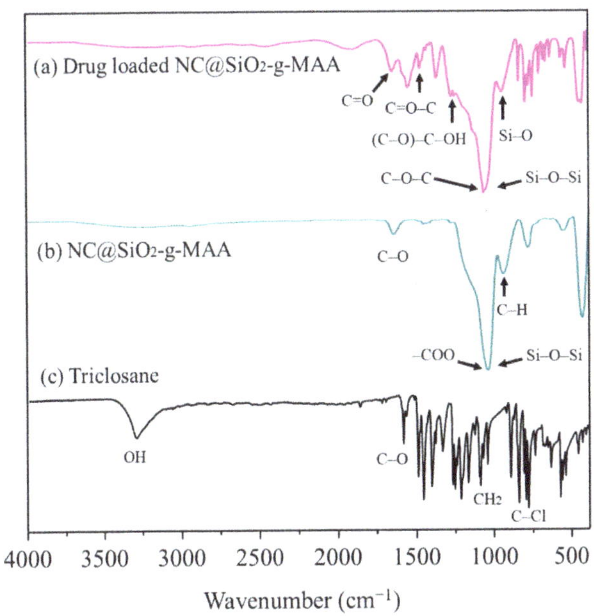

Figure 5. The FT-IR spectra for the (**a**) drug-loaded NCs, (**b**) pristine NCs and (**c**) triclosan (drug).

3.5. Dynamic Swelling Behavior According to pH and Humidity

3.5.1. Mechanism of Controlled Release

Among the polymers whose ionic properties change with pH, MAA, which we used as the ligand-brush shell, can be charged with anions at pH 5 and above. We analyzed the controlled release performance of drug-loaded NCs at pH 4, pH 6 and pH 7, and found that controlled release of the antibacterial drug occurred at pH above 5. Therefore, we designed a humidity- and pH-sensitive self-antimicrobial release system that is very sensitive under acidic conditions. In the release experiment for that agent, the capsules loaded under pH 6 were subjected to release testing at pH 4 and pH 6. As a result, we confirmed that the drug agent was released only above pH 6, as shown in Figure 6, because the anionic MAA ligand brush was pulled out of the anionic drug molecules by electro-repulsive forces as the entangled grafting polymer chains opened above pH 6. The NCs were influenced in different pH ranges due to deprotonation of the carboxylic acid groups, which gave rise to electro-repulsive forces between the carboxylate anions and the anions of the agent. In contrast, the entangled ligand-brush polymer chains did not expand and retained the entangled polymer chains, so the antibacterial drug loaded into the ligand brushes was not released at pH 4.

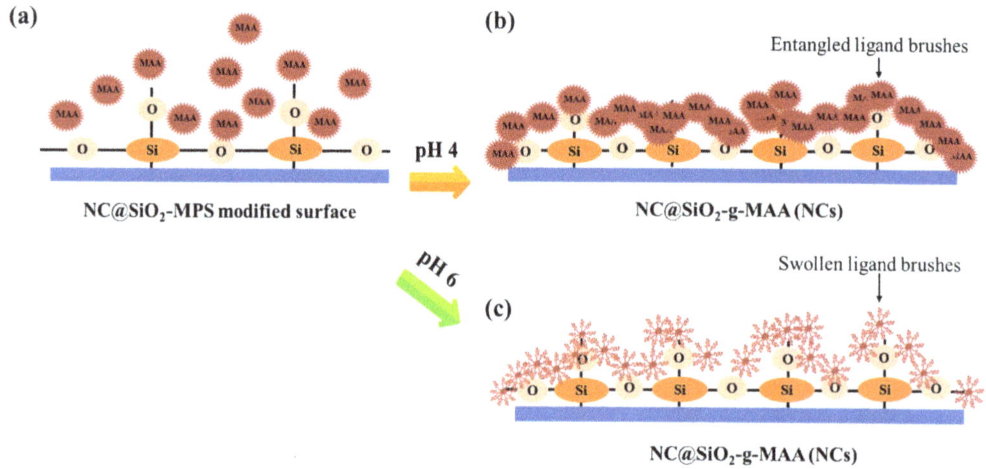

Figure 6. Self-controlled antibacterial drug release mechanism of the on the pristine NCs; (**a**) preparation of ligand brushes on the NC surface, (**b**) pH trigger mechanism of the ligand brushes at pH 4, and (**c**) pH trigger mechanism at pH 6.

This controlled drug release can be explained by MAA in different pH conditions. pK_a indicates the dissociation of an acid, whereas pH indicates the acidity or alkalinity of a system, with a smaller pK_a value indicating a stronger acid [21–24,64]. The pK_a value of the pristine MAA polymer was ~4.8 at neutral pH, which means that it almost entirely deprotonated, making it an anionic polymer. The carboxyl groups are ionized to be electrically negative. A large number of carboxylic acid groups in MAA were deprotonated as carboxylic acid at a pH above 5, resulting in a high negative-charge density in the NCP@SiO_2-g-MAA, which can cause swelling. At pH 6, most of the carboxylic acid groups of MAA were ionized as $-COO^-$ anions and repelled each other, causing the polymer ligand-brush outer shell to swell and the network to open for controlled drug release. By contrast, at pH 4, the outer shell of the NCs did not swell due to the formation of hydrogen bonds among the carboxylic groups of the MAA. Second, the intermolecular interactions between the Si–O–Si groups from the MAA produced an electrostatic attraction or intermolecular hydrogen-bonding interaction among them at pH 4, hindering drug diffusion (Figure 6b). On the contrary, electrostatic repulsion occurred at pH 6 and accelerated the controlled release and diffusion of the agent from the NCs. The controlled release mechanism of the drug is depicted in Figure 6c. As ionization increases, the swelling behavior is governed by the relaxation mechanism of the polymer chain, which explains why the MAA ligand brushes swell above pH 6.0. When the pH is 6.0, q has a value close to 2, indicating that the solvent penetration mechanism is relaxation-controlled transport. On the contrary, when the pK_a value is acidic (pH 4.0 in this experiment), there is no repulsive force between the ligand brushes because ionization of the MAA ionization group does not occur, so the mechanism of controlled release is not affected by Fickian release.

3.5.2. Dynamic Swelling Behavior According to pH and Humidity Response

To examine the dynamic swelling behavior of the hydrogel according to the pH, 500 mg of dry NCs were immersed in pH 4 and pH 6 buffer solutions for 8 h. The swollen NCs were then taken out of the buffer solutions. We measured their mass and calculated the mass swelling ratio, q, using Equation (2).

$$q = \frac{W_s}{W_d} \quad (2)$$

where q is the weight swelling ratio, W_s is the mass of the NCs after wet swelling, and W_d is the dry mass of the NCs before swelling. Table 2 shows the change in the mass swelling ratio (q) of NCs soaked in buffer solutions at pH 4.0 and pH 6.0 for 8 h. The amount of water absorbed by the polymer network was greater when the pH was 6.0 than when the pH was 4.0. After 1 h in pH 4 buffer solution, the q value was 1.000, but in the pH 6 buffer solution, it was 2.011. This result can be explained by the fact that the hydrophilic network is increased by ionization of the carboxylic acid groups in the MAA repeat unit in the hydrogel at a pH higher than pK_a 5, which allows it to absorb a relatively large amount of moisture. Nevertheless, swelling of the weak MAA ligand brushes occurred even at pH 4.

Table 2. The weight swelling ratio (q) of NCs in pH 4 and pH 6 conditions.

Buffer Solution	q Value of NCs							
	1 h	2 h	3 h	4 h	5 h	6 h	7 h	8 h
pH 4	1.000	1.002	1.007	1.009	1.011	1.014	1.015	1.015
pH 6	2.011	2.018	2.035	2.052	2.088	2.105	2.106	2.107

At pH 5 or higher, the electrostatic repulsion between negatively charged groups in the NCs caused the ligand brush portion of the surface to expand rapidly and release the drug agent. The ionization of the carboxyl groups present in the MAA changed with the surrounding pH environment, and so selective drug agent release occurred as the pH changed, as shown in Figure 7. NCs capable of selective drug agent release were prepared using pH-switched polymers by ionizing the carboxyl groups present in MAA according to the surrounding humidity. The pH-sensitive NCs showed hydrophobicity at pH 4; at pH higher than that, the carboxyl groups of the MAA ionized and became electrically negative (pH 5 corresponds to the pK_a of MAA). At pH 6.0, the electrostatic repulsion between the negatively charged groups caused the ligand-brush portion of the MAA to expand rapidly, releasing the loaded antibacterial drug. As a result, the triclosan attached to the ligand was released, and triclosan was identified after its controlled released in both the pH 6 and pH 7 buffer solutions, confirming that controlled release occurred through the influence of pH, as shown in Figure 7a.

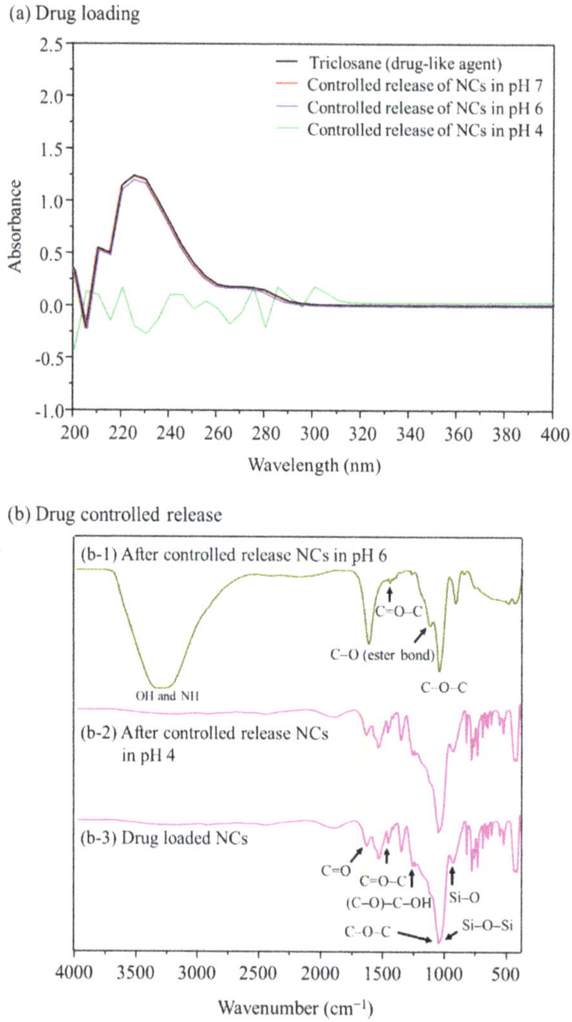

Figure 7. (**a**) Controlled release of triclosan from NCs at different pH values; (**b**) FT-IR spectra of the controlled-release NCs at different pH conditions: (**b-1**) drug-loaded NCs, (**b-2**) after controlled release at pH 4, and (**b-3**) after controlled release at pH 6.

We evaluated the controlled drug release using drug-loaded NCs in pH 4 and pH 6 solutions, as shown in Figure 7b. For the controlled-release NCs, the adsorption of pH 6 DML buffer solution components at 3400 cm^{-1} and 3350 cm^{-1} corresponded to the –OH and –NH groups, respectively. However, the controlled-release NCs in pH 4 buffer preserved the specific peaks of triclosan (Figure 7b), indicating that controlled release did not occur at pH 4. As shown in Figure 7 for the NCs at pH 6, the MAA C=O absorption band was observed at 1709 cm^{-1}, and the bands at 1500 cm^{-1} could be assigned to the stretching vibration absorption of C=O–C and the asymmetric stretching vibration absorption of $–COO^-$ anion groups caused by grafting MAA onto the surface of the NC intermediate (NC@SiO_2-MPS). These results demonstrate the chemical reactions between the (C=O)–OH of MPS and the carboxyl group of MAA that were generated when grafting the ligand brushes. In the controlled-release test at pH 6, the triclosan attached to the ligand was released, and no more drug material was identified, confirming that controlled release occurred through the influence of pH, as show in Figure 7. The drug release curve shows whether the drug was completely released at pH 6 and pH 7.

3.6. Self-Antimicrobial Activity

As described in Figure 8, the entangled MAA brushes grafted onto the hybrid NCs swelled only in certain humidity and pH environments, in which the increase of energy in the hydrogenated capsular surface produced electrostatic repulsion with the anionic drug. The self-antimicrobial activity of the drug-loaded NCs was tested at pH 4, pH 6 and pH 7. At pH 4, no clear zone of *E. coli* formed, as shown in Figure 8a. However, the pH 4–loaded NCs produced a 13-mm clear zone, and the pH 6-loaded NCs produced a 15–19 mm clear zone at 45% humidity, as shown in Figure 8b. In addition, the density change in *E. coli* was evaluated at pH 4, pH 6 and pH 7 and 100% humidity for 24 h. As shown in Figure 8c, the initial *E. coli* density with pristine NCs (without drug loading) was 0.117 (OD), which increased to 23.2 (OD) after incubation for 24 h at pH 6, indicating a greater than 198-fold growth of *E. coli*. The *E. coli* density with the drug-loaded NCs increased from 0.117 to 0.7 (OD) at pH 4, decreased about 3-fold at PH 6, and decreased from 0.105 to 0.1 (OD) at pH 7. Meanwhile, the *E. coli* density with the drug-loaded NCs decreased from 0.117 to 0.08 (OD) at pH 6 and increased slightly, from 0.105 to 0.3 (OD), at pH 7, an approximate 2.9-fold increase. Finally, all *E. coli* bacteria were killed after 24 h of exposure to the drug-loaded NCs in pH 6 and pH 7 conditions with 100% humidity, illustrating controlled release of the anionic antibacterial drug at the target pH. The weak swelling behavior of the MAA ligand brushes caused by moisture resulted in the release of the drug, which inhibited the growth of *E. coli*, a result similar to the swelling behavior seen in the NCs in the pH 4 buffer solution.

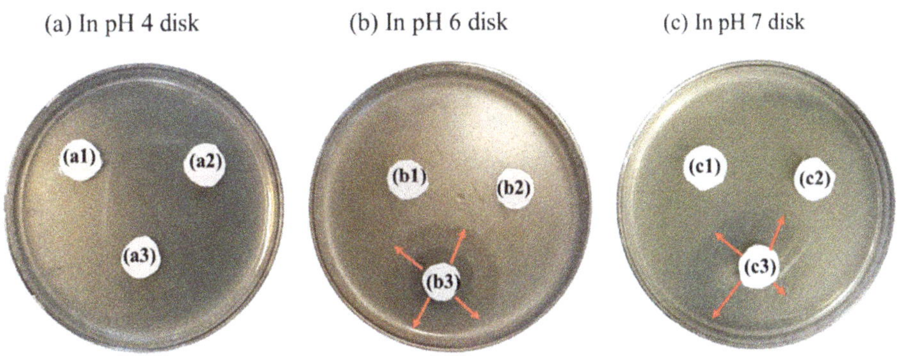

(d) Density change of *E. coli* by self-antibacterial NCs in 100% humidity condition.

Sample	pH 4		pH 6		pH 7	
	Initial density (OD)	After 24 h (OD)	Initial density (OD)	After 24 h (OD)	Initial density (OD)	After 24 h (OD)
Pristine NCs	0.115	15.6	0.117	23.2	0.105	11.2
Drug-loaded NCs in pH 4	0.115	0.9	0.117	0.7	0.105	0.1
Drug-loaded NCs in pH 6	0.115	3.5	0.117	0.08	0.105	0.3

Figure 8. Self-antibacterial activity of ligand brush NCs at different pH conditions with 45% humidity of (**a–c**) pH 4, 6, 7 disks: (**a1**,**b1**,**c1**) control sample, (**a2**,**b2**,**c2**) pristine NCs, (**a3**,**b3**,**c3**) drug-loaded NCs in pH 4, 6, 7; (**d**) density change of *Escherichia coli*.

3.7. Cytotoxicity Testing

We next examined the numbers of live cells after cultivating A549 cells in various samples, as shown in Table 3 and Figure 9. The number of live cells that adhered to the flask before exposure and the number of live cells after cultivation for 24 h were quantified; cytotoxic tests were performed by determining the number of adherent cells and examining images of live cells. In Table 3, the cell adhesion activity was calculated using Equation (3).

$$\text{Cell adhesion activity } (\%) = \left(\frac{\text{Number of cell adhesion after exposing (24 h)}}{\text{Number of cell adhesion before exposing}}\right) \times 100 \quad (3)$$

Table 3. Cytotoxicity of antibacterial NCs to cultured A549 cells.

Sample	Number of Cells Adhered		Cell Adhesion Activity (%)
	0 h	**24 h**	
Control	79	188	237.39
Pristine NCs	70	98	140.8
Drug-loaded NCs	115	162	140

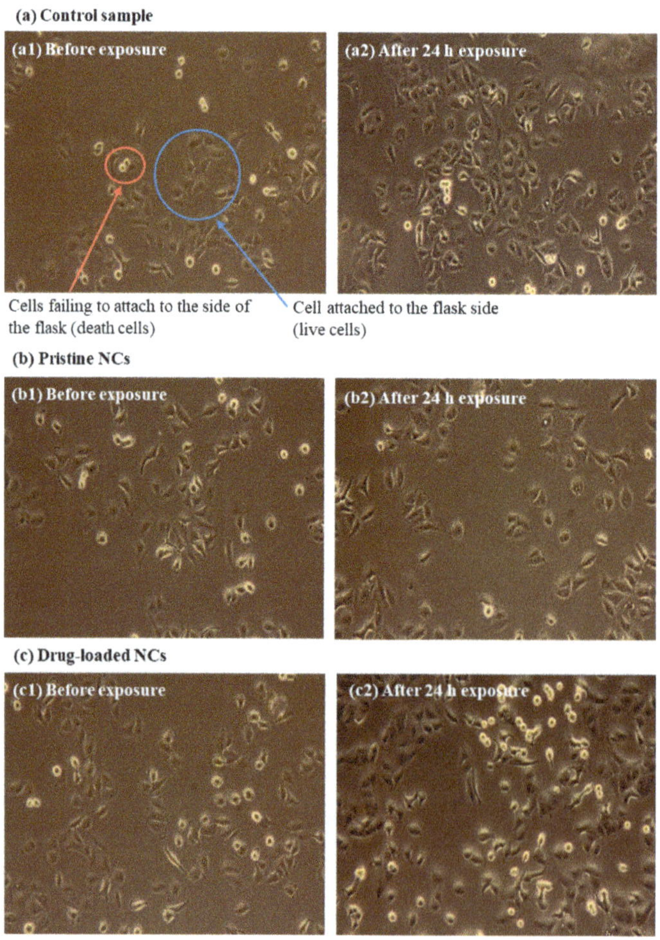

Figure 9. Antibacterial activity of ligand-brush NCs: (**a**) C.S., (**b**) pristine NCs, and (**c**) drug-loaded NCs.

The number of live cells in the control sample was 79, which increased to 188 after 24 h. In experiments using pristine NCs, the number of living cells increased from 70 to 98. With drug-loaded NCs, the number of living cells increased from 115 to 162. With the pristine, non-drug-loaded NCs, the cell proliferation rate was 237%, whereas the cell proliferation rate of the drug-loaded NCs was approximately 140%. Thus, the cell proliferation rate of the drug-loaded NCs was less than 97% of that in the control sample, but it was still 140%, indicating no cytotoxicity. Figure 9 shows images of dead and living cells after exposure to the indicated samples.

4. Conclusions

This work has demonstrated humidity- and pH-responsive NCs prepared with ion changeable polymer brush. The anion-modified NCs were formed as a container and loaded with an anionic antibacterial drug by the anionization of the grafted ligand brushes. The surface area (347.03 cm^2/g) of our NCs was increased using a nanoscale capsule and copolymer ligand brushes. That large surface area contained a high density of the antibacterial drug, and drug release was easily controlled by the swelling of the ligand brushes. The grafted, anionic ligand-brush copolymer has an obvious positive influence on the negatively charged antibacterial drug. The diameter of the substrate NC@SiO_2 was 676.3 ($\pm$119.6) nm, which increased to 887.8 ($\pm$141.2) nm upon grafting of the ligand brushes. We also defined the mechanism of the controlled drug release. The self-antimicrobial NCs showed a controlled release in target pH and humidity conditions. In the antimicrobial experiment using *E. coli*, a clear zone of 13–20 mm formed at pH 6 and 45% humidity, and all the *E. coli* was killed at pH 6 and pH 7 at 100% humidity. The density of *E. coli* decreased from 0.117 to 0 (OD) at pH 6 with 45% humidity and from 0.105 to 0 (OD) at pH 6 with 100% humidity, demonstrating self-antimicrobial activity at target humidity and pH conditions.

Author Contributions: Conceptualization, E.K. and Y.T.L.; methodology, E.K.; software, E.K.; validation, E.K. and Y.T.L.; formal analysis, E.K. and Y.T.L.; investigation, E.K. and Y.T.L.; resources, E.K. and Y.T.L.; data curation, E.K. and Y.T.L.; writing—original draft preparation, E.K.; writing—review and editing, E.K. and Y.T.L.; visualization, E.K.; supervision, Y.T.L.; project administration, E.K.; funding acquisition, E.K. All authors have read and agreed to the published version of the manuscript.

Funding: This work was supported by a National Research Foundation of Korea (NRF) grant funded by the Korea government (MEST) (No. NRF-2019R1A2C1011539).

Institutional Review Board Statement: Not applicable.

Informed Consent Statement: Not applicable.

Conflicts of Interest: The authors have no conflict of interest to declare.

References

1. Zhou, J.; Li, T.; Zhang, C.; Xiao, J.; Cui, D.; Cheng, Y. Charge-switchable nanocapsules with multistage pH-responsive behaviours for enhanced tumour-targeted chemo/photodynamic therapy guided by NIR/MR imaging. *Nanoscale* **2018**, *10*, 9707–9719. [CrossRef] [PubMed]
2. Wang, W.; Hao, X.; Chen, S.; Yang, Z.; Wang, C.; Yan, R.; Guo, Z. pH-responsive Capsaicin@ chitosan nanocapsules for antibiofouling in marine applications. *Polymer* **2018**, *158*, 223–230. [CrossRef]
3. Ali, S.I.; Heuts, J.P.; van Herk, A.M. Vesicle-templated pH-responsive polymeric nanocapsules. *Soft Matter* **2011**, *7*, 5382–5390. [CrossRef]
4. Zhang, Z.; Hadjichristidis, N. Temperature and pH-Dual Responsive AIE-Active Core Crosslinked Polyethylene–Poly (methacrylic acid) Multimiktoarm Star Copolymers. *ACS Macro Lett.* **2018**, *7*, 886–891. [CrossRef]
5. Mei, L.; Xie, R.; Yang, C.; Ju, X.J.; Wang, W.; Wang, J.Y.; Chu, L.Y. pH-responsive Ca-alginate-based capsule membranes with grafted poly (methacrylic acid) brushes for controllable enzyme reaction. *Chem. Eng. J.* **2013**, *232*, 573–581. [CrossRef]
6. Tummala, S.; Kumar, M.S.; Pindiprolu, S.K. Improved anti-tumor activity of oxaliplatin by encapsulating in anti-DR5 targeted gold nanoparticles. *Drug Deliv.* **2016**, *23*, 3505–3519. [CrossRef]
7. Wang, J.; Sui, M.; Fan, W. Nanoparticles for tumor targeted therapies and their pharmacokinetics. *Curr. Drug Metab.* **2010**, *11*, 129–141. [CrossRef]

8. Wang, C.; Liu, X.; Chen, S.; Hu, F.; Sun, J.; Yuan, H. Facile preparation of phospholipid–amorphous calcium carbonate hybrid nanoparticles: Toward controllable burst drug release and enhanced tumor penetration. *Chem. Commun.* **2018**, *54*, 13080–13083. [CrossRef]
9. Hamarat Sanlıer, S.; Yasa, M.; Cihnioglu, A.O.; Abdulhayoglu, M.; Yılmaz, H.; Ak, G. Development of gemcitabine-adsorbed magnetic gelatin nanoparticles for targeted drug delivery in lung cancer. *Artif. Cells Nanomed. Biotechnol.* **2016**, *44*, 943–949. [CrossRef]
10. Wong, T.W.; Dhanawat, M.; Rathbone, M.J. Vaginal drug delivery: Strategies and concerns in polymeric nanoparticle development. *Expert. Opin. Drug Deliv.* **2014**, *11*, 1419–1434. [CrossRef]
11. Zhang, L.; Wang, Y.; Tang, N.; Cheng, P.; Xiang, J.; Du, W.; Wang, X. Bioinspired stability improvement of layer-by-layer microcapsules using a bioadhesive for enzyme encapsulation. *React. Funct. Polym.* **2016**, *99*, 73–79. [CrossRef]
12. Chuang, C.Y.; Don, T.M.; Chiu, W.Y. Synthesis and properties of chitosan-based thermo-and pH-responsive nanoparticles and application in drug release. *J. Polym. Sci. A Poly. Chem.* **2009**, *47*, 2798–2810. [CrossRef]
13. Chuang, C.Y.; Don, T.M.; Chiu, W.Y. Synthesis and characterization of stimuli-responsive porous/hollow nanoparticles by self-assembly of chitosan-based graft copolymers and application in drug release. *J. Polym. Sci. A Poly. Chem.* **2010**, *48*, 2377–2387. [CrossRef]
14. Tamizhrasi, S.; Shukla, A.; Shivkumar, T.; Rathi, V.; Rathi, J.C. Formulation and evaluation of lamivudine loaded polymethacrylic acid nanoparticles. *Int. J. Pharm. Tech. Res.* **2009**, *1*, 411–415.
15. Ghasemi, R.; Abdollahi, M.; Zadeh, E.E.; Khodabakhshi, K.; Badeli, A.; Bagheri, H.; Hosseinkhani, S. mPEG-PLA and PLA-PEG-PLA nanoparticles as new carriers for delivery of recombinant human Growth Hormone (rhGH). *Sci. Rep.* **2018**, *8*, 9854. [CrossRef] [PubMed]
16. Depan, D.; Shah, J.; Misra, R.D.K. Controlled release of drug from folate-decorated and graphene mediated drug delivery system: Synthesis, loading efficiency, and drug release response. *Mater. Sci. Eng. C* **2011**, *31*, 1305–1312. [CrossRef]
17. Thomas, C.R.; Ferris, D.P.; Lee, J.H.; Choi, E.; Cho, M.H.; Kim, E.S.; Zink, J.I. Noninvasive remote-controlled release of drug molecules in vitro using magnetic actuation of mechanized nanoparticles. *J. Am. Chem. Soc.* **2010**, *132*, 10623–10625. [CrossRef]
18. Maver, U.; Godec, A.; Bele, M.; Planinšek, O.; Gaberšček, M.; Srčič, S.; Jamnik, J. Novel hybrid silica xerogels for stabilization and controlled release of drug. *Int. J. Pharm.* **2007**, *330*, 164–174. [CrossRef]
19. Luo, X.; Matranga, C.; Tan, S.; Alba, N.; Cui, X.T. Carbon nanotube nanoreservior for controlled release of anti-inflammatory dexamethasone. *Biomaterials* **2011**, *32*, 6316–6323. [CrossRef]
20. Muppalla, R.; Rana, H.H.; Devi, S.; Jewrajka, S.K. Adsorption of pH-responsive amphiphilic copolymer micelles and gel on membrane surface as an approach for antifouling coating. *Appl. Surf. Sci.* **2013**, *268*, 355–367. [CrossRef]
21. Ball, V. Electrodeposition from Tannic acid-polyamine blends at pH = 5.0 is due to aggregate deposition and oxidation. *Colloids Surf. A Physicochem. Eng. Asp.* **2019**, *578*, 123530. [CrossRef]
22. Azam, F.; Ijaz, H.; Qureshi, J. Functionalized crosslinked interpenetrating polymeric network for pH responsive colonic drug delivery. *Int. J. Polym. Mater. Polym. Biomater.* **2021**, *70*, 646–655. [CrossRef]
23. Tian, B.; Liu, S.; Lu, W.; Jin, L.; Li, Q.; Shi, Y.; Du, Y. Construction of pH-responsive and up-conversion luminescent $NaYF_4$: Yb3+/Er3+@ SiO_2@ PMAA nanocomposite for colon targeted drug delivery. *Sci. Rep.* **2016**, *6*, 21335. [CrossRef] [PubMed]
24. Chang, Y.H.; Chen, Z.M.; Peng, H.Y.; Tsai, Y.J.; Lin, T.B.; Chen, G.D.; Change, S.Y. Extraction of Substance P by Functionalized Iron Oxide Nanoparticles for Matrix-Assisted Laser Desorption/Ionization Mass Spectrometric Analysis. *J. Chin. Chem. Soc.* **2011**, *58*, 457–462. [CrossRef]
25. Wu, Y.; Sadatmousavi, P.; Wang, R.; Lu, S.; Yuan, Y.F.; Chen, P. Self-assembling peptide-based nanoparticles enhance anticancer effect of ellipticine in vitro and in vivo. *Int. J. Nanomed.* **2012**, *7*, 3221.
26. Luo, Y.L.; Yang, X.L.; Xu, F.; Chen, Y.S.; Zhao, X. pH-triggered PMAA-b-HTPB-b-PMAA copolymer micelles: Physicochemical characterization and camptothecin release. *Colloid Polym. Sci.* **2014**, *292*, 1061–1072. [CrossRef]
27. Sun, L.; Chen, Y.; Zhou, Y.; Guo, D.; Fan, Y.; Guo, F.; Chen, W. Preparation of 5-fluorouracil-loaded chitosan nanoparticles and study of the sustained release in vitro and in vivo. *Asian J. Pharm. Sci.* **2017**, *12*, 418–423. [CrossRef]
28. Derakhshandeh, K.; Soheili, M.; Dadashzadeh, S.; Saghiri, R. Preparation and in vitro characterization of 9-nitrocamptothecin-loaded long circulating nanoparticles for delivery in cancer patients. *Int. J. Nanomed.* **2010**, *5*, 463. [CrossRef]
29. Hu, Y.; Gu, X.; Chen, W.; Zou, S.; Wang, C. Macroporous nanocomposite materials prepared by solvent evaporation from pickering emulsion templates. *Macromol. Mater. Eng.* **2014**, *299*, 1070–1080. [CrossRef]
30. Gao, D.; Chang, R.; Lyu, B.; Ma, J.; Duan, X. Preparation of epoxy-acrylate copolymer/nano-silica via Pickering emulsion polymerization and its application as printing binder. *Appl. Surf. Sci.* **2018**, *435*, 195–202. [CrossRef]
31. Chevalier, Y.; Bolzinger, M.A. Emulsions stabilized with solid nanoparticles: Pickering emulsions. *Colloids Surf. A Physicochem. Eng. Asp.* **2013**, *439*, 23–34. [CrossRef]
32. Xiao, M.; Xu, A.; Zhang, T.; Hong, L. Tailoring the wettability of colloidal particles for Pickering emulsions via surface modification and roughness. *Front. Chem.* **2018**, *6*, 225. [CrossRef] [PubMed]
33. Pu, K.Y.; Li, K.; Liu, B. A molecular brush approach to enhance quantum yield and suppress nonspecific interactions of conjugated polyelectrolyte for targeted far-red/near-infrared fluorescence cell imaging. *Adv. Funct. Mater.* **2010**, *20*, 2770–2777. [CrossRef]

34. Liu, X.; Shi, L.; Zhang, Z.; Fan, Q.; Huang, Y.; Su, S.; Huang, W. Monodispersed nanoparticles of conjugated polyelectrolyte brush with high charge density for rapid, specific and label-free detection of tumor marker. *Analyst* **2015**, *140*, 1842–1846. [CrossRef] [PubMed]
35. Zhang, Z.; Fan, Q.; Sun, P.; Liu, L.; Lu, X.; Li, B.; Huang, W. Highly Selective Anionic Counterion-based Fluorescent Sensor for Hg_2^+ by Grafted Conjugated Polyelectrolytes. *Macromol. Rapid Commun.* **2010**, *31*, 2160–2165. [CrossRef] [PubMed]
36. Wang, Y.; Fan, D.; He, J.; Yang, Y. Silica nanoparticle covered with mixed polymer brushes as Janus particles at water/oil interface. *Colloid Polym. Sci.* **2011**, *289*, 1885–1894. [CrossRef]
37. Hatanaka, D.; Takemoto, Y.; Yamamoto, K.; Kadokawa, J.I. Hierarchically self-assembled nanofiber films from amylose-grafted carboxymethyl cellulose. *Fibers* **2014**, *2*, 34–44. [CrossRef]
38. Yu, X.; Huang, S.; Chen, K.; Zhou, Z.; Guo, X.; Li, L. Preparation of functional Janus particles with response to magnetic force. *Ind. Eng. Chem. Res.* **2015**, *54*, 2690–2696. [CrossRef]
39. Giovannini, G.; Gubala, V.; Hall, A.J. Off–on switchable fluorescent probe for prompt and cost-efficient detection of bacteria. *New J. Chem.* **2019**, *43*, 13094–13102. [CrossRef]
40. Tkachov, R.; Senkovskyy, V.; Oertel, U.; Synytska, A.; Horecha, M.; Kiriy, A. Microparticle-Supported Conjugated Polyelectrolyte Brushes Prepared by Surface-Initiated Kumada Catalyst Transfer Polycondensation for Sensor Applications. *Macromol. Rapid Commun.* **2010**, *31*, 2146–2150. [CrossRef]
41. Bollhorst, T.; Rezwan, K.; Maas, M. Colloidal capsules: Nano-and microcapsules with colloidal particle shells. *Chem. Soc. Rev.* **2017**, *46*, 2091–2126. [CrossRef] [PubMed]
42. YeonáKim, J.; BonáYoon, S. Fabrication of nanocapsules with Au particles trapped inside carbon and silica nanoporous shells. *Chem. Commun.* **2003**, *6*, 790–791.
43. Kong, X.; Yu, Q.; Zhang, X.; Du, X.; Gong, H.; Jiang, H. Synthesis and application of surface enhanced Raman scattering (SERS) tags of Ag@ SiO_2 core/shell nanoparticles in protein detection. *J. Mater. Chem.* **2012**, *22*, 7767–7774. [CrossRef]
44. Caruso, M.M.; Blaiszik, B.J.; Jin, H.; Schelkopf, S.R.; Stradley, D.S.; Sottos, N.R.; Moore, J.S. Robust, double-walled microcapsules for self-healing polymeric materials. *ACS Appl. Mater. Interfaces* **2010**, *2*, 1195–1199. [CrossRef]
45. Pan, S.; Chen, M.; Wu, L. Synthesis of raspberry-like polymer/SiO_2 hybrid colloidal spheres grafted by block-copolymer poly (MPC-b-MPS) for underwater superoleophobic anti-biofouling coatings. *J. Colloid Interface Sci.* **2018**, *522*, 20–28. [CrossRef]
46. Shiojima, T.; Inoue, Y.; Kyomoto, M.; Ishihara, K. High-efficiency preparation of poly (2-methacryloyloxyethyl phosphorylcholine) grafting layer on poly (ether ether ketone) by photoinduced and self-initiated graft polymerization in an aqueous solution in the presence of inorganic salt additives. *Acta Biomater.* **2016**, *40*, 38–45. [CrossRef]
47. Xu, P.; Wang, H.; Tong, R.; Du, Q.; Zhong, W. Preparation and morphology of SiO 2/PMMA nanohybrids by microemulsion polymerization. *Colloid Polym. Sci.* **2006**, *284*, 755–762. [CrossRef]
48. Li, H.; Yuan, J.; Qian, H.; Wu, L. Synthesis and properties of SiO_2/P(MMA-BA) core–shell structural latex with siloxanes. *Prog. Org. Coat.* **2016**, *97*, 65–73. [CrossRef]
49. Viscomi, F.N.; Dey, R.K.; Caputo, R.; Cui, B. Enhanced adhesion of electron beam resist by grafted monolayer poly (methylmethacrylate-co-methacrylic acid) brush. *J. Vac. Sci. Technol. B Nanotechnol. Microelectron.* **2015**, *33*, 06FD06. [CrossRef]
50. Neubauer, M.P.; Poehlmann, M.; Fery, A. Microcapsule mechanics: From stability to function. *Adv. Colloid Interface Sci.* **2014**, *207*, 65–80. [CrossRef]
51. Park, J.; Cullen, D.A.; Chen, J.; Polizos, G.; Sharma, J. Same solution synthesis and self-assembly of porous silica nanoparticles into microspheres. *Appl. Surf. Sci.* **2019**, *467*, 634–639. [CrossRef]
52. Chen, H.; He, J.; Tang, H.; Yan, C. Porous silica nanocapsules and nanospheres: Dynamic self-assembly synthesis and application in controlled release. *Chem. Mater.* **2008**, *20*, 5894–5900. [CrossRef]
53. Xu, J.B.; Spittler, D.A.; Bartley, J.P.; Johnson, R.A. Alginic acid–silica hydrogel coatings for the protection of osmotic distillation membranes against wet-out by surface-active agents. *J. Memb. Sci.* **2005**, *2260*, 19–25. [CrossRef]
54. Oshiro Junior, J.A.; Paiva Abuçafy, M.; Berbel Manaia, E.; Lallo da Silva, B.; Chiari-Andréo, B.G.; Aparecida Chiavacci, L. Drug delivery systems obtained from silica based organic-inorganic hybrids. *Polymers* **2016**, *8*, 91. [CrossRef] [PubMed]
55. Scheiner, M.; Dickens, T.J.; Okoli, O. Progress towards self-healing polymers for composite structural applications. *Polymer* **2016**, *83*, 260–282. [CrossRef]
56. Gieseler, G.; Pepelanova, I.; Stuckenberg, L.; Villain, L.; Nölle, V.; Odenthal, U.; Scheper, T. Purification of bone morphogenetic protein-2 from refolding mixtures using mixed-mode membrane chromatography. *Appl. Microbiol. Biotechnol.* **2017**, *101*, 123–130. [CrossRef] [PubMed]
57. Li, W.R.; Xie, X.B.; Shi, Q.S.; Zeng, H.Y.; You-Sheng, O.Y.; Chen, Y.B. Antibacterial activity and mechanism of silver nanoparticles on Escherichia coli. *Appl. Microbiol. Biotechnol.* **2010**, *85*, 1115–1122. [CrossRef]
58. Pal, S.; Tak, Y.K.; Song, J.M. Does the antibacterial activity of silver nanoparticles depend on the shape of the nanoparticle? A study of the gram-negative bacterium Escherichia coli. *Appl. Environ. Microbiol.* **2007**, *73*, 1712–1720. [CrossRef]
59. Raffi, M.; Mehrwan, S.; Bhatti, T.M.; Akhter, J.I.; Hameed, A.; Yawar, W.; ul Hasan, M.M. Investigations into the antibacterial behavior of copper nanoparticles against Escherichia coli. *Ann. Microbiol.* **2010**, *60*, 75–80. [CrossRef]
60. Leung, Y.H.; Ng, A.M.; Xu, X.; Shen, Z.; Gethings, L.A.; Wong, M.T.; Lee, P.K. Mechanisms of antibacterial activity of MgO: Non-ROS mediated toxicity of MgO nanoparticles towards Escherichia coli. *Small* **2014**, *10*, 1171–1183. [CrossRef]

61. Crecente-Campo, J.; Lorenzo-Abalde, S.; Mora, A.; Marzoa, J.; Csaba, N.; Blanco, J.; Alonso, M.J. Bilayer polymeric nanocapsules: A formulation approach for a thermostable and adjuvanted E. coli antigen vaccine. *J. Control. Release* **2018**, *286*, 20–32. [CrossRef] [PubMed]
62. Cho, H.W.; Yoon, K.J. Effect of P(MMA-co-MAA) compatibilizer on the miscibility of nylon 6/PVDF blends. *Eur. Polym. J.* **2003**, *39*, 1249–1265.
63. Bayramoğlu, G.; Yılmaz, M.; Şenel, A.Ü.; Arıca, M.Y. Preparation of nanofibrous polymer grafted magnetic poly (GMA-MMA)-g-MAA beads for immobilization of trypsin via adsorption. *Arıca Biochem. Eng. J.* **2008**, *40*, 262–274. [CrossRef]
64. Guillaume, J.L.; Pichot, C.; Guillot, J. Emulsifier-free emulsion copolymerization of styrene and butyl acrylate. I. Kinetic studies in the absence of surfactant. *J. Polym. Sci. A Polym. Chem.* **1990**, *28*, 119–136. [CrossRef]

MDPI AG
Grosspeteranlage 5
4052 Basel
Switzerland
Tel.: +41 61 683 77 34

Pharmaceutics Editorial Office
E-mail: pharmaceutics@mdpi.com
www.mdpi.com/journal/pharmaceutics

www.ingramcontent.com/pod-product-compliance
Lightning Source LLC
Chambersburg PA
CBHW040746250926
51850CB00007BB/1426
* 9 7 8 3 7 2 5 8 7 9 3 1 1 *